W9-ACD-174

ROCKS FROM SPACE

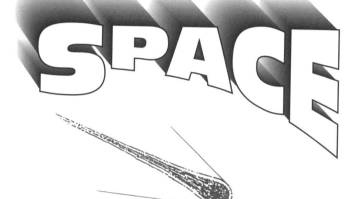

Rocks from Space

Meteorites and Meteorite Hunters

SECOND EDITION

O. Richard Norton

Illustrated by Dorothy S. Norton

Mountain Press Publishing Company
Missoula, Montana
1998

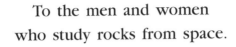

To the men and women
who study rocks from space.

© 1994, 1998
O. Richard Norton

First edition, 1994
Second edition, 1998

Cover illustration © 1994
Dorothy S. Norton

*Cover: A small asteroid enters Earth's atmosphere
at dusk over San Francisco Bay, leaving behind a dust train
and pushing a shock wave through the atmosphere.*

All photographs © 1998
by O. Richard Norton unless otherwise credited

All illustrations © 1998
by Dorothy S. Norton unless otherwise credited

Library of Congress Cataloging-in-Publication Data
Norton, O. Richard.
    Rocks from space : meteorites and meteorite hunters / O.
Richard Norton ; illustrated by Dorothy S. Norton.—2nd ed.
      p.   cm.
    Originally published: 1994.
    Includes bibliographical references and index.
    ISBN 0-87842-373-7 (alk. paper)
    1. Meteorites—Handbooks, manuals, etc. 2. Meteorites—
Pictorial works. 3. Asteroids—Pictorial works. 4. Comets—
Pictorial works.
    I. Title.
QB755.N67   1998
523.5'1—dc21
                                        97-51574
                                          CIP

PRINTED IN THE UNITED STATES OF AMERICA

Mountain Press Publishing Company
P. O. Box 2399 · Missoula, Montana 59806
(406) 728-1900 · FAX (406) 728-1635

Contents

Preface to the Second Edition

I wrote *Rocks from Space* to fill an obvious void. In the early 1990s, I could hold in one hand all the books written on meteorites over the past ten years. Most were badly out of date but still useful. Of these, at least half were no longer available except as increasingly rare jewels in used book stores. A few were intended for the scientifically literate or meteorite researcher and were of little use to enthusiastic but nonacademically trained amateurs. This was an opportune time to fill the gap with a useful and entertaining book.

I had no idea at the time that the science of meteoritics would explode into the public consciousness virtually simultaneously with the publication of the first edition. Public awareness of meteorites continues to increase rapidly. The last four years have seen hundreds, if not thousands, of new meteorite collectors, all vying for the limited amounts of meteoritic material that makes it to the commercial market. The Internet is loaded with meteorite news, meteorite research, meteorite dealers, meteorite collectors, and meteorite book sellers. And the meteorite frenzy goes on. A popular magazine published in New Zealand, *Meteorite!*, has appeared, filled with timely articles by amateurs and professionals. Substantial meteorite collections have been placed on the public auction block at staggering prices. Would you pay $2,000 per gram for a meteorite that scientists say came from Mars?

All this activity should come as no surprise to those in meteorite circles. Much has happened recently to spark an interest in things that appear in the sky and sometimes fall to Earth. Two magnificent comets, Hyakutake and Hale-Bopp, were wonderful spectacles and naturally raised questions of our vulnerability to celestial objects passing close to Earth. Comet Hyakutake passed a mere 9 million miles from Earth. Even as I write this, a controversy rages over a stunning discovery that suggests mini cometlike bodies weighing 20 to 40 tons, each the size of a small house, are continually slamming into Earth's atmosphere at the unbelievable rate of 7,000 to 43,000 per day! As if to support this outrageous idea, scientists have just found that Earth's upper atmosphere, usually thought of as a bone-dry region, actually harbors large amounts of water. Water vapor cannot easily rise from the lower atmosphere to the upper. Instead, it must come from space beyond Earth.

Meteorite scientists justifiably have become concerned that enthusiastic collectors wanting a piece of the action might snatch up the raw materials of their science. What was once an obscure, esoteric subject studied by only a small group has blossomed into prominence, and the scientists themselves have inadvertently provided the greatest stimulus. They find themselves between a meteorite and a hard spot. Some of the most stunning work exciting the world's imagination was the August 1996 announcement by NASA scientists that Mars rock ALH 84001 may contain the remnants of past life on Mars. This was followed within a year by the landing of *Pathfinder* on Mars. There, through *Pathfinder* cameras, all the world saw the bleak, barren Martian landscape strewn with Mars rocks—possible future Mars meteorites *in situ*.

The need for a book like *Rocks from Space* is greater than ever. The many changes in the second edition reflect the events of these past four years. Many new meteorites have been recovered, some as a direct result of the first edition and others falling through the roofs of houses and bouncing on the streets. Scientists now recognize new groups of meteorites, a collector found a lunar meteorite in one of the most desolate spots on Earth, and sifting through lunar soils brought back from the Moon, scientists found a submillimeter-size object identified as a rare enstatite chondrite. Included in this edition are the most current classifications of meteorites with both old and new nomenclature presented. The new edition contains many photographs of rare specimens, including one of Robert Haag's Adamana chondrite with its amazing orientation.

In writing the second edition, I have been fortunate to have the attention and interest of some of the most respected meteoriticists in the United States. They have been invaluable in pointing out errors that inadvertently crept into the first edition. Dr. Alan Rubin of the University of California at Los Angeles has patiently given me his time, providing important materials on recent research that is rapidly changing the face of meteoritics. I have spent many hours corresponding with Dr. Ursula Marvin of the Harvard-Smithsonian Center for Astrophysics discussing the historical side of meteoritics. She is the scientific historian for the international Meteoritical Society and has given me invaluable insights (as well as corrections) into the first several chapters. Dr. Rhian Jones and Dr. Adrian Brearley of the University of New Mexico's Institute of Meteoritics opened their fabulous collection of meteorites to me. Like a kid in a meteorite candy shop, I spent many hours examining and photographing every conceivable meteorite type. Institute personnel were always available whenever I needed help interpreting slab specimens and thin sections, which was more often than I would like to admit. Carol Schwarz of the Johnson Space Center in Houston has on numerous occasions responded to my requests for sometimes obscure images that required considerable effort to exhume from the immense NASA image files. Dr. Dana Johnston, chairman of the Geosciences Department at the University of Oregon, graciously made the department's Zeiss petrologic microscope available to me for photography of thin sections. My good friend and colleague Dr. Bob Reynolds, igneous petrologist at Central Oregon Community College, was always available to help me interpret the sections.

I can always count on meteorite dealers to come up with interesting new falls and finds. At a recent Tucson Gem and Mineral Show, Allan Langheinrich made room for my copy stand, cameras, and lights to photograph some of his rare specimens that I could hardly believe I was seeing. Blaine Reed has more than once provided sample specimens for study. Michael Casper has done the same, as well as giving me free rein to use any of his choice photographs. You can't miss these in the book.

Finally, Dorothy Sigler Norton, my in-house scientific illustrator, editor, and most-severe-but-benevolent critic whom I had the great fortune to marry some years ago, has once again proven her devotion and patience during the writing of this edition. Through the years, she has developed her own personal Rosetta stone to decipher my sometimes obscure sketches, miraculously transforming them into works of art. She has grown to love rocks from space as I do.

Preface to the First Edition

In 1990, David Flaccus, publisher of Mountain Press, visited his daughter at her horse ranch in Bend, Oregon. A neighbor rancher told him about an astronomer living in Bend who might be interested in writing a book on meteorites, a book with little or no scientific prerequisites attached. The only prerequisite would be a more-than-casual interest in rocks, especially rocks from space. I was the one to whom the neighbor referred, and the source of the rumor had been a student of mine.

David contacted me and we scheduled a meeting at the ranch. What Dave didn't know was that he was getting a two-for-one package. My wife, Dorothy Sigler Norton, is a highly competent scientific illustrator with more than a casual interest in astronomy. She would be the illustrator for the book. Having an in-house illustrator is beyond the dreams of most authors. Dave was delighted and hired both of us that day. I never did find out why Dave wanted to publish a book on such an esoteric subject. Years earlier, Mountain Press had become known for their Roadside Geology books, and perhaps Dave thought it would be an interesting extension to that popular series.

After our initial meeting, we set our sights on perhaps a year for writing and illustrating, and another few months to get the manuscript into final form. How naïve we were. The schedule neglected a few important facts. We still had to make a living during the writing period; and like all active fields of science, meteoritics refused to stand still while I wrote the book.

As the months and years rolled by, the scope of the book widened and deepened. What seemed at first to be a rather narrow subject turned out to be extraordinarily broad. Meteoritics touches so many scientific disciplines today that it was impossible to treat it adequately without at least a glimpse into these diversified fields. Who would have guessed that the study of meteorites would involve the geology of glaciers, the orbits of comets and asteroids, the U-2 spy planes, a spacecraft to Jupiter, and the death of the dinosaurs? This is only half the story. Science, with all of its complex ramifications, is a human activity. The history of our pursuit of meteorites is littered with not-too-admirable human traits involving theft, hoaxes, and courtroom trials driven by personal passions, scientific one-upmanship, and personal monetary gain. In contrast, the study of meteorites is intimately involved with the search for our beginnings. This can only be thought of as one of humankind's noblest endeavors. A book about meteorites would not be worth writing or reading without involving the human element.

While writing this book, I was constantly bombarded with new facts and figures, sometimes conflicting with what I had written just days earlier. Certain areas are very fluid. For example, estimates of the frequency of Earth encounters with sizable meteoroids varies from author to author. Days after I returned the edited manuscript to the publisher, I received word that between 1975 and 1992 United States military satellites, part of the early-warning system, detected meteoroids up to several yards across entering the atmosphere from space. During seventeen years of secret observations, 136 explosions high in the atmosphere were detected. This is an average of eight per year, which increases earlier estimates by at least ten times. Three of these yielded the energy equivalent of a 20-kiloton nuclear blast. They were all apparently stony objects, which usually self-destruct high in the atmosphere. Had they been iron, we would have seen some new craters on Earth.

To prepare a book like this requires the efforts and cooperation of many individuals: the scientists who uncover new knowledge at an increasingly prolific rate; the rock hound, the collector, or the average citizen who finds new specimens to help resupply the world's meteorite coffers; and the museums and research centers that disseminate new knowledge to the world. All have contributed to this book through their personal experiences, their research, their writings, their photographs, and their specimens. I take no credit for their contributions but am grateful that they chose to support this work.

Heading the list is Dr. Ursula Marvin, of the Smithsonian Astrophysical Observatory, who kindly sent me photographs of the oldest and newest meteorites on the planet—the Ensisheim meteorite, dating from A.D. 1492, and meteorites she helped to find nearly five hundred years later in 1982 on the Antarctic ice shelf. Linda Schramm, of the Department of Mineral Sciences, Division of Meteorites, at the Smithsonian Institution, provided several of the hard-to-find illustrations, as did the American Museum of Natural History's Department of Library Services. The chapter on Harvey Nininger benefited a great deal from the help of Margaret Huss, Harvey Nininger's daughter. She provided many of the photographs, and what she couldn't provide, Dr. Chuck Lewis, at Arizona State University, could and did. In the same chapter, quotations appear from Nininger's book *Find a Falling Star,* reprinted by permission of Paul S. Eriksson Publishers. Robert A. Haag gave me free rein to photograph his vast collection of meteorites. Photographs of many of his meteorites appear in this book.

There are many others I wish to thank for sending me their valuable photographs and comments, all of which were gratefully received. Photographic searches are grueling experiences, and these people and their respective organizations lightened my task. With my thanks, I acknowledge them here: the National Air Photo Library of Canada; the National Aeronautics and Space Administration; Dr. Donald E. Brownlee, University of Washington; Dr. Linda Watts, Planetary Science Branch, Johnson Space Center, Houston; the National Optical Astronomy Observatory, Tucson, Arizona; Dr. William Livingston, National Optical

Astronomy Observatories, Tucson, Arizona; Mr. Drew Barringer, president of the Barringer Crater Company, Decatur, Georgia; Meteor Crater Enterprises, Flagstaff, Arizona; Dr. Rhian Jones, Institute of Meteoritics, University of New Mexico; Mr. Hal Povenmire, Florida Fireball Patrol; Dr. David Raup, University of Chicago; Dr. Walter Alvarez, University of California, Berkeley; Dr. Tom Gehrels and the Spacewatch project, University of Arizona; Max Planck Institut für Aeronomie, Katlenburg-Lindau, Germany; Walter Zeitschel, Hanau, Germany; Ronald N. Hartman, Mt. San Antonio College, Walnut, California; Allan Parker and James Baker, Visalia, California; Arizona Historical Society; Thomie Davis, Tucson, Arizona; Mr. Gordon Nelson, Tucson, Arizona; and Mark Zalcik, Edmonton, Alberta.

This book was read in various draft stages by people with and without technical expertise in meteoritics. I especially wish to thank Ms. Dolores Hill of the Lunar and Planetary Laboratory, University of Arizona, for reading and providing helpful criticisms on the mineralogy and structure of meteorites. Dr. Roy S. Clarke Jr. of the Smithsonian Institution read the section on the Old Woman meteorite. Mss. Carolyn Horton and Susan Fisher of Bend, Oregon, labored through the first draft and provided exhaustive criticism and suggestions. The second and third drafts were tackled ruthlessly, rigorously, and effectively by the editorial staff at Mountain Press; it is a better book as a result of their guidance and persistence.

Unlike most authors' spouses, who tolerate many months lost to the computer and the reference library, Dorothy enthusiastically jumped in with brushes and pencils. She studied each chapter as it came from the word processor, considering the illustrations that would make the book more readable and enjoyable. We had many conferences and sketching sessions. It was a thrill to see the beautiful cover painting magically appear on her canvas. She showed patience when I didn't and tolerance when I became at times disagreeable. If nothing else, the book is a masterpiece of scientific art and illustration because of her. I cannot thank her enough for her considerable labors. I can only continue to love her.

Finally, I offer this book as a tribute to David Flaccus, founder of Mountain Press Publishing Company. David did not live to see the completion of this, his last book. A month before his death, Dorothy completed the cover painting and hastened to show it to him. Dorothy had provided paintings for many book covers for Mountain Press, but this one was special to him. His smile told her he was pleased. With the completed cover and the manuscript nearing completion, he knew it was well on its way. His original working title, *Rocks from Space*, has remained unchanged.

A rock-strewn plain in Ares Valles on Mars, photographed by the Pathfinder lander within days of its landing on July 4, 1997. –NASA/JPL photo

The asteroid 253 Mathilde photographed by the NEAR spacecraft during its close encounter on June 30, 1997. –NASA photo

Introduction

The alien craft roars down out of the sky, a glowing fireball leaving a trail that marks its path. A heat shield releases and falls thousands of feet to the plains below. Parachutes deploy, slowing the acceleration of the craft. A set of huge beach balls expands around the precious cargo. This would not be a classical landing with roaring rockets and attitude controls. Just beach balls. No one sees it hit the surface of Mars, bouncing once . . . twice . . . sixteen times before rolling to a stop.

It is a frigid, still night. The craft waits for the first light of the Martian dawn to announce that humans have once again returned to Mars, the first time since *Viking* landed more than twenty years ago. The beach balls deflate and retract. Three folded petals unfurl like a flower opening around the lander, and ramps position themselves. In the center sits a little toylike roving vehicle called Sojourner, ready to roll down the ramp and out onto the surface of Mars. The mother ship called *Pathfinder* sends a signal toward a blue star 135 million miles away. "I'm okay," it tells Earth.

Within hours of the landing, people around the world turn on their computers and access the Internet. On the screens before us appears the rock-strewn surface of Mars; we are participants in the exploration, along with scientists at the Jet Propulsion Laboratories in Pasadena. We watch as Sojourner traverses the landing site, making the first historic in situ analysis of Mars rocks.

Completely overshadowed by the spectacular *Pathfinder* landing on Mars was another wonderful achievement. Months earlier, a small spacecraft called *NEAR,* an acronym for *Near-Earth Asteroid Rendezvous,* was launched toward the asteroid belt between Mars and Jupiter. Its destination was the main belt asteroid 433 Eros, a cylindrical-shaped hunk of rock about 25 miles long and 8 miles wide. Early on its three-year journey to Eros, it passed by another asteroid called 253 Mathilde, and made over five hundred images of the battered and fragmented rock. This is the fourth asteroid directly photographed by spacecraft.

Why highlight a Mars landing and an asteroid flyby in a book about meteorites? Simply because we are probing the sources of the meteorites. This search actually began more than twenty-five years ago during the Apollo Moon program when astronauts collected 800 pounds of Moon rocks from the lunar surface. Using these authentic examples, scientists have identified terrestrial Moon rocks that made their way to Earth as meteorites long before humans reached the Moon. In essence, we are journeying out to visit the meteorites, instead of waiting for meteorites to come to us. In meteorite collections on Earth, there are about a dozen rocks we believe came from Mars. We can prove it by remotely analyzing Mars rocks using probes on the Sojourner rover. It is the next best thing to being on Mars and examining the rocks ourselves.

The case for including asteroids in *Rocks from Space* is quite clear. All the asteroids pictured in this book are fragments of larger parent bodies. They are giant chips. The impacts that produced these fragments also produced smaller chips that somehow make their way to Earth as meteorites. The asteroid belt is the home of most of the meteorites. Asteroid 253 Mathilde, especially interesting, is a primitive C-class asteroid thought to be composed of carbonaceous matter and hydrated minerals similar to carbonaceous chondrite meteorites. The other two asteroids photographed by the *Galileo* spacecraft on its way to Jupiter are S-class, similar in composition to ordinary chondrites that comprise the great majority of all our meteorites.

To hold in your hand something alien, created before Earth existed, is irresistible. Once, while giving a lecture to a group of vacationers on a cruise ship, I casually handed a meteorite to someone in the first row and asked that it be passed around so everyone would have a chance to hold a rock from space older than Earth. The reaction was electric. People clambered to touch it and a riot nearly ensued. The allure of these ancient rocks is universal. They speak in all languages. To hold something older than Earth is to return to creation itself. Some meteorites are examples of the earliest solid bodies, primitive and pristine. They have not changed appreciably in the past 4.5 billion years. Locked within them are clues to the formation and early evolution of the solar system, Earth, and indeed, life itself. Other meteorites show indisputable histories of heating, metamorphism, and recrystallization. These in turn tell of parent bodies, subplanetary masses that must have existed (and perhaps still do) somewhere in the asteroid belt, where no major planet ever formed.

As we learn more about the constituents of the solar system, we better understand the interrelationships between the various parts. When Thomas Jefferson was president, scientists of the day considered phenomena such as meteors and fireballs and objects such as asteroids, meteorites, and comets to be completely independent. Only after nearly two hundred years of study can we say with certainty that they are all related, and one cannot be studied effectively without introducing the others. This book reflects that notion. Although it is primarily a book about meteorites, it also deals with their asteroid "parents" and their comet "cousins."

This book tells the story of meteorites and our attempts to read their histories. The notion that rocks fall to Earth from space was slow in coming, but once we accepted these stones as true celestial bodies, the stage was set for the science of meteoritics. The struggle to understand meteors, fireballs, and meteorites makes entertaining reading, and you'll find bits of that history scattered throughout this book. We really could not make much headway until the tools of the trade were developed. We needed a working knowledge of basic sciences such as chemistry and physics, especially as they relate to mineralogy, crystallography, and petrology, for the study of meteorites is, in essence, the study of rocks and minerals applied to celestial bodies. Today, meteoritics has become an interdisciplinary science, borrowing from sciences that, in the past, seemed to have little in common.

The modern study of meteorites has led us down different paths. That not all celestial objects are benign has been a real eye-opener. We now realize that large meteorites, asteroids, and comets have impacted Earth in the past, sometimes with devastating results. Since the time of Charles Darwin, scientists have considered the evolution of species to be a slow, gradual change brought about by natural selection as animals and plants respond to the outside environment. Within the biota, whatever was best fit to deal with the environment would survive. This gradualism formed the cornerstone of not only biology and paleontology but also geology. Geological processes are also governed by slow change, not easily seen in many human lifetimes. Such changes occur on a different time scale. Now we know that cosmic impacts are yet another geologic process but one that occurs virtually instantaneously.

Cosmic impacts not only change the face of the land but, more important, have led to the extinction of countless species of plants and animals over geologic time, playing a vital role in the evolution of life. Like the process that produces them, these extinctions occur rapidly and catastrophically. A growing body of evidence suggests that the rise of mammals, distant precursors of you and me, was made possible by the impact of one or more giant meteorites at the end of the Mesozoic era, 65 million years ago. That impact eliminated the early mammals' competitors, including the dinosaurs.

The impact of Jupiter by Comet Shoemaker-Levy in 1994 confirms that catastrophic events are still taking place in the solar system. And there are more flying mountains and rocky icebergs out there. Big ones. They could change the face of life on Earth today or in the future as others did in the past. Is anyone watching?

Part I

FALLS, FINDS, AND CRATERS

Comet Hale-Bopp as it appeared on March 31, 1997, displaying a broad, diffuse dust tail and a narrow straight ion tail. Comet dust is the principle component of interplanetary dust particles found along the solar system plane. Hale-Bopp put out 200 times more dust than Comet Halley did during its appearance in 1986.

–Lynn Carroll photo

ONE

Shooting Stars, Meteor Showers, and Fireballs

"Empty" Space

The vacuum of space is a myth. Space is never empty. There is always something occupying it, bounding it. Even in the farthest reaches of our Milky Way Galaxy, space is thinly veiled in cold gas and tiny dust particles. Recent observations have even revealed gaseous matter in extragalactic space.

Zodiacal Light

In our solar system, space is occupied by dust-sized particles and the flood of charged atomic particles streaming from the Sun called the solar wind. Astronomers have long recognized a soft pyramidal glow of light often seen extending from the western horizon long after sunset on moonless nights from early February to early March. The same pyramid of light is visible again above the eastern horizon well before dawn from late September through October. Astronomers have named it the zodiacal light, since the glow occupies that part of the sky in which the familiar constellations of the zodiac are found. That is the plane of our solar system, or the ecliptic plane.

Greek astronomers recognized the phenomenon more than two thousand years ago. Arab astronomers were also aware of it. They called it the tall twilight, implying that it extended far above the dawn and twilight glow of the Sun. In the *Rubaiyat* (A.D. 1100), Persian poet and astronomer Omar Khayyám refers to the morning zodiacal light as a "false dawn." The first clear description of zodiacal light in the Western world was published in the mid-seventeenth century in London. By the 1680s, French astronomers correctly theorized that zodiacal light was sunlight reflecting off dust-sized particles scattered along the ecliptic plane.

I clearly remember the first time I became aware of the zodiacal light, at Kitt Peak National Observatory, 40 miles west of Tucson, Arizona. I was preparing to use one of the telescopes that night and opened the dome shutters at sunset. It was late February and I could see the winter star Aldebaran shining brightly in the fading blue sky. Astronomers usually begin observing as soon as twilight fades, and I waited impatiently for night to come and the sky to darken. It didn't. After more than an hour of waiting I began to wonder if the observatory clock was wrong. At that latitude, astronomical twilight should begin about an hour after sunset. The opened dome was facing west and I could see a yellowish light filling the opening. I stepped outside, looked to the west, and saw something I had never noticed before.

There, rising from the western horizon almost to the zenith was a broad cone of yellowish light that was definitely not the glow of twilight. The glow was brightest closest to the sunset point and gradually faded with increasing altitude above the western horizon. At its broadest point it extended about 10 degrees on either side of the ecliptic but tapered rapidly higher above the horizon. It was nearly as bright as the summer Milky Way and it lay along the zodiac, which from my latitude of 32 degrees was tilted somewhat to the south at that time of year.

The zodiacal light, photographed about one hour after sunset on February 14, 1996, extends to 50 degrees above the western horizon. The light wedge straddles the ecliptic plane seen tilted to the left. The planet Venus is the bright object near the center of the wedge, and Saturn, also in the light, is near the horizon.

Interplanetary Dust

Seventeenth-century French astronomers correctly explained the zodiacal light as reflected sunlight. Concentrated along the ecliptic plane within the planetary orbits are countless billions of micron-sized particles in orbit around the Sun—one micron is one-thousandth of a millimeter. These particles, called interplanetary dust particles, are the solar system's smallest meteoroids. They are composed of silicate minerals like those found in Earth's rocks. Astronomers once thought of them as remnant material left over from the formation of the planets. About a century ago, however, astronomers finally recognized that micron-sized particles could not remain in orbit around the Sun for long. The pressure of the Sun's radiation is sufficient to push the tiniest particles out of the solar system. The larger particles are slowed in their orbits and spiral gradually into the Sun. Thus, the pressure of solar radiation could sweep the solar system clean of micrometeoroids within a few hundred thousand years, a brief period in the solar system's 4.6-billion-year history. The exact length of time depends on the masses of the particles. Further contributing to the depletion of micrometeoroids are the planets, which act like giant vacuum cleaners, constantly sweeping them up. So there must be a source of interplanetary dust particles to replace those that are continuously being lost.

There are two possible sources. Asteroids are one source. The asteroid belt between the orbits of Mars and Jupiter is a zone roughly 100 million miles wide, in which thousands of rocky bodies orbit the Sun. They range in diameter from a few hundred feet to more than 500 miles. Their orbits cross and collisions occasionally occur. These collisions break the asteroids into smaller pieces, which in turn collide, creating ever smaller bodies. This continual fragmentation of asteroids creates micron-sized and larger chunks, adding to the particles responsible for the zodiacal light.

A second source for micrometeoroids is comets. They trail dusty tails as they rush toward the Sun, and most comet orbits take these icy bodies far outside the plane of the solar system. Comets with highly inclined orbits cannot add much to the dust that reflects the zodiacal light, but periodic comets whose paths stay close to the ecliptic plane probably do account for substantial amounts of zodiacal dust.

Interplanetary dust particles scatter sunlight efficiently, especially if they are dense. We have all seen the effects of sunlight striking a frosted glass window—the window seems to light up as sunlight is scattered forward through the glass. This forward scattering is precisely the same mechanism causing the glow of zodiacal light when particles are between us and the Sun. The glow is brightest closest to the sunset point and fades rapidly with distance from that point.

Another ephemeral light in the eastern sky is visible from Earth. Considerably fainter than the zodiacal light, it also involves these tiny

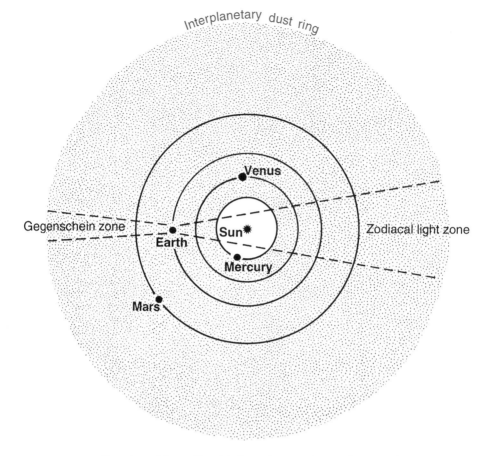

Location of the zodiacal light and gegenschein, relative to the Sun and Earth, viewed from above the ecliptic plane.

members of the solar system. This time the particles scatter sunlight along the ecliptic, reflecting light from the Sun back toward us. This back-scattering of sunlight occurs well beyond Earth's orbit because Earth is between the particles and the Sun. This fainter light is called gegenschein and is an extension of the zodiacal light along the zodiac. Under the darkest, clearest skies, some observers have seen a faint band connecting the zodiacal light to the gegenschein.

The best time to see zodiacal light is early spring or early fall; the ecliptic projects higher in the sky above the western horizon in the spring and the eastern horizon in the fall. In the Northern Hemisphere, it is best to observe the zodiacal light in latitudes between 35 degrees and 25 degrees. From these latitudes, the ecliptic stands most nearly vertical and the zodiacal light appears to extend straight up from the horizon. A clear sky with little haze or dust and no moon is required. Under the best conditions, you should be able to see the light extending about 10 degrees on either side of the ecliptic, and the glow should reach nearly to the zenith.

The *Pegasus* satellite, launched in 1965, was the first spacecraft to detect particles of interplanetary dust directly. It was equipped with enormous winglike structures to sweep space along its orbit. As the satellite orbited Earth, detectors in the wings sensed impacting particles. Three satellites were launched that year to study the distribution, mass, and velocity of the particles for applications to the American manned space program. *Pegasus* detected many dust-sized particles, but far fewer than expected. Interplanetary particles proved to be no hazard to astronauts, either in Earth orbit or during manned lunar missions. Other spacecraft, such as the Vikings that reached Mars in 1976 and the Voyagers that traveled to the outer planets a decade later, suffered no serious impacts with the dust of space.

The most definitive study of the distribution of interplanetary particles came in late January 1983, when a new kind of satellite greatly extended our ability to detect the dust visually. The *Infrared Astronomical*

Infrared Astronomical Satellite (IRAS) *in Earth orbit.* –NASA

Satellite (IRAS) was designed to detect heat radiating off relatively cold bodies. Armed with a battery of infrared detectors, *IRAS* was so sensitive that it could detect heat from a single dust speck 2 miles away! These particles absorb light and reradiate much of the energy in the infrared wavelengths. In the first few weeks of operation from its 560-mile-high orbit, *IRAS* sensed heat radiating from dust along the ecliptic

11

within the inner solar system, close to Earth. This was the dust responsible for the zodiacal light. *IRAS* also detected heat from a donut-shaped ring of dust between 200 and 300 million miles from the Sun. That placed the dust ring well within the asteroid belt, confirming our suspicions that colliding asteroids are the source of this space dust. In addition to asteroid dust, several distinct dust bands found in this ring seem to correspond to the known periodic comets whose paths take them close to the ecliptic plane.

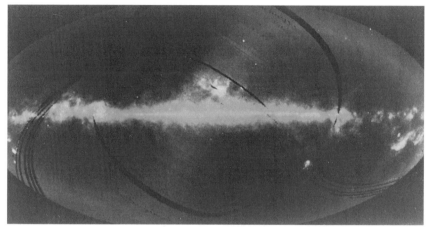

IRAS *view of the entire sky. The bright central band is the result of clouds of dust along the plane of the galaxy. The faint S-shaped band above and below the plane is composed of dust along the plane of the solar system.* –NASA

Interplanetary Dust Rains on Earth

After years of observing light and heat from these dust specks, scientists have finally succeeded in collecting them. The first samples were taken in 1970 using collectors on balloons. In 1974, NASA began using high-flying U-2 aircraft to capture samples filtering into the upper atmosphere from space. These tiny particles, only a few hundredths of a millimeter across, are composed of delicate, fluffy aggregates of tiny mineral crystals. They are carbon-rich silicates similar in composition to primitive stony meteorites but different enough to place them in a separate category. They may be cometary dust.

A different type of cosmic dust particle is retrieved from the ocean floor. These were first collected more than a century ago from sea-floor sediments and separated out with magnets. Now, magnetic collectors are towed across the ocean floor, and they have recovered curious spherical particles, the remains of micrometeoroids that melted in the

Brownlee particle photographed by a scanning electron microscope. The particle, 9 microns in diameter, is composed of tiny interlocking crystals, with an elemental composition similar to that of carbonaceous chondrite meteorites.
—Courtesy of Donald E. Brownlee, University of Washington, and NASA

atmosphere. These are the very specks that produce the flashes of light we see shooting across the sky at night. The melted particles recondense as tiny spherules that slowly drift to Earth. They are made of tightly packed crystals of olivine, magnetite, and glass, and measure from less than 1 millimeter to 3 millimeters in diameter (0.039-0.117 inch).

Astronomer Donald E. Brownlee from the University of Washington is a leading authority on cosmic dust. Today these dust specks bear his name: Brownlee particles. He has recovered thousands of dust particles both from the deep ocean floor and from high in the stratosphere. These samples of the smallest meteoroids in the solar system are preserved in a growing collection at the Lyndon B. Johnson Space Center in Houston, Texas, where the nation's lunar rocks are stored. Researchers from around the world may request samples for study from the Cosmic Dust Catalogue, which pictures hundreds of these particles photographed through a scanning electron microscope.

From these studies, scientists estimate that only a few hundred particles per cubic mile exist near Earth. Even at this surprisingly low density, Earth collects between 35,000 and 100,000 tons of this dust annually.

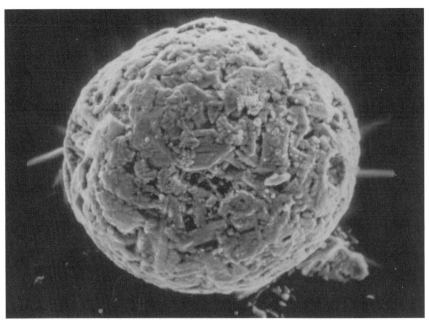

A 300-micron-diameter sphere collected from the mid–Pacific Ocean floor. The tightly packed crystals have elemental compositions dominated by iron, magnesium, and silicon. Sulfur and carbon were probably lost when the meteoroid melted in the atmosphere. –Courtesy of Donald E. Brownlee, University of Washington

Meteors

People often confuse the terms meteor, meteoroid, and meteorite. The solid body that survives its fall to Earth is a meteorite. While still in space, the body is a meteoroid. A meteor is an atmospheric phenomenon produced by a tiny particle, probably comet dust, that enters Earth's atmosphere at high speed, burns briefly, and then disappears. Individually, the dust speck and the brief streak of light it produces are commonly though incorrectly used synonymously with the word meteor.

Meteors were recognized as atmospheric phenomena in Aristotle's day (350 B.C.). The word comes from the Greek word *meteoros,* or *meteora,* which means things that are lifted high in the air. The Greeks, however, did not associate meteors with astronomical objects. They thought astronomical bodies could never make physical contact with imperfect Earth, since the heavens represented celestial perfection, which could not exist on Earth.

These tiny cometary particles enter Earth's atmosphere at speeds commonly exceeding 26 miles per second. In a matter of seconds, the small bodies collide with millions of air molecules. The high-speed collisions rapidly heat up the particles. Reaching temperatures of over 2,000 degrees Fahrenheit, the particles quickly burn up, leaving a momentary streak of light in the night sky. These small particles last only a second or two before vaporizing.

For an average meteor to be seen from Earth, it must be within 100 miles of the observer. Most of this distance is in altitude. Meteors first appear when the meteoroid is approximately 60 miles above Earth, where the atmospheric density is great enough to heat the meteoroid to incandescence. Using this distance as an average, we can calculate the total number of meteors potentially visible to observers worldwide: On any given day, the number probably exceeds 25 million.

It is a surprising fact that most meteors cannot be seen by the unaided eye. Often while observing through telescopes, I have seen faint meteors rushing through the field of view. Their yellow color indicates that the bodies producing them have much smaller masses and thus heat to lower temperatures than meteors visible to the human eye, which typically glow a bright white. If we include the use of binoculars or small telescopes, which allow us to see meteors one hundred times fainter than the faintest star visible to the unaided eye, the observable number of meteors on any given day increases dramatically to 400 billion.

Most meteoroids range in size from a few thousandths of an inch in diameter to the size of a pea. The average meteoroid is about the size of a grain of sand. Particles of this size generally burn up in the atmosphere. The pea-sized meteoroids have a better chance of surviving their fiery passage but usually they, too, are consumed. Their larger size extends their longevity, producing some of the brightest meteors. They go out in a blaze of glory.

Curiously, the smallest micrometeoroids are the most likely to survive their atmospheric passage. Their size, though small by meteoroid standards, is large compared with their total mass, and a large surface area makes a very efficient heat radiator. They can radiate the heat of entry rapidly enough to prevent their small masses from vaporizing. Once they have lost their cosmic speed, they may remain suspended in Earth's upper atmosphere for years before filtering down. While there, they may act as condensation nuclei around which water vapor freezes. The frozen particles produce eerie bluish clouds some 50 miles high, so high that they capture the light of the Sun after sunset. These clouds, called noctilucent clouds, are best seen in the summer near the Arctic Circle (See color plate I).

Sporadic Meteors

Sporadic meteors are so common that you can easily see three or four an hour almost any evening under a clear, moonless sky. They may appear from any part of the sky and are not predictable.

A sporadic meteor near the north celestial pole. Meteor shows several explosive events in its train. Earth's rotation causes stars to appear as short arcs.
—Peter Zana photo, courtesy of Hal Povenmire, Florida Fireball Patrol

Unlike planets and most asteroids, these tiny meteoroids do not show a preference for the ecliptic plane. They can approach Earth from any direction and usually travel at speeds far greater than Earth's orbital speed, 18 miles per second. Speeds greater than 26 miles per second are common. Their entry speeds vary according to the direction they are traveling.

If we could look down on Earth's North Pole, we would see it traveling around the Sun in a counterclockwise direction while Earth is rotating on its axis. From noon to midnight we are on the trailing side of Earth, facing away from its orbital direction. Meteoroids traveling at 26 miles per second in the direction of Earth's motion must catch up to Earth during this time. They strike Earth at 26 miles per second less Earth's 18 miles per second orbital speed, or 8 miles per second—a relatively slow speed. As the night wears on, our position on the rotating Earth slowly changes. By midnight, we begin to enter the morning sky and face the direction of Earth's orbital motion, at which point the meteor situation changes. In the early evening position, the great bulk of Earth shielded us from meteoroids traveling in a direction opposite our own. But after midnight, our side of Earth is in full view

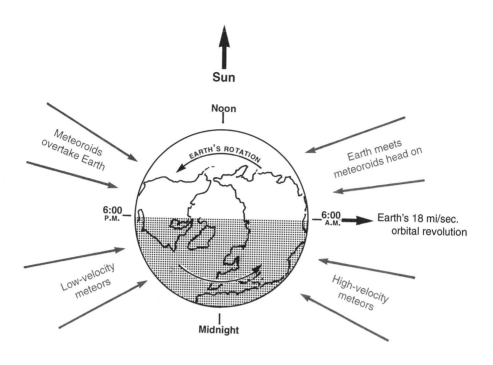

Meteoroids encounter Earth at different speeds, depending upon the meteor's path relative to Earth's leading and trailing sides.

of particles encountering it face-on. The meteoroid velocities must then be added to Earth's orbital motion, that is, 18 miles per second plus 26 miles per second—an incredible 44 miles per second. Meteoroids striking Earth's atmosphere at these high speeds account for the brighter meteors, usually seen from midnight to dawn (no one ever said stargazing was convenient!).

More meteors are visible between midnight and dawn because at that time two sources exist: those meteoroids striking Earth head-on, which we did not see in the early evening, and those meteoroids crossing Earth's orbit at an angle, allowing Earth to catch up. You can observe sporadic meteors any time of year from midnight until dawn.

Meteor Showers

For centuries, observers have noticed a periodicity to the occurrence of some meteors, which appear at predictable times of the year. Unlike sporadic meteors, which may originate from any direction in the sky at any time, predictable meteors appear to radiate from a specific region of the sky on specific evenings of the year. They also occur in greater numbers than sporadic meteors. It is not unusual to count a dozen or more every hour. Occasionally, thousands of meteors rain from the sky in an hour, giving credence to another popular term, meteor storm.

Enter the Comets

A great moment in nineteenth-century astronomy came when mathematicians and science historians joined astronomers to solve the mystery of periodic meteor showers.

Observations of comets and calculation of their orbits were big business in nineteenth-century astronomy. More than a century earlier, Isaac Newton had derived the mathematics and gravitational laws necessary to calculate the orbits of planets, asteroids, and comets. With this powerful mathematical tool, astronomers could make observations of newly discovered comets and calculate their orbits around the Sun.

Astronomers proved that most comets have extremely elongated elliptical orbits, which send them out well beyond the last planet after they pass around the Sun, not to return again for thousands of years. A few comets have much shorter elliptical orbits, giving them relatively short periods of revolution that allow them to return to the Sun within a periodic cycle of a few years to a century or so. These are the periodic comets.

That fact has a direct bearing on meteor showers. Comets contain frozen water, carbon dioxide, carbon monoxide, and other volatiles mixed with stony debris. They have been called dirty snowballs, which

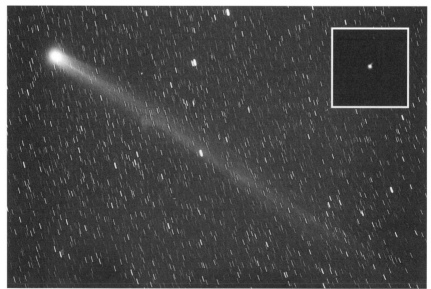

Comet Hyakutake at its closest approach to Earth, less than 10 million miles, on March 24, 1996. The straight ion tail seen here stretches over 10 million miles. Inset: A high-resolution image near the 1-mile-diameter icy nucleus shows a single jet or geyser spewing out gas and dust, continually adding to the growing tail.
–Lynn Carroll photo

The icy nucleus of Comet Halley, photographed by the Halley Multicolor Camera aboard the European Space Agency's Giotto *spacecraft from a distance of about 6,000 miles, March 13, 1986. Two active geysers spew gas and dust into space. The dark surface shows several impact craters.*
–Dr. H. U. Keller photo,
Max Planck Institut für Aeronomie,
Lindau Harz, Germany

well describes them. Comets spend most of their time in the frozen reaches of deep space. Most comets entering the inner solar system remain for only a few months. Sharp-eyed amateur astronomers usually spot them near the orbit of Mars. As a comet rushes toward the Sun, its solid body warms, and its ices sublimate to gases. This releases the fine, dusty material locked in the frozen mass since the comet's birth. Cracks develop in the comet's icy nucleus, and jets of sublimating gases transport tons of stony dust-sized particles into space.

March 13, 1986, was a historic day. The *Giotto* spacecraft, sent by the European Space Agency to rendezvous with Comet Halley, finally reached its frozen target. It was scheduled to come within 600 miles of the icy surface. This was unknown territory, as no telescope or spacecraft had ever revealed a close-up of the surface of a comet. While still several thousand miles away, *Giotto* was struck by tiny ice particles from the comet, which disoriented the space probe and turned its cameras away from the comet, but not before the probe had beamed real-time television images of Halley's surface back to Earth. The images revealed a peanut-shaped body covered with charcoal-gray material– probably the very dust we have been discussing. Two large geysers actively spewed out jets of water vapor and dust. The solar wind pushes much of this water vapor and dusty material behind the comet as it

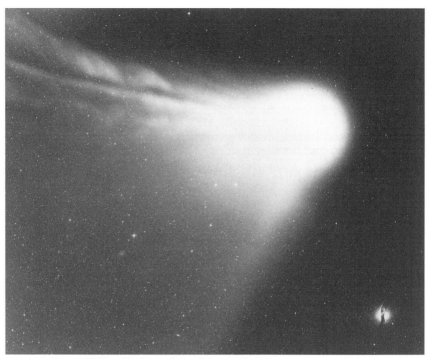

Comet Halley displays a broad dust tail as it passes across the Southern Hemisphere on April 15, 1986. The galaxy Centaurus A is visible in the lower right. –National Optical Astronomy Observatory

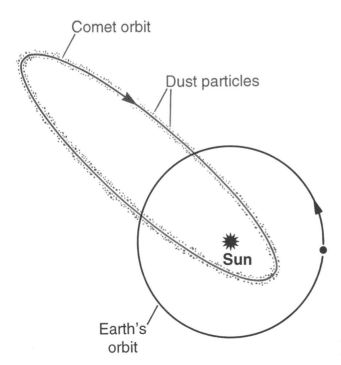

Comet orbit

Dust particles

Sun

Earth's orbit

A swarm of meteoroids distributed uniformly along a comet's orbit. They orbit the Sun with precisely the orbital period of the comet.

approaches the Sun, forming a diaphanous tail that stretches for millions of miles across the inner solar system.

This dusty material scatters more or less uniformly along the comet's orbit. If the comet has an extremely elongated elliptical orbit, it will not linger but will pass out of the inner solar system within a few months after rounding the Sun. The debris left behind will become part of the dusty material that pervades the solar system today, little of it reaching Earth. If the comet has a relatively small, elliptical orbit like Comet Halley, the solid material will remain in the solar system and continue to move along the orbit long after the comet has passed by the Sun.

If a comet enters the inner solar system and crosses Earth's orbit, there is a good chance the particles will be seen every year at the same time. Comet Halley, for example, encounters Earth's orbit twice during

Comet Halley's orbit crosses Earth's orbit twice each year. Earth meets the outgoing dust particles on May 4 and the incoming particles on October 21, resulting in moderate meteor showers on those two nights.

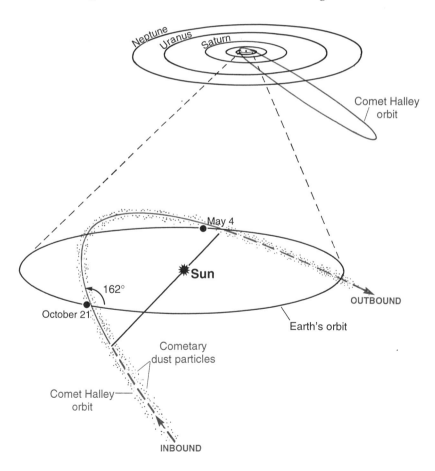

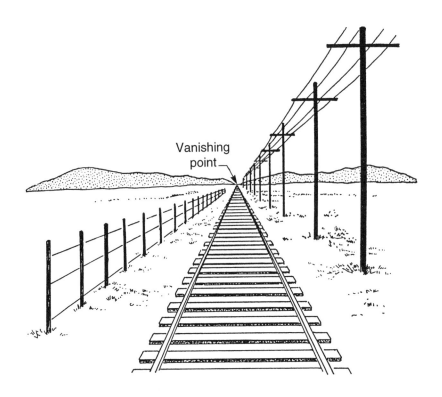

*The optical illusion of railroad tracks appearing to converge at a
vanishing point is analogous to the radiant point of a meteor shower.*

its passage around the Sun. The comet's orbit is inclined to the plane
of Earth's orbit by 162 degrees. It crosses Earth's orbit traveling south
to north when approaching the Sun and north to south when receding.
On October 21 each year, Earth passes through the inbound stream of
Halley's tail particles, and on May 4 the following year, Earth passes
through the outbound particle stream. Twice a year, nearly six months
apart, Earth encounters tiny meteoroids from the tail of Comet Halley,
which appear as fast-moving meteors, with an expected frequency of
between twenty and twenty-five per hour. Comet Halley orbits the Sun
in a retrograde direction, that is, in a direction opposite the motion of
Earth and other planets. For this reason, Earth encounters Halley's
particles head-on, so the speed of Halley's meteors is quite high—nearly
40 miles per second.

Where to Look for Meteors

Shower meteors come from specific regions of the sky. The particles are traveling together in space in the same direction along a
common orbit. They strike Earth's atmosphere at nearly the same position along Earth's orbit, and consequently we see their meteor trains

each year against the same background of stars. And though the meteoroids are traveling parallel to each other, their meteor trains appear to diverge from a small area of the sky, called the radiant point.

This phenomenon is only an illusion, similar to the familiar convergence of railroad tracks. The tracks are parallel but they appear to come together (or vanish) at a point in the distance. Similarly, meteor trails extend from the radiant point and diverge toward the observer. The radiant point is about 60 miles above Earth, where the meteoroids encounter denser atmosphere and burn up. Since the radiant point is always in the same small area of the sky for a given meteor shower, the shower is named for the constellation in which it appears. Thus, on May 4 the radiant point of Comet Halley meteors is in the zodiacal constellation of Aquarius near the star Eta Aquarii, and the meteor shower is called the Eta Aquarids. On October 20–21, Halley meteors appear to radiate from the constellation Orion, and are called the Orionids.

If a comet is very old, having orbited the Sun thousands of times, its debris particles will gradually spread out uniformly along the comet's orbit, as with Comet Halley. If the comet is new, having only recently begun to orbit the Sun, the particles left behind in the comet's orbit are more likely to be bunched into a compact swarm. The swarm will continue to orbit the Sun in the comet's orbit. Earth could encounter

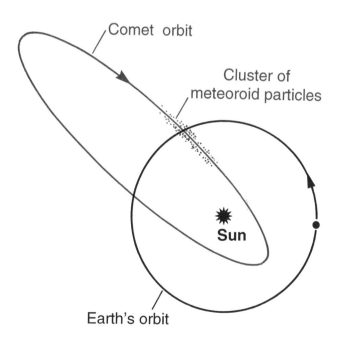

Meteoroid particles clustered along a comet's orbit can produce meteor storms if Earth encounters them.

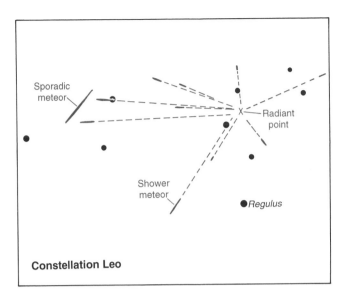

Meteors appear to diverge from the radiant point in the constellation Leo during the Leonid meteor shower, November 16–17, 1966.
—National Optical Astronomy Observatory

Woodcut of the great Leonid meteor shower, 1833.

thousands of meteoroids within a matter of days or weeks after the comet crosses Earth's orbit. Meteoroids clustering along a comet's orbit have produced the great meteor showers witnessed over the centuries.

The Great Meteor Showers

In the twentieth century, North Americans witnessed two spectacular meteor showers, one in 1946 and the other in 1966. The famous Leonid meteor shower occurred on November 16–17, 1966. It appeared to radiate from a point within the sickle-shaped head of Leo, the Lion, a constellation of the spring sky. The Leonids have a long history.

On November 17, 1833, the night sky provided the backdrop for one of the most spectacular meteor showers ever recorded. A delightful woodcut of the great Leonid meteor shower at first glance seems fanciful, surely inaccurate. But judging from eyewitness accounts, the illustration is probably not exaggerated. Within a few hours, well over 200,000 meteors were recorded from one location alone. The radiant point appeared on the zenith, where the head of Leo is found at dawn on November 17. Astronomical historians investigating ancient meteor showers discovered that between A.D. 902 and 1833, a period of 931 years, there were apparently twenty-eight noteworthy showers. If these showers were actually twenty-eight periodic episodes of the same shower, the Leonids were recurrent every 33.25 years. Less spectacular showers occurred a year or two before the main event, showing that it took the meteor swarm two or more years to pass through the intersection point on Earth's orbit. The critical question was: Would another great meteor storm occur thirty-three years after 1833? Astronomers the world over awaited the Leonids of 1866 to confirm their predictions. Right on schedule the Leonids appeared again; the shower was not as spectacular as in 1833, but thousands of meteors once again rained down on happy observers. Another more modest shower occurred in 1867 as the remainder of the swarm passed across Earth's orbit.

After the 1866 Leonid shower, astronomers armed with a more precise position of the radiant point began to calculate a more accurate orbit for the Leonid meteors. They noticed that Comet 1866 I, the first comet to be discovered in 1866 (Comet Tempel), had an orbit identical with the calculated orbit for the Leonid meteor swarm. At last the connection between comets and meteor showers was clear: Comets spawn swarms of tiny meteoroids, producing meteor showers in their wakes. Astronomers investigated other known meteor showers. Comet Halley was recognized as the source of two meteor showers. The famous Perseid meteor shower occurring between August 9 and 12 was linked with Comet 1862 III, Comet Swift-Tuttle, and another November shower was connected to Comet 1852 III, Biela's Comet. Several yearly

YEARLY METEOR SHOWERS

Shower	Maximum Activity	Hourly Count	Velocity (miles/sec.)	Associated Comet
Quadrantids	Jan. 3–4	100	25.0	———
Lyrids	April 21–22	12	28.5	Thatcher (1861 I)
Eta Aquarids	May 3–5	20	39.3	Halley
Delta Aquarids	July 29–30	30	24.8	———
Perseids	Aug. 11–12	60	35.6	Swift-Tuttle
Draconids	Oct. 8–9	variable	12.2	Giacobini-Zinner
Orionids	Oct. 20–21	25	39.8	Halley
Taurids	Nov. 7–8	12	17.5	Encke
Leonids	Nov. 16–17	variable	42.4	Tempel-Tuttle
Geminids	Dec. 13–14	60	20.6	Asteroid Phaeton
Ursids	Dec. 22	10	20.0	Mechain-Tuttle

The 1966 Leonid meteor shower, during which numerous meteors appeared in the course of a few minutes. –National Optical Astronomy Observatory

meteor showers occur, caused by evenly distributed comet particles. Two have major swarms: the Leonids and the Draconids.

In 1899, astronomers anticipated the return of the Leonids. This time they were disappointed. Apparently, Jupiter's gravity had pulled the swarm from Earth's orbital plane. Thirty-three years later, in 1932, the Leonids again failed to appear. They seemed to be lost. But as early as the mid-1950s, astronomers began to note a slow but steady increase in the number of Leonids appearing every November 16–17. By the early 1960s, a few astronomers in America believed the swarm was returning. A bold prediction was made that on the night of November 16, 1966, Earth would again witness the Leonids' spectacular return.

It was a night I will never forget. Every hour from sundown to midnight, I stepped outside to see whether the Leonids had returned. The later it got, the more dubious I became. At midnight, I gave up and decided to turn in; but on the chance the belated shower might occur in the early morning hours, I set my alarm for one o'clock. No meteors. At two, I reluctantly struggled out into the backyard to give the Leonids one last chance, and found the sky filled with "falling stars." Everywhere I looked I saw fast-moving meteors. So many were falling that it was impossible to count them. Nothing like this had occurred since the shower of 1866. There were estimates of over 150,000 meteors falling in the three hours or so before dawn. The Leonids had returned; the 1833 woodcut was no exaggeration.

27

Will the Leonids return in 1999? There is a good chance the meteor swarm will return in that year or a year later, barring perturbations by Jupiter. Observations may be possible one year prior to and one year after 1999. Mark your calendar for November 16–17, 1999, and set your clock for the best show on Earth!

The 1946 Draconid Meteor Shower

Another fine meteor shower occurred in 1946, though not the great spectacle of the Leonids twenty years later. The periodic comet Giacobini-Zinner had crossed Earth's orbit, leaving behind a clump of cometary particles guaranteeing a shower. Earth passed through that clump on October 9. Though a full moon lighted the sky that night, observers in the western United States counted as many as two hundred meteors every hour.

The meteors seemed to radiate from the head of the circumpolar constellation Draco, so they are named the Draconid meteors. Comet Giacobini-Zinner orbits the Sun every 6.6 years, creating the potential for a repeat of the shower every other complete orbit of the comet. In 1933, Earth encountered the comet's debris, but 6.6 years later, when the comet again passed across Earth's orbit, Earth was opposite that point in its orbit. It needed another 6 months to reach the intersection point. By that time, the debris had thinned, so no shower was seen. In 1946, 6.6 years later, the comet and Earth arrived at the intersection of the two orbits at the same time, resulting in the Draconid shower.

It seems reasonable to suppose that on every other orbit of the comet, every thirteen years plus a few weeks, Earth should expect a return of the Draconids. But they did not appear in 1959 or in 1972 or 1985. The comet still exists, but apparently its debris no longer intersects Earth's orbit. So few meteors are seen on October 9 each year that few books now list the Draconids.

The Faithful Perseids

If you want to observe a respectable meteor shower, plan to watch the Perseids. This shower stems from the debris of Comet Swift-Tuttle scattered uniformly along its orbit. Every year, between August 9 and August 12, Earth encounters the particles. If no moon lights the sky and if you are far away from the glare of city lights, you may see as many as sixty meteors per hour, especially between midnight and dawn. The tiny meteoroids are moving as fast as the Halley meteoroids, around 36 miles per second, and leave short, bright trails. They extend from the constellation Perseus, which rises in the northeast just before midnight. A reclining garden chair, a blanket, something warm to drink, and patience are all you need to attend the show. Face northeast and concentrate on the radiant point about 45 degrees above the horizon. The point climbs as the night proceeds into the morning hours and by

A Perseid meteor. –Courtesy of Hal Povenmire, Florida Fireball Patrol

dawn is just northwest of the zenith. It is common to see meteors streaking as much as halfway across the sky from the radiant point, but if you trace their paths backward, they should all intersect in Perseus.

Comet Swift-Tuttle came close to Earth in 1992, adding to the particles along its orbit. This was expected to rejuvenate the shower in 1993. So certain were astronomers, that NASA postponed its shuttle launch and repositioned the Hubble space telescope away from the predicted particle paths. Europeans were given the best chance to see a good shower, since astronomers predicted the swarm would arrive while North America was still in daylight. The world waited in great anticipation and was disappointed. Astronomers were perplexed. The anticipated meteor storm failed to appear. Europeans saw a brief increase to two hundred meteors per hour, which soon subsided. In the United States, the numbers remained around sixty per hour. Even while the shower was in progress, new calculations showed that the storm would actually occur on August 12, 1994, this time favoring viewing from the United States.

The night of August 11–12, 1994, was clear in central Oregon, where several of us settled down for a night of meteor watching. The early hours brought only an occasional meteor. Midnight passed with

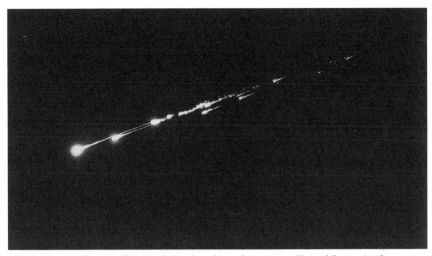

The Peekskill fireball lighted the sky along the eastern United States in the early evening of October 9, 1992. The fireball broke into numerous fragments as it neared the end of its atmospheric flight. Many of these pieces probably reached the ground, scattering over a 50-mile path. –Courtesy of Sara Eichmiller

no significant increase. At about 1:00 A.M., we briefly abandoned our posts and went indoors for some late refreshments and much-needed hot coffee. During our short break, the peak began. It was definitely show time when we returned to the night sky. Over the next hour, we counted at least 125 meteors, at least twice the normal number, among them several brilliant ones typical of Perseids. Some observers recorded a peak of around 250 by 3:00 A.M. It was exciting to see, but not the meteor storm some had predicted.

Although bright moonlight interfered with observations in 1995, meteor counts told a discouraging story. The shower was weakening, and the time of peak activity had narrowed to only an hour or so. During the peak, counts briefly reached 160 per hour. By 1996, the number had fallen to 120. The peaks over the last several years were undoubtedly caused by material deposited in clumps along the orbit of Comet Swift-Tuttle during its near-Earth passage in 1992. For a few years, the Perseids had two peaks; the normal 60 per hour that lasts for a few days and the distinct but diminishing peak that lasted for only an hour or so.

Observers in 1997 reported only about 75 meteors per hour. Meteor peaks are notoriously difficult to predict, and some experts believe that clumps of material are still awaiting passage through Earth's orbital plane. Some predictions now say that new peaks will appear after the turn of the century. This is enough encouragement to keep the dedicated Perseid meteor watchers returning every August, indefinitely, waiting. . . .

Fireballs and Rocks from Space

Meteors, even the bright ones, should not be confused with the rare, intense fireballs that bring lots of phone calls to local police dispatchers. Fireballs are produced by walnut-size to several-foot-diameter chunks of rock and iron that were knocked off asteroid parent bodies by collisions. Particles in an average meteor shower never survive to reach Earth's surface, but most meteoroids that produce fireballs are massive enough to survive atmospheric passage, often explosively disintegrating into several smaller pieces before reaching Earth's surface.

When a sizeable meteoroid hits the atmosphere at cosmic speeds it quickly heats up, melting the stony surface material into a glasslike substance. At the same time, atmospheric gases around the heated mass glow brightly, making the fireball appear much larger and closer than it really is.

The atmosphere acts as a brake and effectively reduces the meteoroid's velocity. Thus, the melting phase typically lasts only a few seconds, not enough time to conduct heat into its interior. When these meteoroids reach the ground as meteorites, they are seldom even warm to the touch and never hot. Frequently, they are quite cold. The relatively poor heat conduction of the meteorite's rocky material coupled with the all-too-brief heating phase assure the survival of the meteorite in a near pristine condition. With these survivors from space, it is finally possible to touch a "falling star."

A daylight fireball streaks over Crater Lake, Oregon, near sundown in the summer of 1992. –Dorothy S. Norton painting

One-ton Norton County meteorite on display at the Institute of Meteoritics, University of New Mexico. –Ken Nichols photo

Redrawn from the first of four woodcuts, this scene depicts the fall of the Ensisheim meteorite. The original print, made in 1492, is at the Universitäts bibliothek in Tübingen, Germany.

TWO

Rocks from Space? Impossible!

February 18, 1948, was a memorable day for those who saw the fiery intruder. Shortly before 4:00 P.M., a great fireball appeared in the cloudless afternoon sky. It was traveling at high speed from southwest to northeast. Two pilots flying a B-29 over east-central Colorado may have been the first to see it. Reports quickly came in from northwest Kansas and southwest Nebraska. A huge ball of fire trailing a long, thin smoke cloud was seen traveling toward the Kansas-Nebraska border. So fast was it moving that no one reported seeing it for longer than five seconds before it passed behind buildings or simply disappeared beyond the horizon. Several people reported hearing explosions followed by a roaring sound similar to the noise of a jet engine. Near Norton, Kansas, the explosion was especially loud, shaking structures at the airport. Observers a few miles west saw the smoke train punctuated by spherical puffs, apparently marking places where the meteoritic body had disintegrated in flight.

A search of the area eventually resulted in the recovery of several large masses of a rare type of stony meteorite. The largest mass, weighing 2,360 pounds, was recovered six months later from a wheat field just north of the Kansas-Nebraska state line. The meteorite had penetrated 10 feet into the soft earth, and a farmer nearly drove his tractor into the hole. For twenty-eight years this meteorite from the now-famous Norton County, Kansas, fall remained the largest single stony meteorite ever recovered, and the largest meteorite recovered from a witnessed fall. (On March 8, 1976, explosions and a fireball preceded the fall of 4 tons of stony meteorites near Jilin, China. Among them was a stone weighing 1.7 tons. It surpassed the Norton County discovery as the largest stone meteorite recovered from a witnessed fall.)

In the Beginning—Superstition and Myth

Hundreds of people witnessed the fall of the Norton County, Kansas, meteorite. Jarring explosions and sounds like the roar of a jet plane followed close behind the intensely brilliant ball of fire; many reports said it was brighter than the Sun. The recovery of large pieces of the meteorite, which disintegrated along its trajectory, provides convincing proof that stones fall from the sky, right? Wrong! Dramatic scenes like this have been reenacted countless times throughout history, but as recently as two hundred years ago, scientists still were not convinced.

Curiously, many early people do not seem to have made the connection between fireballs and stones from the sky, yet in some historical periods the worship of meteorites was common. Images of sacred stones, possibly meteorites, appear on many Greco-Roman coins. The best-known sacred stone was in the temple of Apollo at Delphi. In Greek mythology, Kronos (Saturn), a Titan and ruler of the universe, had the disquieting habit of swallowing his children as soon as they were born, in an effort to avoid his predicted destiny of one day being overthrown by one of his offspring. At the birth of Zeus, Kronos's wife, Rhea, tricked him by giving him a stone to swallow (Kronos must have been very nearsighted), thus saving the infant. When Zeus later conquered the Titans, Kronos regurgitated the stone and it fell to the center of the world. The temple eventually built there became the site of the most revered oracle of the ancient world. As late as A.D. 180, the Roman Pausanias wrote that he had seen the stone there, "A stone of no great size which the priests of Delphi anoint every day with oil."

Many temples were reputed to contain "heavenly stones." Unfortunately, no proof of them exists today—with the possible exception of the holy stone of Mecca located in the Kaaba, the most sacred Moslem shrine. Tradition says this stone was given to Abraham by the angel Gabriel. When meteorites were finally recognized as astronomical bodies, in the early nineteenth century, many scientists speculated that the stone of the Kaaba was a meteorite. Few Western scholars and scientists have examined the stone. Those who have tell us the stone is black with a light gray interior; the black exterior could be a fusion crust, a rind that forms when a meteorite burns in Earth's atmosphere. The interior of stone meteorites are usually light gray. So far, so good. Some report that the stone is unusually heavy, which again sounds good. Others claim it floats on water. Meteorites definitely do not float. It may be impactite, a shock-melted glass produced by the impact of a large meteorite. Impactite, like vesicular lava, frequently contains hollow cavities formed by gas bubbles when the rock was liquid, which could cause the glassy rock to float like pumice. Perhaps it's not a coincidence that in central Saudi Arabia there are several meteorite craters in which impactite glass and meteorites have been found.

The Ensisheim Meteorite

On November 7, 1492, near noon, a loud explosion announced the fall of a 280-pound stone meteorite in a wheat field near the village of Ensisheim in the province of Alsace, France, which at the time was part of Germany. An old woodcut depicting the scene shows the fall witnessed by two people emerging from a forest. Actually, a young boy was the only eyewitness. He led the townspeople to the field, where the meteorite lay in a hole 3 feet deep. After it was retrieved, people began to chip off pieces for souvenirs until stopped by the town magistrate. The townspeople accepted the stone as an object of supernatural origin. King Maximilian of Germany, passing through Ensisheim three weeks later, examined the meteorite and proclaimed it a sign of God's wrath against the French, who were warring with the Holy Roman Empire. Maximilian ordered the great stone displayed in the parish church in Ensisheim as a reminder of God's intervention in the conflict. There it remained until the French revolution, when it was confiscated from the church by French revolutionaries and placed on display in a new national museum in nearby Colmar. In the museum, French scientists removed numerous pieces from the main mass for study. Many of those pieces eventually wound up in museums around the world. Ten years later, it was returned to the church in Ensisheim. The remaining specimen, then weighing only 122 pounds and nearly without crust, was once again removed from the church and taken to its final resting place, the town hall in Ensisheim. And there, much diminished, but occupying a position of honor, it remains to this day (see color plate I).

Skepticism and Disbelief

The Ensisheim meteorite is the oldest witnessed fall in the Western Hemisphere from which specimens are preserved. (The world's oldest surviving meteorite fell at Nogata, Japan, on May 19 in the year A.D. 861.) There is no doubt of its authenticity. Yet memory seems to have faded as the centuries passed. There was no credible scientific explanation for the stones falling from heaven, so most "authorities" before 1700 claimed that meteorites somehow formed in the atmosphere during violent thunderstorms. Perhaps earthly particles inside clouds, consolidated by the heat of lightning and being heavier than air, would fall from the cloud as thunderstones. Typically, these ideas were vague at best. Aristotle taught that stones could not fall from the realm of the heavenly bodies, since this would violate the doctrine of heavenly perfection. Nor could they be allowed to form in the atmosphere. Yet, about 467 B.C., people saw a large stone fall in broad daylight in Thrace, near Aegospotami, which must have been something of an embarrassment to Aristotle. Since he believed that stones could only form within Earth's interior, he was challenged to explain the phenomenon. He finally concluded that indeed it did form on Earth but was subsequently lifted into the atmosphere by strong winds. Strong winds, indeed!

By the eighteenth century, European science was slowly maturing. The great scientists of the preceding century had made the first critical observations of the world as it really was, not as the imaginative mind would have liked it to be. Galileo's telescope had revealed for the first time that planets were not divine lights moving with perfect circular motion around the center of the universe (Earth). Instead, the telescope showed planets to be other physical worlds—some, perhaps, like Earth itself. Jupiter was attended by moons unknown to Aristotle's followers. Johannes Kepler, Galileo's contemporary in Germany, had painstakingly derived empirical laws of planetary motion that described how these bodies moved and predicted their distances from the Sun. Newton followed with his laws of motion and the concept of gravity, giving astronomers powerful mathematical tools for determining the orbits of planetary bodies and comets.

With the seventeenth century came the slow infusion of intellectual honesty into investigations of nature, spearheaded by these great moments of insight. No longer was the world described according to whims of egocentric fancies; the world was simply the way it was—and without our help. The budding European scientific community took it upon itself to make careful observations of nature without the influence of personal beliefs. This led to a skepticism of any observations that could not be substantiated by scientific fact or tested through independent observation. The result was the development of modern scientific method.

The Enlightenment, as this period is called, produced a backlash that set the study of meteorites back for a century or more. Fireballs and meteorites had always been associated with superstition and religion, unsubstantiated by scientific observations. The beginning of the eighteenth century saw this enlightened attitude in full swing. No one in the scientific community had ever seen a "thunderstone" form in the atmosphere nor could anyone provide a convincing argument explaining how such a thing could happen. The fledgling sciences of chemistry and mineralogy could provide no explanation. Consequently, in the minds of most scientists, stones simply could not fall to Earth from space. Because eyewitness accounts of meteorite falls were tainted by fantastic stories of evil omens, personal disaster, and divine intervention, none of which could be substantiated, it is no wonder the reports were not believed. The stories were dismissed as folklore. Still, reports continued to come in from the countryside. (I once heard a volcanologist express with some frustration an axiom in geology: "Volcanoes never erupt in the presence of a geologist." Likewise, a meteorite never falls in the presence of a meteoriticist.)

Scientific skepticism continued until the end of the eighteenth century. Around 6:30 P.M. on December 14, 1807, a fireball exploded over Weston, Connecticut, and dropped several stones near the town. A search of the fall path by scientists from Yale College yielded about 330 pounds of stony material. This was the best, if not the first, docu-

A 1628 German engraving depicts artillery shells from embattled celestial legions falling to Earth as meteorites. The reclining farmer seems to have been hit, but there are no authenticated reports of any human victims of falling meteorites.

mented fireball in the young nation, and many witnesses observed the stones falling. The population of the United States was sparse, so the sighting of the fireball over the few towns along its path was unusual and exciting. Specimens from this fall formed the nucleus of the first substantial collection of meteorites in the United States, which is still housed at Yale University.

When he heard about this fall, Thomas Jefferson supposedly said, "I would more easily believe that two Yankee professors would lie than that stones would fall from heaven." Jefferson was a competent scientist, known especially for his studies in paleontology and agronomy. Like most scientists of his day, he was skeptical of the old ideas that stones were somehow produced in the atmosphere or by volcanic eruptions on the Moon. He was certainly aware of numerous European reports of falling meteorites from the previous thirty years, particularly those from France. This was the first serious report from the United States. On the subject of the Weston meteorite, he wrote:

> A cautious mind will weigh the opposition of the phenomenon to everything hitherto observed, the strength of the testimony by which it is supported, and the error and misconceptions to which even our senses are liable. It may be difficult to explain how the stone you possess came into the position in which it was found. But is it easier to explain how it got into the clouds from whence it is supposed to have fallen? The actual fact however is the thing to be established.

Jefferson was skeptical that stones could be created in the atmosphere. It was not the fall of stones itself he doubted, as he had no personal experience or strong opinion about them. If he did make the remark about the Yankee professors, he was speaking as a Virginia politician about the veracity of New Englanders in general. As there is no source to confirm the alleged quotation, we should consider it folklore.

Stones from Heaven

Before the Weston, Connecticut, fall, American scientists like Jefferson had been somewhat ambivalent about European reports of falling stones. Over a century later, one lone American scientist would lead the world in the recovery and study of these curiosities.

Several well-witnessed fireballs that dropped stones in Europe in the 1790s finally began to stem the tide of disbelief. A shower of stones fell near the town of Barbotan in southwest France in 1790, immediately after the appearance of a fireball. Depositions from more than three hundred eyewitnesses were collected, much to the surprise of skeptical French scientists. Four years later, in the early evening of June 19, 1794, stones rained down from the sky on the eastern and southern outskirts of the city of Siena, in Tuscany, Italy. Over two hundred specimens were picked up by the local inhabitants and tourists. This fall produced considerable debate among Italian scientists. Their arguments followed the same well-worn path. Most insisted the stones came from condensations in "igneous clouds" or perhaps from the concentration of volcanic ash from Mount Vesuvius. But at least there was no denying that real stones had fallen from the sky that day. The debate was now not the reality of the stones but their origin.

The conclusion of a lone German scientist, Ernst Friedrich Chladni, was that meteorites are celestial bodies from space. True to the meteoriticist's axiom, Chladni did not observe fireballs nor did he witness meteorites falling from the sky. He simply reexamined reports of fireballs and falling meteorites made by laymen over many centuries beginning as early as the first century A.D. He noted striking similarities in these reports regarding the phenomenon of fall. Fireballs seemed to travel at great speed, suggesting a cosmic origin. Their growth in size and brightness as they dropped into the lower atmosphere and the explosions they made immediately before the fall of stones strongly suggested that fireballs and meteorites were related and that meteorites were simply the charred residue of fireballs.

In 1794, Chladni summarized his investigations in a sixty-three-page book with the unassuming title, *On the Origin of the Mass of Iron Discovered by Pallas and Others Similar to It, and on Some Natural Phenomena Related to Them*. Pallas is a stony-iron meteorite found in 1749 on Mount Emir in Siberia, about 145 miles south of the city of Krasnojarsk. The 1,600-pound meteorite was first examined by the German scientist Peter Simon Pallas in 1772 and later transported to Krasnojarsk. Today, the 1,133-pound main mass of the Pallas, or Krasnojarsk, iron is in the Academy of Sciences in Moscow. Amazingly, Chladni never studied any actual meteorites while writing his epic book. He only gathered reports of their falls. He did not study the Pallas meteorite itself, even though it occupies a prominent place in the title of his book.

A sample of the Krasnojarsk pallasite with its original 1749 label. Much of the olivine has weathered out, leaving numerous voids within the iron-nickel network. –Allan Langheinrich collection

Chladni's book confronts all the old arguments, from fireballs as atmospheric ignitions to meteorites as volcanic rocks. Three important conclusions set the stage for the skeptics' defeat: stones and irons did fall, stones and irons were the remnants of fireballs (though fireballs were not always observed), and meteorites had their origin in space. This third conclusion is not exactly what it seems. Chladni noted that meteors and fireballs had velocities far greater than would be expected if Earth's gravity alone acted on the bodies. They must therefore have a "cosmic" origin with high independent velocities. Although he speculated that meteorites might be interplanetary debris, their apparent random directions entering Earth's atmosphere suggested to him an origin not tied to the Earth or Sun. He envisioned meteorites forming out of aggregates of material in interstellar space. The idea of interstellar meteorites persisted well into the mid-twentieth century.

The first two conclusions were already finding general acceptance by many scientists. The third, and most important, fell on deaf ears. Members of the respected French Academy of Sciences were most resistant to the idea of stones falling from the sky. Then at midday on April 26, 1803, as if to mock the doubting scientists, meteorites literally rained down on L'Aigle in Normandy, France. More than three thousand stones were gathered within days of the fall. Jean-Baptiste Biot, a renowned French scientist, investigated and presented overwhelming evidence to the Institut National de France. Biot found that the meteorites had fallen within an elliptical area. He was the first to describe the shape of the strewn field commonly taken by meteoritic bodies as they scatter along the flight path of the incoming fireball. After Biot's careful and thorough investigation of the L'Aigle fall, French scientists finally accepted the overwhelming evidence, and most of them agreed with Biot's conclusions. Rocks did, indeed, fall from space.

Origins of the Celestial Rocks

Early chemical analysis revealed the celestial origin of meteorites. English chemist Edward Charles Howard provided the long-awaited proof. In 1802, he showed that the chemical composition of four different stone meteorites was almost identical. These were found in four widely separated localities in Europe and India with different kinds of country rock. Thus they were not related to the terrestrial rock of each location. This suggested a single origin not on Earth.

Two kinds of meteorites seemed to fall from the sky: stones and irons—and sometimes a mixture of the two. Their compositions appeared so different that, at first, they did not seem related. Howard again applied his skills to the problem. He analyzed the metallic iron in both types of meteorites and found it alloyed with nickel, a metal rarely found in iron ores on Earth. This was strong evidence that the two types of meteorite were indeed related. Today we know that iron in meteorites is never found without the presence of nickel and that virtually all meteorites contain nickel in some form.

It was known that fireballs traveled with cosmic speeds and somehow came from outside Earth's atmosphere. If meteorites were bodies like planets that orbit the Sun, they would be held in their orbits by gravitational forces. It seemed unlikely that these stones could break loose from their prescribed orbits and find their way to Earth. But there was some middle ground—the Moon. The existence of craters on the Moon had been known since Galileo's day. Astronomers continually debated the nature of these craters. Were they volcanic or the result of

Impact craters on the far side of the Moon. More than 30,000 craters are visible on the side facing Earth. Many more exist on the side facing away from Earth, as this Apollo 10 *photo verifies.* –NASA

impacts by celestial bodies? Most astronomers of the day favored the idea of lunar volcanoes. In fact, the great English astronomer William Herschel reported observing active volcanoes on the Moon in 1787. Other astronomers quickly followed with reports of their own observations of lunar volcanism. Perhaps the stones that entered Earth's atmosphere came from lunar eruptions. After all, violent volcanic eruptions on Earth were known to have hurled rocks for distances of several miles. Astronomers recognized that rocks from the Moon had to be expelled at about a mile and a half per second to escape the Moon's gravitational pull; and the orbits of these escapees had to intersect Earth's orbit so that Earth's gravity would pull them in. The probability of meeting these conditions seemed small indeed.

Other evidence, however, also seemed to point to the Moon. All the stony meteorites were generally similar in composition, suggesting a single origin. The chemical link between the irons and stones was not yet understood. A solar system origin would suggest that stones can come from all over the solar system; that is, they would have several origins and therefore a variety of compositions. Neither the stones nor irons showed evidence of oxidation, so they had to originate on a body with no atmosphere. The Moon seemed to be the perfect place, since it was known to have no atmosphere. Furthermore, calculations of the density of the Moon's crust matched the average density of stone meteorites. These arguments influenced Chladni profoundly, so much so that he changed his mind about an interstellar origin and accepted a probable lunar origin in 1805.

On January 1, 1801, the first asteroid was discovered, followed a year later by the discovery of a second. By 1807 four had been found, all occupying orbits between Mars and Jupiter. Many scientists were convinced that asteroids were pieces of one larger planet that somehow had been disrupted. Originally, Chladni thought meteorites might be chunks of rock and iron from a destroyed planet, so perhaps meteorites were simply pieces of asteroids that had found their way to Earth. Several years later, returning again to the problem of high velocities of fireballs but this time with asteroids in mind, Chladni rejected the lunar origin theory and returned to his original idea that meteorites were independent bodies, perhaps pieces of a broken planet coming from outside the solar system.

We can only marvel at the conclusions these early pioneers reached. The lunar origin idea was largely rejected before the mid-nineteenth century, but 150 years later lunar rocks were found in Antarctica. Though meteorites generally do not come from the Moon, these few specimens were indeed blasted off the lunar surface, not by volcanoes but by other impacting meteorites. Some meteorites do seem to be remnants of much larger parent bodies from the region between Mars and Jupiter. Others, more primitive, were never part of any larger body.

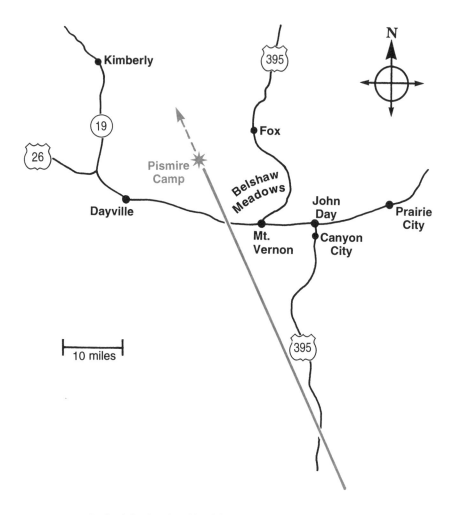

Path of the October 23, 1987, meteoroid in the John Day area in central Oregon.

THREE

Effects of a Falling Meteoroid

At 2:37 on the afternoon of October 23, 1987, a bright daylight fireball blazed across the sky in central Oregon, traveling south to north. It was reported that many people in the John Day area, east of my home in Bend, had witnessed the event. I saw an opportunity to investigate firsthand the fall of what must have been a sizable meteorite. Larry Pratt, a local amateur astronomer, happened to be outdoors a few miles south of Bend, and spotted the fireball low in the northeast sky, traveling in a northerly direction. He estimated its altitude and azimuth at 20 degrees and 70 degrees, respectively. Later we learned of other sightings as far west as the Willamette Valley, as far north as Yakima, Washington, and as far east as western Idaho. Three days later, Larry, my wife, and I headed east to the town of John Day. We thought many people must have seen the fireball nearly overhead at that location. Stopping at the *John Day Weekly News* office, we made inquiries.

Shortly after the fireball occurred, the newspaper had printed a story about a plane crash in nearby Belshaw Meadows. The sheriff's office had quickly investigated but found no downed plane in the area. The editor told us that several people had either observed the fireball or heard it explode. All day we traveled along Highway 26 from John Day to Mount Vernon to Dayville, talking to people who had witnessed the event. The stories we heard from those closest to the falling mass were strikingly similar:

> Bob Maddox was 11 miles west of the town of Mount Vernon. At approximately 2:30 P.M., he heard a sonic boom, then observed a flash like lightning. He looked up and saw nearly overhead a compact, bluish cloud that slowly changed to white. The boom continued to rumble like thunder for about one minute. Maddox commented that he thought a jet plane had broken up and exploded.

> Bob Pugh was about a half mile from home, just off Highway 26 in Mount Vernon, when he heard a "whomping" sound much like the sound made by helicopter rotor blades. There was an explosion, followed by a continuation of the "whomping" sound for about three seconds. Looking up, he saw a blue-white cloud nearly overhead, but no fireball.

Many people we questioned mentioned the "whomping" sound. Even after the explosion, the sound persisted for a few seconds. Pugh likened it to the sound an acetylene torch might make if swung around your head. We tentatively interpreted this to mean that the meteoroid was rotating as it traveled, and that it was not round but elongated:

> Mrs. Born was in Fox Valley, north of Mount Vernon, when she heard a sound like the "whomping of a flat tire." She said it was followed by an explosion. Racing outside, she saw a small smoke cloud about 60 degrees up in the southwest. It seemed to be drifting over the ridge of mountains marking the Malheur National Forest between Fox Valley and Highway 26.

Mrs. Born's observation was especially important, since she saw the fireball from the north. This told us that the meteorite must have impacted somewhere between Highway 26 and Fox Valley, the most probable location being somewhere on the separating ridge.

Most people could not give precise compass bearings nor could they accurately estimate the altitude of the fireball in degrees, both of which were important for correctly determining its flight path. One particular sighting, however, was superb:

> Dave Corliss, a Forest Service archaeologist, was taking compass-bearing readings on a proposed logging site in the Ochoco National Forest near Wolf Mountain, 27 miles southwest of Mount Vernon, when he suddenly saw the fireball. It punched a hole in the clouds just before it exploded and produced a whitish, puffy cloud. He immediately took a compass bearing on the cloud and measured the cloud's altitude: 52 degrees NE; 30 degrees altitude. At the time the fireball was clearly visible, Corliss heard crackling sounds, which he associated with the fireball.

Corliss's observation established that the meteoroid was only about 19,000 feet up when it exploded. A check at the weather bureau had established the cloud deck at 20,000 feet. The cloud reported by Corliss and most of the other observers was produced by the body as it explosively disintegrated. The explosion of a fireball is the terminal event in its flight through the atmosphere. At the moment of the explosion, what remains of its initial velocity is reduced almost to zero, and the remaining solid body simply falls to Earth by gravity alone. We used Corliss's altitude observation to plot the meteoroid's point of fall. His observations, combined with those of others, established that it must have fallen on or near the ridge north of Mount Vernon. If it exploded over the 4,600-foot ridge, then the meteoroid was less than 3 miles up at the time.

One significant and startling report came from a group of loggers working at Pismire Camp up on a ridge northwest of Mount Vernon:

Rod McClain was working next to a loader that was busy loading freshly cut logs. Suddenly he noticed a shadow, "like an aircraft shadow moving on the ground," he said. His immediate thought was that it was a jet plane coming through the area at treetop level. He also heard the "whomping" sound like a helicopter and then an explosion. The sound was extraordinarily loud, since he heard it even over the loud whine of the loader. The rapidly moving shadow reminded one logger working near McClain of the moving shadows made by falling trees, and he immediately "hit the dirt" for fear of being struck by falling timber. The object passed almost exactly overhead and exploded at that point. All the men who witnessed the event confessed immediate fear and confusion. As one logger told it, "I knew it was the hand of God come to take me home."

The loggers were standing right under the flight path of the falling meteoroid and experienced the most powerful shocks from the exploding mass. Their position marked the only single point on the true flight path we were able to determine from all the reports. The fireball cast a shadow on the ground as it passed over the trees. This is testimony to its brightness because it was a sunny afternoon. The moving shadows were followed by loud booms and a "puff" cloud overhead. The meteorite must have fallen only a few miles north of their position. Unfortunately, the celestial visitor picked a most inhospitable place to make landfall. The forests there are thick with brush and the slopes leading off the ridge are steep. Scattered everywhere are black basalt rocks that mimic the black fusion crust of a meteorite, so a close examination of every rock would be necessary. We knew our chances of finding the meteorite were slim under those conditions. A preliminary search left us empty-handed. We have visited the area several times since 1987, but have not found a trace of the meteorite that must be there—somewhere.

This story illustrates the factors involved in the passage of a fireball and the subsequent hunt for the fallen meteorite. Today we know a great deal about the phenomena accompanying a falling meteoroid. We all have a chance of witnessing a great fireball sometime in our lives, so it is useful to know what to expect in the few seconds from the first sighting of the burning mass to its final plunge to Earth.

The Meteoroid "Test"

All meteoroids are subjected to a rigorous test before they can become meteorites: They must survive passage through Earth's atmosphere. Of the millions of particles that enter the atmosphere each year, about 10,000 tons of meteoritic debris make it to Earth. Most of this tonnage is composed of meteoritic dust that filters unscathed through the atmosphere. Particles from less than 0.0001 inch (dust size) to about one-eighth inch in diameter are totally consumed during their atmospheric passage. Larger bodies can make it through if their velocities are not too high.

Characteristics of Fall

All falling meteorites announce themselves with light and sound. The light from the fireball is usually seen over a wide area, often hundreds of miles from the spot where it comes to rest on the ground. Though it is possible for the sound accompanying the fireball to be heard over as wide an area, it usually is heard only by people close to the impact point, within 25 or 30 miles.

The height at which a burning meteoroid first becomes visible as a fireball depends on its initial mass and velocity when it enters the uppermost part of the atmosphere. All meteoroids in space are moving at cosmic velocities. Since they have highly elliptical orbits that carry them from the asteroid belt to points well inside Earth's orbit, their cosmic velocities can be as high as 26 miles per second in the vicinity of Earth. Usually, meteoroids cross Earth's orbit obliquely, either forward of or behind the moving Earth, so their speeds fall somewhere between 10 miles per second and 26 miles per second.

It's All a Matter of Physics

A meteoroid falling to Earth is a dramatic event, with predictable characteristics. All bodies in motion possess inertia, a body's resistance to a change in its motion. The force required to change its motion—to speed it up or slow it down or change its direction—depends upon the body's momentum. Momentum is a product of the body's mass, the amount of material it contains, and its speed or velocity. The more mass the body has or the faster it is traveling, the greater the force necessary to change its state of motion. A 1-ton meteoroid hitting Earth's atmosphere at 25 miles per second possesses enormous momentum. The force necessary to slow it is equally enormous, and is produced in the atmosphere. The atmosphere reduces the meteoroid's momentum by creating a drag on it, thus slowing its velocity. It further reduces its momentum by heating the meteoroid until it begins to lose mass by the ablation process.

A moving mass also has energy associated with its motion. This is called kinetic energy, which also involves the body's mass and velocity, but this time the quantities are different. Mathematically, it looks like this: $KE = 1/2mv^2$.

Like momentum, the greater a meteoroid's mass and velocity, the greater its kinetic energy. Unless a meteoroid possesses an enormous mass—millions of tons—velocity is the more important factor. The important thing to notice about the kinetic energy equation is that it varies with the square of its velocity but only with the first power of the mass. For example, if two meteoroids have the same velocity but one has twice the mass of the other, the more massive body will have twice the kinetic energy. For two meteoroids of the same mass but with velocities differing by a factor of two, the meteoroid with the higher velocity will have four times (v^2) the kinetic energy. Generally, the mass of a moving body is considered a constant, but for a meteoroid passing through the atmosphere, there is considerable mass loss. The mass of the body must be considered a variable in this special case.

The kinetic energy of a falling meteoroid can be converted to other forms of energy; kinetic energy means everything to its survival. The higher its kinetic energy when it meets the upper atmosphere, the higher the atmospheric drag on the meteoroid and the more rapidly it heats up, as atmospheric resistance acts to convert this energy of motion to heat, light, and sound.

Light Phenomena

A fireball's light is generated by two distinctly different mechanisms: the burning of the solid body itself and the incandescence of the atmosphere immediately around the burning mass. A meteoroid usually does not become luminous, so is not noticed until it reaches an altitude of between 60 and 90 miles. Above that altitude, the air is too thin to produce much drag on a meteoroid (low Earth-orbiting satellites have orbited at 90 miles). Below this altitude the atmosphere becomes dense enough to begin to produce a substantial resistance to a meteoroid's motion. At this point, a meteoroid begins to convert some of its kinetic energy to heat, which begins to melt the outer parts of the meteoroid, and it becomes incandescent. Its temperature reaches more than 3,000 degrees Fahrenheit; material begins to slough off as the rock surface literally liquefies; and almost immediately, the ablated material vaporizes. Though catastrophic for the meteoroid, this alone would not make

A large meteoroid breaks up 50,000 feet above the Nevada desert while still aglow from its fiery passage through the upper atmosphere.
–Dorothy S. Norton painting

it very noticeable at such a high altitude; but as heating continues, the air around the incandescent rock begins to ionize (lose electrons). As atmospheric gases recapture electrons they release light, causing the air itself to become incandescent. Though the rock may not be more than a foot or two in diameter when first entering the atmosphere, the glowing air mass surrounding it may measure hundreds of feet across: This is the fireball seen from Earth—a mass of glowing hot gas.

Most people who observe fireballs report seeing colors. The colors derive from two sources: the composition of the vaporizing material and the composition of the luminous air surrounding the hot meteoroid. Many elements give off diagnostic colors when vaporized. For example, sodium produces an intense yellow color, nickel a green color, and magnesium a blue-white color, and these elements are found in meteorites. Atmospheric gases also produce distinctive colors when excited into fluorescence. During displays of the aurorae, greens and reds produced by nitrogen and oxygen are common. A fireball often changes color as it progresses along its path. Typically, a fireball will turn from brilliant white to bright red immediately before it is extinguished. Or it may resemble a Fourth of July sparkler, displaying a spectrum of colors.

The Song of the Fireball

Fireballs that produce meteorites also produce a variety of cacophonous sounds. A jet plane traveling at or above the speed of sound generates a pressure wave, or shock wave, in front of the plane. On the ground this wave is detected as a loud explosive sound, the familiar sonic boom. Meteoroids generate similar shock waves, but since they travel much faster than sound, they can precede the sound by several seconds or as much as a minute or two. Those who observe fireballs often report no sound at all while the fireball is visible. Seconds or minutes after the fireball has vanished, the pressure wave passes by the observer and several loud reports are heard. If the meteoroid breaks into several pieces in flight, each can generate its own shock wave, which will be heard as a continuous rumble like distant thunder as the shock waves overlap each other. One sound many observers report hearing while the fireball is visible is a hissing noise like radio static. Dave Corliss reported hearing this sound during the sighting of the fireball in central Oregon. Because sound waves travel much more slowly than light waves, the two phenomena could not occur simultaneously. The sound is also selective, as some observers report hearing it while others viewing the same fireball from different locations fail to hear the sound. Australian physicist Colin S. L. Keay finally solved this three-hundred-year-old mystery with a phenomenon called electrophonics. He showed that the plasma trails of large fireballs generate several kilowatts of power in the form of 1- to 10-kilohertz, low-frequency radio emission. These radio transmissions, traveling at the speed of light, are converted to sound when they interact with natural or man-made objects located near the observer. Power poles, antennae, trees, and tall buildings act as natural transducers converting the electromagnetic energy into sound. If there is no transducer nearby, the observer does not hear the sound simultaneously with his observation of the fireball.

People have also heard whistling sounds close to the impact point. This could be the actual sound of the meteoroid passing through the air at a low speed just before impact. The "whomping" sound described by many who observed the daylight fireball over central Oregon in 1987 is not unique. Many eyewitnesses say it resembles the noise of a flat tire on a moving car. It may be the sound of an irregular body in rotation.

The Ablation Process

A meteoroid in a state of incandescence rapidly loses material by ablation, and the ablated material carries away the accumulating surface heat. First the material vaporizes, then it condenses again as it cools. It is deposited behind the fireball as a long, narrow dust or smoke train. This train of cloudy material can be most impressive. The actual amount of material lost depends to some extent on the composition of the meteoroid. The more friable stony meteoroids lose more mass than the irons. Even so, estimates of material in the dust train of the iron meteorite that fell in the Sikhote-Alin Mountains in far-eastern Siberia in 1947 range as high as 200 tons. This is even more amazing considering that the remaining falling body was estimated to weigh only 70 tons, and of this only about 25 tons of meteoritic material have been found at the impact site. Most of the meteorite, then, was vaporized and recondensed to form the train. This material is often luminous at night and usually remains in the atmosphere for hours. High-altitude winds scatter it over a great distance, and eventually it falls to Earth as tiny, roughly spherical micrometeorites.

Mass Loss

Rapid changes take place in a meteoroid as kinetic energy is converted to heat and light. Meteoroids with high initial velocities will be subjected to far greater atmospheric resistance than slower-moving meteoroids and therefore suffer greater ablative effects. This is easily seen in the accompanying graph, in which entry velocities of a 1-ton iron meteorite are related to the percent of the initial mass retained on reaching Earth's surface. Initial velocities of 12 miles per second and 24 miles per second at an entry angle of 45 degrees are compared. In both cases maximum ablation takes place between altitudes of 10 and 20 miles. The higher-velocity body suffers much greater mass loss,

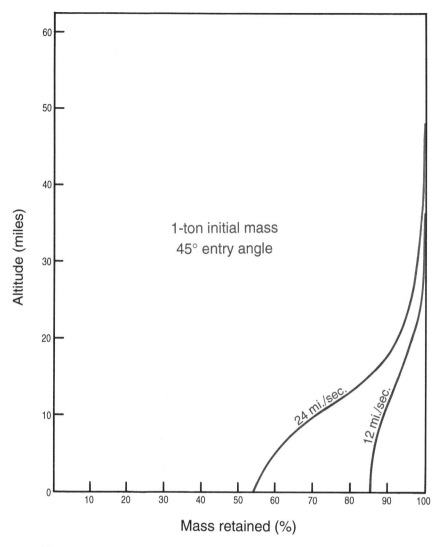

The initial velocity of a meteoroid determines how much material will be ablated during its atmospheric passage. −Data from Heide 1964

retaining only 55 percent of its initial mass, while the lower-velocity mass retains 86 percent. Doubling the entry velocity more than triples the loss of mass. Mass loss would be even greater for higher entry velocities. This assures the destruction of many meteoroids before they can reach Earth. Most observed fireballs that resulted in surviving meteorites had velocities of around 12 to 15 miles per second.

Many meteoroids enter the atmosphere at considerably higher speeds than these and survive. Meteoroids encounter enormous forces from the pressure of the air resisting their motion. Frequently, the force on a stony meteoroid, especially on the forward end, is sufficient to cause the mass to break into several smaller pieces. These pieces have a better chance of survival than the larger single mass, since smaller stones with less mass are more readily slowed in the atmosphere, therefore suffering less heating and ablation. That may allow them to survive even when the main mass has a high initial velocity. Sometimes fireballs brighten along their trajectories as they fragment into many smaller fireballs. Most meteorite showers, in which hundreds or even thousands of meteorites fall over a wide area, are undoubtedly the result of multiple fragmenting of a single mass.

Effects of Impact

Many meteoroids lose all their cosmic velocity at some altitude, so their final plunge to Earth is due to gravity alone. The velocity at which a meteorite strikes Earth depends in a complex way upon its initial cosmic velocity, its initial mass, and its size and shape. A meteoroid's

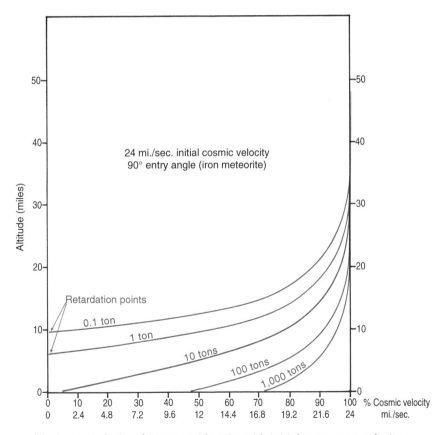

The impact velocity of a meteoroid varies with initial mass, entry velocity, and entry angle. Most meteoroids lose their cosmic velocity several miles above Earth, but meteoroids of greater than 10 tons strike the ground with a percentage of their cosmic velocity intact. –Data from Heide 1964

51

velocity determines its air resistance, and its shape and size determine how much drag it experiences. The angle at which it enters the atmosphere, from vertical to nearly horizontal, also affects drag, a shallow angle of entry giving it a longer path through the atmosphere. Thus a shallow angle of entry results in a higher ablation rate and a higher rate of mass loss.

With so many variables to consider, determining the speed of an impacting meteoroid is complicated. We can get an idea of impact velocities by selecting a reasonable set of conditions for falling meteoroids of various masses. The graph on page 51 shows the rate at which a meteoroid weighing from 200 pounds to hundreds of tons will lose its initial cosmic velocity because of atmospheric drag. For a meteoroid entering the atmosphere vertically, at 24 miles per second, the atmosphere will be remarkably effective in slowing its fall. A meteoroid weighing from a few hundred pounds to more than a ton loses all its cosmic velocity while still several miles above Earth's surface. At that altitude, called the retardation point, it now begins to speed up, because of gravity, at the rate of 32 feet per second every second. Although it accelerates at this familiar rate, the atmosphere provides resistance to free-falling bodies. Within a few seconds of free fall, the meteoroid reaches a terminal velocity of between 200 and 400 miles per hour, depending upon its remaining mass, and does not accelerate beyond that speed.

The Hoba iron meteorite, the largest intact meteorite in the world, near Grootfontein, Namibia. –Courtesy of Frederick Pough

52

Once equilibrium has been established between air resistance and the force of Earth's gravity, a meteoroid's free-fall speed is far too slow to generate further melting or to ionize the air around it, so the meteoroid ceases to give off light. It falls as a dark body. At that relatively slow speed, a meteoroid (now, technically, a meteorite) the size of a baseball or basketball may not even penetrate the ground when it strikes Earth. Larger meteorites, up to several tons, create holes equal to or a bit larger than their diameters and may penetrate the ground up to several feet. The penetration depth depends to a large extent on the nature of the surface they hit. A stony meteorite will tend to shatter if it strikes a hard rock surface at more than 200 miles per hour, and many have. Iron meteorites, however, being less brittle and with higher tensile strength, often survive impacts at those low velocities. Large iron meteorites weighing several tons have been found intact, resting on hard surfaces with little evidence of impact other than a shallow penetration hole.

Incoming meteoroids are more likely to be destroyed when their masses are greater than about 10 tons. A 10-ton meteorite retains about 6 percent of its cosmic velocity. If it originally traveled at 24 miles per second, it would have a residual cosmic velocity of 1.4 miles per second, which is an incredible 5,200 miles per hour when it hits the ground. The impact speed goes up rapidly with increasing mass. A 100-ton meteorite retains nearly half its cosmic speed, or 12 miles per second.

Fortunately, meteorites in the 100-ton class are exceedingly rare. None are known to exist on Earth today. There is good reason to believe that meteorites of 100 tons or more do not survive impact. Meteorites that weigh hundreds or thousands of tons are not retarded much by the atmosphere, retaining a large fraction of their cosmic velocities. They self-destruct on impact with Earth, exploding violently and leaving an impact crater behind as a mark of their moment of glory.

The largest known meteorite on Earth was discovered in 1920 on a farm 12 miles west of Grootfontein, Namibia, in southwestern Africa. This meteorite rests where it was found on the Hoba farm, and for good reason: The remarkably rectangular iron meteorite measures 9 feet by 9 feet by 3.5 feet, and its estimated weight is more than 60 tons. It rests in a depression in limestone rock only about 5 feet deep. This is an ancient meteorite that has undergone considerable weathering. The cavity it made at impact was probably much larger and deeper than the depression in which it rests today, and weathering of the limestone over the centuries may have "weathered out" the once-buried meteorite.

French political cartoonist Honoré Daumier drew this cartoon of an elderly man looking for the Comet of 1847 while his wife frantically tries to direct his attention to a brilliant fireball.

FOUR

Tracking Fallen Meteorites

At about 9:00 P.M. on March 25, 1969, a huge fireball lit up the sky over western Nevada. It was traveling northwest to southeast, closely paralleling Highway 95, which runs down the western side of the state. Hundreds of eyewitnesses jammed the telephone lines along the fireball's route. Being the only astronomer in Nevada at the time, I became the state expert on all things celestial, and the local sheriff in Reno contacted me for ready answers I didn't have. I hadn't seen the fireball, but the community expected an intelligent response from its resident astronomer. I called the local television station and requested they ask those people who had witnessed the fireball to contact me with their reports. For the next several days the telephone rang constantly. The people of western Nevada were wonderfully cooperative. Their response was enthusiastic and interest was high, but the reports were generally vague in detail and sometimes greatly exaggerated in estimates of altitude and position. Many reports reflected the excitement of the moment, but fell far short of drawing an accurate conclusion:

> My wife and I were traveling south on Highway 395 about 80 miles north of Reno. We had just passed through Susanville when suddenly the whole landscape lit up around us. We looked high through the car windshield and here was a brilliant ball of fire coming down right in front of us. It was sparkling like a Fourth of July sparkler and it turned from white to a reddish color before it hit the ground. It was so close to us that I slowed down because I was afraid it might hit us. My wife thought it landed about a mile away, behind a group of hills just to the east of the highway, but I think it must have landed further away than that, more like five miles down the highway. We could show you just where we think it landed if you like.

The report was useful because the observers could tell the direction of the fireball by referring to groups of hills. They knew where they were on the highway when they saw the fireball, which was also helpful. But their estimates of just where the object might have fallen were greatly underestimated. Most witnesses I talked to in the days after the sighting gave estimates of how close the falling meteorite had been.

People in Reno saw it fall just south of town, perhaps in the hills near Virginia City. People in Carson City were sure it had landed in some nearby hills just off Highway 50. Residents of Lee Vining, California, 200 miles south of Reno, thought it had landed south of them. People in Las Vegas, 400 miles south of Reno, saw it moving southward and thought it might have landed somewhere around Hoover Dam, or maybe even closer. In fact, the fireball was witnessed all the way to the Mexican border, where it was still heading south. If the meteoroid made it down safely, it might rest today in the Gulf of California.

Fireballs do seem to be much closer than they really are. This is true even for experienced observers. Consider the problem: A fireball is usually visible for only three or four seconds before it vanishes. In that all-too-brief span of time, investigators know they cannot reasonably expect the average person to compile a perfect set of accurate observations: the meteor's direction of motion, preferably in azimuth coordinates; its altitude in degrees; the total number of seconds observed, and the local time; its magnitude or brightness compared with other celestial objects; its color and any color changes; any sounds it might make; the length of any smoke train. And while you're at it, a picture would be helpful, too!

When a fireball suddenly appears (there's never any warning), it is such an astonishing moment that most people are spellbound. Who would think of estimating its altitude in degrees? But with enough reports, as inaccurate as many are bound to be, scientists can often glean enough solid information to eliminate inaccuracies and, with a bit of luck and lots of footwork, locate the meteorite if it has survived.

The best approach to observing a fireball is to be prepared for the inevitable. In the course of a lifetime, anyone can expect to experience one or more fine fireballs, some of which may produce meteorites. Knowing what to look for when that moment arrives could put you one step closer to the recovery of a valuable meteorite. The fireball represents the meteoriticist's hope—the recovered meteorite, his goal.

What Time of Year Do Fireballs Occur?

Although meteoroids may enter Earth's atmosphere at any time during the year, there seem to be times when they have a better chance of surviving their fiery rites of passage.

The first hint of a statistical preference for certain times of the year comes from the mid-nineteenth century, but the data are meager, representing too few falls to be meaningful. A hundred years later, with the rising interest in meteorites, many more falls were recorded. Of 581 falls documented between 1800 and 1960, the maximum number of falls occurred in May and June. The falls drop to a minimum in December and January, with a further dip in March.

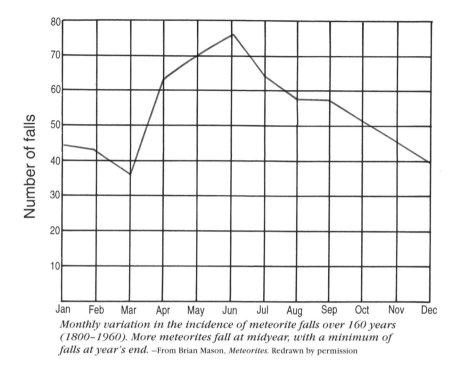

Monthly variation in the incidence of meteorite falls over 160 years (1800–1960). More meteorites fall at midyear, with a minimum of falls at year's end. –From Brian Mason, *Meteorites.* Redrawn by permission

Since most of these meteorite falls were observed in the Northern Hemisphere, these data must have been influenced by the Northern Hemisphere's winter season, when there are fewer people outside on cold winter nights. However, if the fall of meteorites is uniform throughout the year, then the number of reports during the spring and fall months should be essentially the same, but they are not. Most meteoriticists conclude that these curves are telling us something about the distribution of meteoroid bodies in space. There may indeed be a greater concentration of meteoroids along Earth's orbit during the Northern Hemisphere's summer months than during the winter months.

At What Time of Day Do Fireballs Occur?

If there is a favored month or season in which meteorites fall, it seems reasonable to assume a favored time of day as well. Analyses made in the mid-nineteenth century suggested a statistical preference. Again, a modern analysis is used to support or deny these early claims. Using 469 recovered meteorites whose falls were observed, the American meteoriticist Frederick C. Leonard plotted a graph showing an

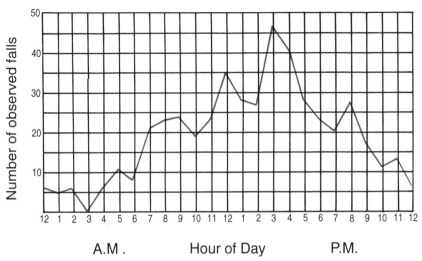

Number of observed meteorite falls plotted against the time of day. Most meteorites are recovered in afternoon and early evening hours. From Brian Mason, *Meteorites*. Redrawn by permission

obvious peak, around 3:00 P.M., and an equally obvious minimum twelve hours later, at 3:00 A.M.

At first the cause of maximum and minimum occurrence seems obvious. Human beings are diurnal animals. Few people are up and about at 3:00 A.M., especially out-of-doors. Three o'clock in the afternoon is a more active time for people and many of us are outside then. So the peak seems artificially based on human levels of activity. This effect is undoubtedly present in the curve, but most meteoriticists suggest an additional interpretation. Chapter 2 discusses the effects of Earth's orbital motion on the meteoroid's speed as it enters the atmosphere. From noon to midnight, meteoroids that strike the Earth are traveling in the same direction as Earth and therefore encounter it at minimum speed. From midnight to noon the opposite is true, when Earth encounters meteoroids head-on and they are striking the atmosphere at high speeds. Because the survival rate of meteoroids goes down as their atmospheric entry velocity goes up, meteoroids observed in the early morning hours are less likely to survive than meteoroids observed in the late afternoon hours. So survival appears to be the key factor here, rather than the nonrandom fall rate the statistics seem to suggest.

Fireball Flight Paths

Certain observations are crucial to locating a meteorite. The most important observations of meteors occur during the fireball stage, when the true trajectory, or path, of the fireball can be determined if the height of the meteor can be measured for at least two points along its path. This will fix both the direction of travel and the angle of

Determining the True Path of a Fireball

Two observers separated by several miles simultaneously note the fireball's path against background stars. Since the meteor is only a few dozen miles above them, it will appear to occupy different positions for the observers, depending upon their own positions on Earth. Lines traced back to each observer will intersect at both ends of the fireball's real path.

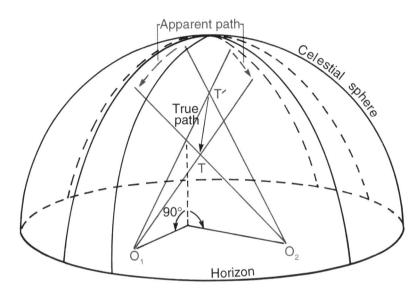

The true path of a meteor (TT´) can be calculated by using two or more ground observations (O_1 and O_2) of its apparent path against the celestial sphere.

descent. When a fireball moves across the sky, an observer doesn't see its true path but, instead, its apparent path from the viewer's position on the ground. Two such observations made simultaneously make it possible to locate the actual trajectory and determine the meteoroid's true direction and approximate angle of descent.

How High Is the Fireball?

After determining the flight path of a fireball, the next step is to ascertain its height. The true position of a meteoroid in space cannot be determined by a single observation alone; observations from at least

Determining the Height of a Meteor

Once the direction and altitude of the retardation point (R) have been measured, it is possible to sketch the observations from both points. For example, an observer at point A sees the meteor fade out a hand's width above some distant trees. Investigators later measure the azimuth of the trees as 155 degrees (65 degrees south of east) and an altitude of 45 degrees. A second observer, at point B several miles away, views the fireball at the same instant and directly estimates its azimuth as 50 degrees and its altitude as 40 degrees. The intersection point of the two triangles marks the real retardation point. A line from that point to the ground represents the real altitude of the meteor at the retardation point.

Triangles ACR and BCR share the common side H, the meteor's height. In principle, these triangles are used to calculate the height, or the height may be scaled by geometric construction. Retardation point altitudes vary over a wide range. Usually they are between 10 and 15 miles above Earth's surface, although heights as low as 2.5 miles and as high as 25 miles or more, well into the stratosphere, are not uncommon.

If at least two azimuths and two altitudes are obtained along a fireball's path, the heights can be calculated for each position. From these heights, the angle of descent can be constructed. If the lapsed time between the two points can be determined in seconds, the rate of descent can be calculated, and from that, the meteor's velocity.

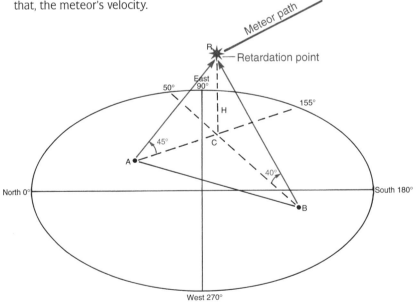

The true height of a meteor can be calculated by using two observation points several miles apart (A and B).

two positions are needed. Ideally, several observers estimate the altitude and azimuth of the meteor in degrees. If this is not possible, each observer should carefully mark his position and the position of the fireball with respect to some feature such as a distant mountain, nearby building, or tree. The true azimuth and altitude can then be estimated by an investigator at that location sometime later.

Generally, viewers do not notice a fireball at the beginning of its luminous flight path but somewhere between the point where it first becomes luminous and its retardation point (or point of extinction, where the light appears to go out). The retardation point is the most important point to determine, since that establishes where the meteoroid loses cosmic velocity and falls freely to Earth. The corresponding point on the ground directly below the retardation point can be determined either mathematically or by geometric construction. Usually, the meteoroid falls to Earth only a few miles from that point along the azimuth bearing (compass direction) of the original flight path.

Area of Fall

End-point observations are crucial to the recovery of a meteorite. The end point is near or at the retardation point, where the fireball quickly disappears and the meteoroid falls on a steep curve to Earth.

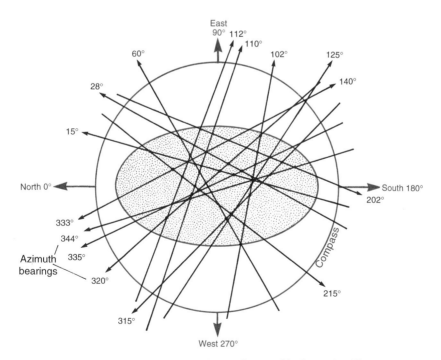

The fall area of the meteorite(s) can be roughly determined by plotting on a map end-point azimuth estimates made by many observers several miles apart.

Observations to Make Before Impact

1. When did the meteor appear? Record the date and the time.

2. Describe your location in longitude and latitude (your local airport or library can provide this information).

3. Note your exact position, that is, where you made your observation. Describe your position using landmarks, or mark your position with a stake for later reference.

4. In what direction was the fireball traveling? You can use a compass bearing or note its direction of travel with respect to a landmark (tree, house, power pole, etc.).

5. What was the altitude of the meteor when you first saw it? When it stopped glowing? Once again, you can gauge its altitude by using familiar landmarks during the day.

6. If the meteor occurred at night and if you are familiar with some of the brighter constellations, you can answer questions 4 and 5 by describing the meteor's path through those constellations. If so, it is particularly important to note the time, since the altitude and azimuth of the stars are constantly changing.

7. How long did the fireball last? The most reliable way to record the duration of the event is to count seconds from the time you first see the fireball until it vanishes.

8. How bright was the fireball? At night, you can compare it with the Moon, bright stars, or planets. If it is an exceptional fireball, it will be brighter than all of these. It will easily cast a shadow and may be comparable to the Sun in brightness.

9. How large was the fireball? You could compare its size with that of the Sun or Moon, which are both one-half degree of arc in the sky. Or you can compare it with some familiar object on the landscape or to your fist or thumb held at arm's length.

10. Did the fireball change color, size, or shape? Sketch its shape and note any sparks, explosions, or breakup of the main body.

11. Did the fireball leave a smoke train? If you saw it at night, was it luminous? How long did the train persist in the sky? Draw a sketch, or better yet, take a photograph of the train. This should be done quickly, since winds aloft will distort the train within minutes. Note also the direction of the train with respect to some landmark.

12. Did you hear any sounds? What was the quality of the sounds? Rumble? Pops? Hiss? How long after sighting the fireball were the sounds audible? If you heard "whomping" or whining sounds, the meteorite passed very close to you.

13. Did you see any dark bodies falling from the point of extinction? If so, how many and in what direction?

Observations to Make After Impact

1. Did the meteorite strike anything (trees, houses, etc.) before it hit the ground?

2. What was the soil type? Sand, clay, rocks? Did the meteorite produce a hole of any size or did it come to rest on the surface? What size hole? How deep? Was the hole vertical or at an angle to the vertical? Estimate this angle.

3. How much time elapsed between the impact and your recovery of the meteorite?

4. Was the meteorite warm or cold to the touch when found?

5. Was more than one meteorite found? If so, carefully mark the location of each meteorite for future reference.

   A photographic record of any aspect of the fall is an invaluable record of the event. Photographs of the meteorite in situ should be taken before the meteorite is removed from the ground. These pictures should be made from several angles. After the meteorite has been removed, photographs of the resulting hole, if any, should be made.

Azimuth estimates of these end points plotted on a map will reveal the most likely location of the fall area. If a dozen or more different azimuth determinations of the end point made by observers at locations several miles apart are drawn on a map, each line will intersect the others at various points. The intersection points lie within a roughly elliptical area. We would look for the meteorite(s) there. That is a rough estimate of the possible fall area; a more accurate fall area could be plotted after recovering specimens.

What to Watch for When Observing a Fireball

I have emphasized the importance of accurately observing the direction and altitude of a fireball. This information is used to figure out the true trajectory of a meteor, its height, and the rough area where a meteorite may have landed. Other observations may help determine fall characteristics and even some physical characteristics of the meteorite. Memories tend to be fleeting after such a brief encounter. If you ever witness a fireball, the accuracy of your observations may determine whether the meteorite will be recovered. Write down as much as you can remember as soon as possible.

Who Owns the Meteorite?

Ownership of a meteorite is sometimes a sticky question and has resulted in many a lawsuit in the United States. If a meteorite landed on private land, the owner of the property owns the meteorite regardless of who found it. You should ask the landowner for permission to search, and if you find a meteorite, offer a fair price for the specimen. An agreement with the landowner made before the search will avoid any misunderstandings after the fact.

If the meteorite is on public land, such as National Forest, Bureau of Land Management (BLM), or state land, the meteorite rightfully belongs to the governing body. People have often gone to a great deal of trouble to recover a meteorite only to have it taken away by a government agency. If the meteorite is found on federal land, the Smithsonian Institution in Washington, D.C., should be contacted. The meteorite legally belongs to the people of the United States regardless of who finds it. The finder might request a piece of the specimen (which probably will be cut for classification and study purposes) or be able to negotiate a reward for finding it. It may be prudent to conclude any negotiations before revealing the location of the specimen in order to protect one's interest in the meteorite. If possible, a valid written agreement should be signed among the parties.

The reward system works. H. H. Nininger certainly proved that in his pursuit of meteorites. People are much more willing to turn over meteorites they find to a governmental body or laboratory if they are offered a fair price for their find, regardless of the laws covering such finds. Bear in mind that most new meteorite finds are made by amateurs (the Antarctic meteorites excepted). Scientists depend upon the army of amateur meteorite hunters to find new specimens. Without the amateur, there would be far fewer specimens available for study. (This is true not only of meteorites but also of fossils.) Freshly fallen meteorites are valuable to scientists, and anyone finding a new meteorite should contact a laboratory or university where meteorites are studied (see Appendix A). The meteoriticists will be glad to test the specimen. If it is a meteorite, they may offer to buy it or they may request a piece of it. The decision is yours. A new meteorite find can be worth a considerable amount of money, depending on how much material of that particular type is already available. The specimen may sell for anywhere between $0.50 per gram and $10 or more per gram (0.035 ounce). Universities are usually hard-pressed for funds, so you may wish to consider a donation of all or part of the specimen.

Selling the Solar System—How Much Is a Meteorite Worth?

How do you put a price on a celestial body? They are valuable tools for acquiring knowledge—perhaps their long-term payoff. But they also have significant commercial value. Check the Internet and you will discover that meteorites are hot items on the commercial market. What used to be a small hobby enjoyed by a few "far out" collectors served by equally few commercial mineral dealers with limited inventories has mushroomed into a worldwide feeding frenzy. New collectors are appearing almost daily, and the trend shows no sign of slowing. Much to the distress of scientists who study meteorites and museum curators who collect them, the world has discovered that meteorites have remarkable commercial value.

Ironically, scientists have been in part responsible for the rapidly increasing prices of meteorites over the past decade or so. When an announcement was made that a particular stony meteorite contained grains from a nearby star that exploded long before the solar system came into being, the price of the specimen rose sharply. Collectors quickly discovered the high cost of stardust. Fortunately, there was plenty of this particular meteorite (Murchison) available, which lowered the asking price somewhat. Then, even while ripples from this announcement were still spreading through the meteorite world, scientists studying SNC (pronounced *snick*) meteorites announced that these rare specimens came from Mars. Collectors clambered to acquire what little was available commercially, sending prices halfway to the red planet. The most recent price stimulus came in 1996 when NASA scientists announced the discovery of possible remnants of Martian life within the SNC meteorite ALH 84001 (see chapter 11). Once again the rush was on.

Recently, well-publicized auctions by major auction houses in the United States began to offer entire private meteorite collections, including many rare types. This attracted the attention of investors with little interest in meteorites beyond their commercial value. Today, more collectors and dealers than ever provide ever rarer specimens at ever higher prices—an inflationary spiral not present in the rest of the economy.

With some private collections surpassing those held by many museums, major collectors now present formidable competition to research institutions. The rarest meteorite types, coveted by scientists and collectors alike, have the dubious honor of being some of the most expensive commodities per gram weight in the world. Some scientists fear significant, privately owned specimens will be lost to science forever. Skyrocketing prices have driven many researchers and smaller collectors out of the market, and specimens needed for study are disappearing into private vaults. Fortunately, most reputable dealers share new finds with scientists. They need the scientists to document and study new specimens, which dealers recognize can only make the meteorites more valuable.

What eventually happens to private collections? Many wind up in research institutions or museums. A few are donated; smaller collections are often sold intact to the highest bidder. The larger collections are so expensive that they are usually sold piecemeal to dealers who in turn will make them available to collectors in yet another recycling.

What can a collector expect to pay for meteorites today? Since the market is in continuous flux, it is always best to check with a reputable dealer. (A list of dealers appears in Appendix B. Before contacting a dealer, the prospective buyer should learn some of the criteria dealers use to establish price. Read chapters 9 through 13 carefully, as they describe the various meteorite types and develop the rudiments of the technical language used by dealers and scientists.) Prices vary enormously, depending upon the type of meteorite (irons, stones, or stony-irons) and their availability. Meteorites are usually sold by the gram. Irons are the cheapest by far, usually sold for under $1.00 per gram. The most readily available and largest irons are often sold by the pound. For these, as the weight goes up the price per pound or gram goes down. The best and most available iron today is called Gibeon and comes from a huge strewn field in Namibia, Africa. This meteorite is remarkably resistant to rusting. Besides sheer weight and size, the collector looks for interesting shapes. If it has a "sexy" or "exotic" shape, it sells for a higher price. Interestingly, most people agree upon what an exotic meteorite shape is. You'll know it when you see it. Collectors can still obtain other common irons such as Odessa (Texas), Canyon Diablo (Meteor Crater, Arizona), Toluca (Mexico), and Sikhote-Alin (Siberia) for prices a bit higher than Gibeon.

Most new collectors begin with irons but soon gravitate to stones. The ordinary chondrites—not "ordinary" in the literal sense—are by far the most common meteorites to fall to Earth. They usually fall in clusters called multiple falls. As with irons, the more plentiful the specimens, the cheaper they are. Don't expect any stony meteorites to cost less than $1.00 per gram; even the most commonly available stones run higher. And buyers may choose broken fragments or individual stones with nearly 100 percent dark fusion crust. Broken fragments have visible interiors and may cost a bit less than complete individuals. The least expensive stones, Gao, come from Upper Volta, Africa. The average 1998 price of about $2.00 per gram will probably increase as supplies dwindle. Other available ordinary chondrites such as Plainview (Texas), Dimmitt (Texas), Tenham (Australia), and Holbrook (Arizona) run considerably higher. The carbonaceous chondrites—the most primitive and unaltered of the chondrites—cost from $8.00 per gram to several hundred dollars per gram. Collectors usually start with Allende (Mexico), the most plentiful and cheapest carbonaceous chondrite, and graduate to more exotic specimens such as Murchison (Australia). Beyond Murchison, the prices rise steeply.

Stony-iron meteorites are sought by collectors because of their unique and beautiful appearance. In particular, collectors seek the pallasites, which have yellow to olive green olivine tucked within the interstices of an iron-nickel network. Three are relatively plentiful: Brenham (Kansas), Imilac (Chile), and Esquel (Patagonia). They are all comparably priced and fall among the medium priced ordinary chondrites.

The achondrites are much rarer than ordinary chondrites and collectors can obtain few for less than several dollars per gram. The eucrite Millbillillie from Australia is the only one still moderately priced. Of the SNC meteorites from Mars, only Zagami (Nigeria) is readily available and costs many hundreds of dollars per gram. This meteorite has sold for as much as $2,000 per gram. Occasionally, 1- to 2-gram fragments of rare SNC meteorites such as Nakhla (Egypt) become available at comparable prices. These rarest meteorites usually come from museums through trades with commercial dealers.

When considering the purchase of expensive meteorites, you must put the prices into understandable terms. In 1997, gold was relatively stable, selling in the precious metals market for around $350.00 per ounce, or $12.50 per gram. The Zagami SNC at $2,000 per gram is over 160 times the price of gold. The only stones worth more are Moon rocks. To my knowledge, only one privately owned lunar meteorite was available in 1998, weighing under 10 grams. It's price? Over one million dollars. In March 1997, a new meteorite from the lunar highlands was recovered from the Libyan Sahara desert and will probably hit the market sometime in the future.

If these prices seem high today, they would seem preposterous by the standards of the early twentieth century. Yet in 1913, the Field Museum in Chicago offered a howardite, a rare achondrite, to the Smithsonian Institution for $5.00 per gram. After heated negotiations, the price was finally lowered to $3.50 per gram, nearly three times the 1913 price of gold, but the buyer backed out of the deal. Today, that same stone would be a steal at $100.00 per gram.

This drawing, based on a quaint etching by Henry Robinson, shows the fireball of August 18, 1783, near Newark-on-Trent, England. The meteoroid appears to have broken into numerous pieces while retaining cosmic velocity, which may have resulted in a meteorite shower. Compare this to the photograph of the Peekskill fireball on page 30.

Noctilucent cloud display photographed from Namao, Alberta, Canada, on July 5, 1988. Bright object above center is the planet Jupiter. —Courtesy of Mark Zalcik

The Ensisheim meteorite in France at the Ensisheim Museum. The meteorite has become rounded through centuries of sample removal. It now weighs only 122 pounds and measures 12.6 inches wide. Remnants of the original shiny black fusion crust can be seen on the top. —Thomas C. Marvin photo, by permission of the Museum of Ensisheim. Reprinted from *Meteoritics*

An Allende carbonaceous chondrite from the Chihuahua, Mexico, fall in 1969. The purple-gray crust is typical of Allende individuals. Note the round chondrules and white inclusions. Specimen is the size of a small apple.

An iron meteorite fragment from the Sikhote-Alin fall. Note the jagged edges produced when the rock was torn from the main body upon impact. Specimen is 4 inches long and weighs 1.2 pounds.

Plate I

Aerial view of Meteor Crater, looking northwest with the San Francisco Mountains in the background, 40 miles away. The crater is nearly a mile wide and 600 feet deep. —David J. Roddy photo, courtesy of Meteor Crater Enterprises

Bedouin guide Haban ibn 'Ali al-Jarbu contemplates the magnificent 4,500-pound Wabar meteorite at the time of its discovery. The cone shape with flat bottom reveals the meteorite's oriented flight. —James P. Mandaville Jr. photo, courtesy of Aramco World

Wabar Crater rim in the Empty Quarter of Saudi Arabia. The black-and-white material scattered on the rim is fused sand formed by the meteorite's heat of impact.
—James P. Mandaville Jr. photo, courtesy of Aramco World

Light-brown fusion crust of the Norton County achondrite. Specimen measures 4 inches across. —Robert A. Haag collection

A 154-pound specimen from Mundrabilla, Western Australia. Ablation of iron sulfide nodules produced the deep pits.
—Robert A. Haag collection

Plate II

Typical glassy black fusion crust on a eucrite achondrite from Camel Donga, Western Australia. Specimen measures 4 inches in its longest dimension.
—Robert A. Haag collection

Two iron-bearing stony meteorites. Both fell at the same time but show different crust colors. The lighter brown specimen shows much greater chemical weathering. The black crust on the specimen at left appears quite fresh.

Three surfaces of a weathered iron meteorite from Canyon Diablo: left: *terrestrial rust;* center: *terrestrial magnetite;* right: *terrestrial magnetite removed by tumbling to reveal nickel-iron.*

Slice of a New Concord stony meteorite, showing abundant nickel-iron grains with rust stains around each, demonstrating that oxidation continues despite careful storage and handling.

Plate III

Beads of ferric hydroxide and hydrochloric acid exude from between the Widmanstätten plates of this iron meteorite from Toluca, Mexico, and expand to form concentric rings of rusting metal. This reaction takes place in the presence of the mineral lawrencite.

Lawrencite reaction attacking a Brenham, Kansas, pallasite. Rust has worked its way under the enclosed olivine crystals, expanding and forcing the olivine from its iron bed.

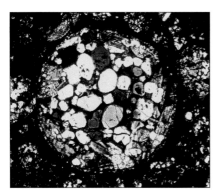

Photomicrograph showing porphyritic texture in a chondrule of Isna CO3 carbonaceous chondrite, showing individual crystals of olivine and pyroxene. Chondrule measures 0.02 inch.
—Courtesy of Rhian Jones, Institute of Meteoritics, University of New Mexico

Photomicrograph showing a chondrule with barred olivine texture from Smarkona LL3 chondrite. Chondrule measures 0.04 inch.
—Courtesy of Rhian Jones, Institute of Meteoritics, University of New Mexico

Plate IV

Slab of Allende CV3, showing well-formed chondrules from 0.008 inch to 0.078 inch in diameter. Several of the large chondrules are surrounded by black magnetite borders. Numerous white, irregular, calcium-aluminum inclusions are present. An unusual field of extremely small chondrules is included within the normal chondrule field. This may be a fragment of a CO chondrite.

A cut and polished slab from a brecciated diogenite that fell near Johnstown, Colorado, in 1924. Large fragments of olive-green hypersthene and small, brown bronzite crystals are set in a pulverized groundmass of pyroxene. Specimen is about 2 inches long.

Widmanstätten figures on an etched slab of an octahedrite from Smithville, Tennessee. Specimen is 4.4 inches at its base.

Cut slab of a Mundrabilla iron from Nullarbor Plain, Western Australia, showing numerous bronze-colored troilite inclusions. Specimen is 9 inches at its longest dimension.
—Robert A. Haag collection

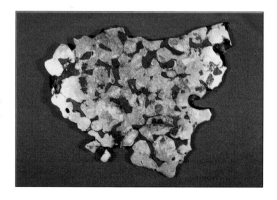

Plate V

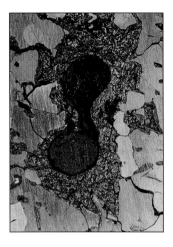

Bronze-colored troilite nodule in a Canyon Diablo iron. Nodule is bordered by graphite (black), and surrounded by schreibersite (silvery).

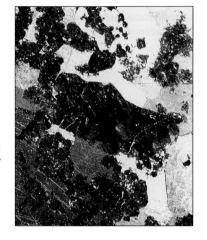

Photomacrograph of a pyroxene crystal in an Odessa, Texas, iron. All of the dark, granular inclusions are silicates.

Section and close-up view of a cut slab from the Brenham, Kansas, pallasite. A continuous network of nickel-iron encloses fractured grains of yellow olivine. Specimen is 12.7 inches across.

A hand-sized pallasite from Imilac, Atacama Desert, northern Chile. The light areas are olivine surrounded by darker nickel-iron, which has shown remarkable resistance to rusting.
—Robert A. Haag collection

Plate VI

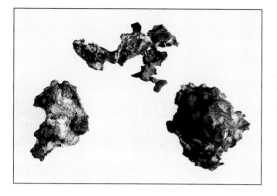

Imilac pallasites found on the surface of the Atacama Desert. Weathering has removed the olivine, leaving strangely shaped nickel-iron remnants behind. —Courtesy Peter Larson, Black Hills Institute of Geological Research, Hill City, South Dakota

Ursula Marvin happily displays a eucrite (ALHA 81006) in its collection bag immediately after she discovered the rare achondrite during the 1981–1982 season. —Robert Fudali photo, courtesy of Ursula Marvin, Smithsonian Astrophysical Observatory

Hubble Space Telescope high-resolution image of the center of the Eagle Nebula. The dark columns are molecular clouds where stars are forming. The red region is fluorescing hydrogen gas. In the blue area, solid dust grains reflect the blue light of the embedded stars. —Courtesy J. Hester, P. Scowen and NASA

The rotating solar nebula flattens into a disk as its center gains mass and collapses to a protosun. —Dorothy S. Norton painting

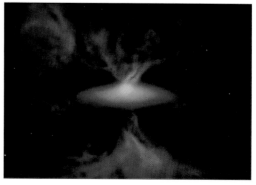

Plate VII

Accretion process leading to the formation of protoplanets. Bodies with tails are icy comets. —Dorothy S. Norton painting

The great daylight fireball of August 10, 1972, over the Grand Tetons and Jackson Lake, Wyoming. —James M. Baker photo

K-T boundary clay near Gubbio, Italy. The clay layer (brown) is to the right of the white layer. Tertiary layers are to the left of the boundary clay, Cretaceous layers to the right. Clay layer is 1/2 inch thick. —Walter Alvarez, Lawrence Berkeley Laboratory, University of California

This image of Comet Shoemaker-Levy 9, made in January 1994 with the newly repaired Hubble Space Telescope, shows sixteen individual pieces of the comet, each with a trailing dust tail. The relative motions of the pieces were watched closely as the impact date with Jupiter in mid-July approached. —Courtesy Dr. H. A. Weaver, T. E. Smith, and NASA

Plate VIII

FIVE

Meteorite Showers

My personal list of the most exciting and rare natural events on Earth would have to include total solar eclipses and volcanic eruptions, but the fall of a swarm of meteorites would be at the top of the list. Surprisingly, meteorite showers occur about as often as solar eclipses. Between the years 1901 and 2000, exactly fifteen annular and total solar eclipses will have been visible from the continental United States. In the same one hundred years there have been as many meteorite showers over the same area. Solar eclipses, being predictable, are more widely experienced because people have ample time to prepare for them. Meteoroids are subject to the same laws of motion as Earth and the Moon, but they are not predictable because they are too small to be observed in space. We become aware of them only when they begin their fiery atmospheric passage. And though a fireball may be seen passing over a half-dozen states, a meteorite shower will probably cover only a dozen or so square miles. Total solar eclipses can be observed only within the limits of an eclipse path, which may be 100 miles wide and hundreds of miles long, but almost everyone has a solar eclipse story. Far fewer people have a good fireball story and almost no one has a meteorite shower tale to tell. When was the last time you heard the beat of celestial stones on your roof?

Meteoroid Fragmentation

The forces acting on a falling meteoroid tend to fragment it during its atmospheric flight. Larger bodies, being subject to greater forces, are more subject to fragmentation. At about 10 miles altitude, a body traveling just under Earth's escape velocity (7 miles per second) is subjected to pressures of about 1,375 pounds per square inch. Fragmentation usually takes place at an altitude of between 7 and 17 miles. At the moment of fragmentation, cosmic velocity rapidly falls to zero and the fragments fall by gravity along a steep arc. Since the fragments share a common trajectory, they tend to remain close together and fall as a group.

Of the three basic meteorite types—irons, stones, and stony-irons—stones tend to fragment most easily. Irons and stony-irons often reach Earth's surface intact. Large iron meteorites are more likely to break on impact than in flight. Are shower meteorites pieces of one large body that broke up in the atmosphere? Or did they travel together through space? Opinions differ. Studies show that specimens from a given meteorite shower are remarkably similar in composition, suggesting they are fragments of one larger mass. Many show an angularity suggesting breakage, although this could be the result of impact with other parent bodies or asteroids while in space.

All meteorites have exterior fusion crusts, the result of heating during atmospheric flight. If a meteoroid breaks in flight, a secondary, less-well-formed fusion crust often forms on the broken face, if it still has sufficient speed after fragmentation. This does not preclude the existence (though unlikely) of meteoroid swarms in space that could enter the atmosphere together, resulting in a possible multiple fall. An example of a possible multiple fall is the Sikhote-Alin, Siberia, fall of large iron meteorites in 1947. Although thousands of fragments were recovered in and around craters formed by the impact of several large iron meteorites, many smaller specimens showed characteristics of complete individuals. This could mean that a large iron meteoroid was traveling in space in the company of many smaller iron bodies before encountering Earth. Multiple falls of individual meteoroids are much rarer phenomena, and fragmentation remains a far more plausible explanation for most meteorite showers. Until we can prove the existence of meteoroid swarms in space, the jury must remain out on this question.

The Distribution Ellipse

Meteorites in a multiple fall or shower tend to fall along an elliptically shaped strewn field, called the distribution, or dispersion, ellipse. The long axis of the ellipse lies along the cluster's direction of motion and the short axis defines the lateral scatter off the axis.

The distribution ellipse of the Homestead, Iowa, meteorite shower has a long axis that extends about 6 miles and is oriented north-south. There is a remarkable sorting of meteorites in the distribution pattern, with the largest and smallest specimens on opposing ends of the ellipse as a result of the kinetic energy of the individual meteorites as they fell. All meteoroids traveling together in space—regardless of their masses—have the same cosmic velocity when they enter the atmosphere. Likewise, if one large meteoroid breaks up in the atmosphere, the individual pieces at first share the same velocity. The mass of the meteoroid will ultimately determine where it will fall. The more massive meteoroids, with their greater kinetic energy, will penetrate the atmosphere to a greater depth before losing their cosmic velocity. Since they have greater momentum, they require more atmospheric braking before

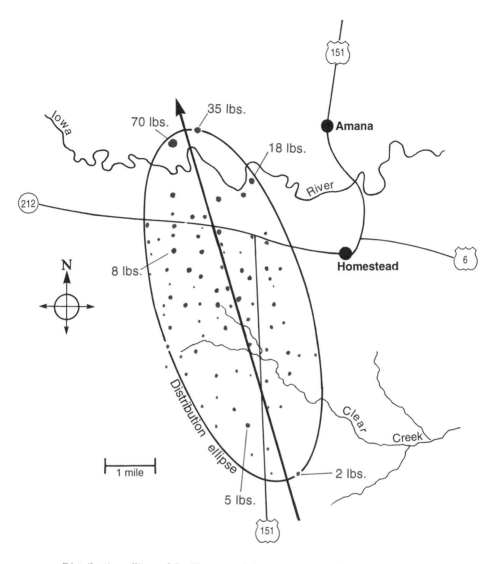

Distribution ellipse of the Homestead, Iowa, meteorite shower (1875).
Note the larger masses at the far end of the ellipse.

they reach the retardation point and fall due to gravity alone. The result is that larger meteorites travel farther down the major axis of the distribution ellipse before impacting Earth.

When meteorites reach their retardation point, their direction changes dramatically, from an angle relatively low to the horizon to a nearly vertical fall. Larger masses, with greater momentum, tend to fall at an angle to the vertical of about 30 degrees, while smaller masses fall more steeply, at about 20 degrees. This difference also results in the distribution of larger masses farther down the major axis of the ellipse.

The job of the meteorite hunter is to figure out the direction of the major axis of the ellipse and to locate the scatter within the strewn field. Studying the distribution of meteorites in the strewn field will reveal the shower's original direction. Like most things in nature, shower meteorites seldom follow textbook descriptions. The distribution ellipse of the Bruderheim, Canada, meteorite shower shows a more typical distribution in the strewn field. Here most of the larger meteorites are on the far end of the distribution ellipse, but a few are mixed in with smaller specimens. Large meteorites are often found with smaller ones that have broken off the main masses on impact, and distribution ellipses are usually several miles long and cover a dozen or more square miles. Finding meteorites in such a large area is a formidable task, even for the most ardent meteorite hunter.

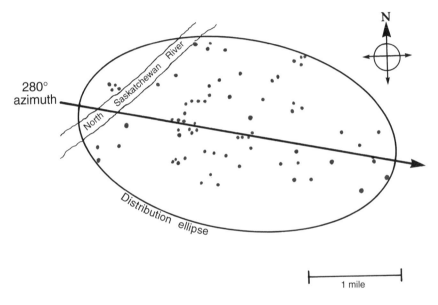

Distribution ellipse of the Bruderheim, Canada, meteorite shower (1960). –Reproduced with permission of the Smithsonian Institution

Meteorite showers have been part of Earth history throughout geologic time, yet their existence has been recognized for only about two hundred years. Curious medieval woodcuts seem to depict the fall of stones from the sky. Unfortunately, the actual events represented in these woodcuts are not documented with observational data, so we cannot know if the woodcut is depicting a real occurrence. Only by the nineteenth century do records become available that are complete enough to be useful, reflecting an increasing awareness among scientists that stones could and did rain down from the skies.

Woodcut from Conrad Lycosthenes' Chronicles of Prodigies *(1557), in which he describes alleged natural disasters, from earthquakes to the lethal rain of stones from the sky.*

Nineteenth-Century Meteorite Falls

L'Aigle. The nineteenth century was blessed with many spectacular meteorite showers. The century began with the fall of three thousand stones near the town of L'Aigle (Orne) in Normandy, France, on April 26, 1803. They were distributed over an ellipse 6 miles long and 2.5 miles wide, with its major axis oriented to the northwest. Jean Baptiste Biot, a member of the National Institute of France, investigated the fall exhaustively. His investigation resulted in the first recognition of a distribution ellipse and the recovery of about 17 pounds of small stones. The largest weighed about 4 pounds. Though not the most spectacular shower on record, historically it remains one of the most important, since Biot's investigation established beyond doubt the reality of meteorite showers.

Pultusk. The largest meteorite shower on record occurred in the early evening of January 30, 1868. Over one hundred thousand stones (possibly as many as three hundred thousand) rained down near the Polish town of Pultusk, a few miles north of Warsaw. Eyewitnesses reported seeing the meteoroid shortly after it entered Earth's upper atmosphere. It became visible at an altitude of almost 200 miles! This is surely the highest fireball on record spotted by the naked eye. Its terminal altitude—the point where it no longer generated light—was also unusually high, about 25 miles. Most of the meteorites, probably all fragments of a single, large stone, fell over an elliptical area measuring about 5 miles by 1 mile, between Pultusk and Ostrolenka. So high was the fall path that meteorites of similar composition were found as far away as Madagascar, Italy, and Russia, all believed to be part of the Pultusk fall. Thousands of tiny meteorites were collectively given the name "Pultusk peas." Larger specimens ranged from about a half-pound to over 4 pounds. All are covered by a black fusion crust contrasting with the light concrete-gray interior.

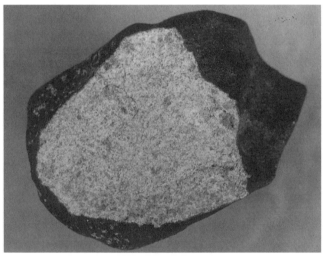

A stony meteorite from the Pultusk, Poland, fall shows dark fusion crust and light interior. This specimen measures about 2 inches in its longest dimension and weighs about 2.5 ounces. Most specimens from this fall are much smaller.

Orgueil Fall

A most significant fall occurred in Orgueil (Montauban), a small town in southwest France, on May 14, 1864. As only about twenty stones were recovered, the Orgueil (or-GAY) shower was a small one. But this meteorite turned out to be a very rare water-bearing Type CI carbonaceous chondrite, containing many complex organic compounds. In 1961, two scientists presented evidence for the presence of tiny structures within the Orgueil meteorite that resemble fossil algae. This

Carbonaceous chondrite from the Orgueil meteorite fall, 1864. –Robert A. Haag collection

most provocative conclusion has been the source of considerable debate ever since. The Orgueil meteorite contains many amino acids fundamental to life, suggesting that the essential materials of life can be, and in fact have been, formed outside the environment of Earth.

The Iowa Three

Homestead, Iowa. Three of the largest nineteenth-century meteorite showers in the United States fell in Iowa, all within the remarkably short period of fifteen years. The first occurred near the tiny town of Homestead, about 20 miles south of Cedar Rapids and 3 miles south of the Amish community of Amana. The meteorites are sometimes called the Amana meteorites. About 10:30 on the night of February 12, 1875, the cold, clear winter sky was punctuated by a great fireball that lit the

A broken fragment from the Homestead, Iowa, stony meteorite. White flecks of nickel-iron and a thick fusion crust are evident.

75

landscape below. It was brighter than the full Moon. Loud, almost continuous rumbles resembling the crash and roll of thunder accompanied the meteor. Two miles above the ground, the meteoroid exploded, sending to Earth more than one hundred pieces of stone, the largest weighing 74 pounds. The meteorites were scattered over a snowy area of 18 square miles from Amana north to Boltonville. The first stone was found lying in the snow where it had landed after bouncing from its original position. Over the next few years about 500 pounds of stones were recovered. Many were subsequently sold by their finders for the unheard-of sum of $2 a pound. Even at this early period in American meteorite history, "meteorite brokers" were discovering that this manna from heaven had a definite monetary value. The American free enterprise system established its place in the meteorite trade and it is still operating nicely today.

Estherville, Iowa. The second Iowa shower occurred at 5:00 P.M. on May 10, 1879, just four years after the Homestead fall. It fell 10 miles south of the Minnesota-Iowa border in northwestern Iowa near the town of Estherville. The fireworks accompanying the fall of the mete-

Stony-iron from the Estherville, Iowa, meteorite shower. Specimen measures 5.5 inches in its longest dimension.
–Robert A. Haag collection

orite were spectacular, stunning enough to stop an argument occurring at that moment in an Estherville baseball game. The Estherville shower produced only a few specimens, but they are truly remarkable specimens! The largest, described by its finder as jagged and rough to the touch, weighed 437 pounds. This was no ordinary stone meteorite. It was a much rarer stony-iron type called a mesosiderite. This specimen experienced a brief history in the hands of several short-term owners. Shortly after its recovery, the meteorite was put on display and admission was charged to view it. In the course of an ownership dispute, it was taken from the display and briefly buried in a cornfield. Later it was exhumed and confiscated by the local sheriff. Just how an attorney got his hands on the specimen is unclear. The attorney, no doubt recognizing its value, quickly transported it out of Emmet County before the local authorities could stop him. The meteorite eventually turned up in the Mississippi River town of Keokuk, Iowa. Charles P. Birge of Keokuk was now in possession of the mesosiderite and he contacted museums in Vienna, London, and Paris, offering to sell the specimen to the highest bidder. On November 30, 1879, the curator of minerals at the British Museum, M. H. Nevil Story-Maskelyne, purchased the main Estherville meteorite from Birge for £450 (about $1,000 in U.S. currency). The meteorite was then cut (with great difficulty) into three pieces, and distributed to museums in London, Paris, and Vienna, where they remain today. The British Museum retained the largest piece, which weighs 229 pounds.

Such meteorite "piracy" does not end there. The second-largest specimen from the Estherville shower, a 151-pound mass, was hidden in a cave on the Abner Ridley farm shortly after it was found on the nearby Amos Pingrey farm. The State of Minnesota purchased it and placed it under the care of the University of Minnesota. An element of intrigue entered the story in 1971 when the specimen was sent to the Smithsonian Institution for verification. When it was cut and examined, it suspiciously resembled the iron meteorites recovered at Meteor Crater in northern Arizona. The original 151-pound Estherville meteorite is still at large. The Estherville meteorite fall was unique. Along with a dozen or so meteorites in the 1-pound weight class, more than five thousand marble-sized iron nuggets fell. They must have accompanied the main masses, but it is unlikely they would have been traveling with the main mass in space. Examination of the interior of the larger specimens reveals a curious structure. Marble-sized lumps of iron are embedded in a ground mass of stony material. The late American meteoriticist H. H. Nininger, after examining hundreds of these nuggets in the Yale University collection, concluded that these small specimens originated from the breakup of the main mass. The stony material around the iron nodules disintegrated and fell away, leaving the stronger iron inclusions to break free and fall as individuals. A hailstorm of meteorite pellets fell that day in Estherville along an elliptical path some 7 miles in length. A farm boy witnessed the rain of small irons as they splashed into a lake, causing the cattle he was leading home to stampede. The iron in these specimens is malleable, and local Estherville artisans shaped many nuggets into jewelry as mementos of this great shower.

Forest City, Iowa. Forest City, Iowa, is only 65 miles east of Estherville. Eleven years after the Estherville fall, almost to the day, stones fell over a 1-mile-by-2-mile area. The date was May 2, 1890, the time 5:15 P.M. This was not like the impressive fall experienced at Estherville, in which over 700 pounds of stony-iron material were recovered. Only five large stones were found, weighing from 4 pounds to 80 pounds. Along with these fell about 1,800 small stones. In total, about 268 pounds were recovered.

Notable Twentieth-Century Meteorite Showers

Holbrook, Arizona. The twentieth century was slow to produce a meteorite shower of note. The first great shower occurred in a little railroad town in northeastern Arizona. The shower of meteorites that fell early in the evening of July 19, 1912, is still Holbrook's chief claim to fame. At about 7:00 P.M., the residents heard a series of loud booming sounds that lasted about a minute. One person said it sounded like a farm wagon being driven rapidly over a washboard road. Several explosions followed in rapid succession. One unusual aspect of the fall was

Meteorites from among the thousands of pea-sized stones that fell over Holbrook, Arizona, in 1912.

the lack of light phenomena, though a smoke train did form as the meteoroid traveled from west to east. Since the Sun was still up at the time, the bright sky may have prevented the fireball from being seen. One young boy living near the railroad yard ran into his house yelling, "It's raining rocks out there!" Thousands of tiny stones were pelting the ground, bouncing off metal roofs in the railroad yard and stirring up dust on the ground where they hit. They fell over an elliptical area measuring about a mile by one-half mile. The largest stone weighed over 14 pounds, but the unusual feature of this shower was the small size of most of the recovered meteorites. Using magnets to attract the tiny rocks (they contained some elemental iron), stone meteorites as small as 2 to 3 millimeters in diameter (0.078 to 0.117 inch) and weighing about 0.01 gram were found. These minuscule specimens win the prize for the smallest meteorites ever recovered. H. H. Nininger found many of them years later in a reexamination of the strewn field. Large red ants had removed the tiny celestial bodies from their original resting places and piled them into ant hills. Nininger dragged his magnet through the ant hills and recovered the world's smallest meteorites from the world's smallest meteorite collectors. In all, more than 16,000 individual stones were found, most no larger than buckshot. Most of the meteorites fell in Aztec near the old railroad yard about 6 miles east of Holbrook. Neither Aztec nor the railroad yard exists today, but if you are very patient and search along the railroad tracks you can still find small meteorites.

Sikhote-Alin, Siberia. Most meteorite showers are falls of stones. Undoubtedly, this is because stony meteorites are more easily broken up in the atmosphere than are irons. They are more brittle and are probably fractured even before they reach Earth. When iron meteorites fall, there is usually far less fracturing and disintegration of the main body.

The Sikhote-Alin fall didn't follow the rules, primarily because it was from an enormous iron meteoroid. Originally it may have been in the 200- to 300-ton class. The great iron mass broke apart in flight and fell in the Sikhote-Alin mountains near Novopoltavka, eastern Siberia, at 10:30 A.M., February 12, 1947. This fall was a monster. More than 25 tons of iron meteorites were recovered in and around 106 craters made by the impacting irons. The raining irons played havoc with the forest, cutting down entire trees and burying themselves in others. Except for the shower of iron meteorites found around Arizona's Meteor Crater, which fell in prehistoric times, the Sikhote-Alin shower has produced more meteorite tonnage than any other in recent history. Individual specimens and fragments number in the thousands.

1969—The Year the Rare Ones Came

Allende, Mexico. In the summer of 1969, I was running a small planetarium on the University of Nevada campus in Reno. Late one afternoon, a visitor came to my office and introduced himself as Frederick

Pough. He asked if I would be interested in buying about 100 pounds of carbonaceous chondrite meteorites, a most unusual question. In all the museums of the world there weren't a hundred pounds of carbonaceous chondrites. I had read about these strange carbon-based meteorites and had seen pictures of them, but I had never actually seen a carbonaceous meteorite. They were so rare that you had to visit a major museum like the Smithsonian or Chicago's Field Museum to see a small bit of a specimen. Now here was this stranger telling me he had a hundred pounds of carbonaceous meteorites in the trunk of his car and they were for sale! I was immediately curious about this man, who looked and acted like a Heidelberg professor. People often bring rocks they have found to the planetarium for identification. In my many years' experience, they had all proven to be "meteor-wrongs." No one had ever brought in a suspected meteorite with such confidence as to call it an iron or stone, much less a carbonaceous chondrite. Yet this "Heidelberg professor" had not only brought in suspected meteorites but identified them without so much as letting me see them. No ordinary man, this Frederick Pough.

We walked out to the parking lot. The stones were in the trunk of a vintage car that had seen many rough off-road miles. Pough lifted the trunk lid and there, carelessly wrapped in old El Paso newspapers and nestled in an open cardboard box, were dozens of the world's ugliest-looking rocks. They were dark rocks covered with what appeared to be a black crust. Where they had been broken, I could see something of the gray interior structure. Sandwiched between the black crust and gray interior was a purple layer. There were small, roughly spherical inclusions protruding from the gray matrix. (See color plate I.) Pough had a look of triumph on his face, which must have contrasted with my own expression of confused denial and doubt.

"Look, I'll have to grind a flat face on one of these rocks before I can say anything about them," I told him. "This will take some time, so why don't you leave them here and check back with me tomorrow?" He looked at me with some surprise and said in a resigned but patient tone, "Oh, that's all right. I can wait while you grind it." He obviously knew something about these rocks that I didn't.

We carried the world's ugliest rocks down to my laboratory, where I selected the specimen with the flattest face and began grinding. Stony meteorites are made of silicate minerals that are quite hard. It usually takes an hour or more to grind a flat face onto one. I touched the specimen to the grinding wheel and immediately a cloud of black, sooty dust rose from the grinding surface. This rock ground flat about as fast as you can sharpen a pencil in an electric pencil sharpener. After an appropriate "Oh my God," I immediately stopped grinding and went to the sink to wash off the gray powder covering the flat face. I carefully let a few drops of water drip onto the freshly ground surface (a wet

Cut face of an Allende meteorite shows densely packed round chondrules set in a gray matrix. Irregular white inclusions are scattered randomly through the meteorite. Specimen measures almost 2.5 inches in its longest dimension.

surface reveals its structure far better than a dry surface). To my amazement, the rock soaked up the water like a sponge! It was dry in an instant. Further, there were dozens of round inclusions all over the fresh surface. I could tell from Pough's look of amusement that he was having great fun with my astonishment and blundering attempts to identify these stones. I quickly excused myself and raced back to my office and my small library of meteorite books. I soon discovered that I didn't know much about carbonaceous chondrites. These were carbonaceous meteorites all right—the round inclusions were the chondrules typical of chondritic meteorites. And the man who gently introduced me to the solar system's oldest rocks was Dr. Frederick Pough, a world-renowned mineralogist and author of a well-known field guide to rocks and minerals. Was it a coincidence that this very field guide was sitting on my bookshelf next to the meteorite books?

And the meteorites? These were freshly fallen specimens from the now-famous Allende, Mexico, fall that had occurred just six months earlier. I did indeed purchase the trunk-load of Allende specimens from Dr. Pough. Where else could I find rocks older than Earth?

The year 1969 was an extraordinary year for meteorite showers of rare vintage. It all began in the dead of night over the little north-central

Mexican town of Pueblito de Allende in the state of Chihuahua. At 1:05 A.M. on February 8, a brilliant fireball appeared, lighting the landscape for miles. The single body producing the fireball apparently disrupted high in the atmosphere and continued to fall, still burning, as many individual fragments. (Many of the specimens display a secondary fusion crust, indicating breakage while still incandescent.) After reaching their retardation points, they fell by gravity and were subject to prevailing winds. They fell over a strewn field with its long axis extending for 30 miles along a south-southwest to north-northeast direction, possibly exceeding 180 square miles in area. This is the largest strewn

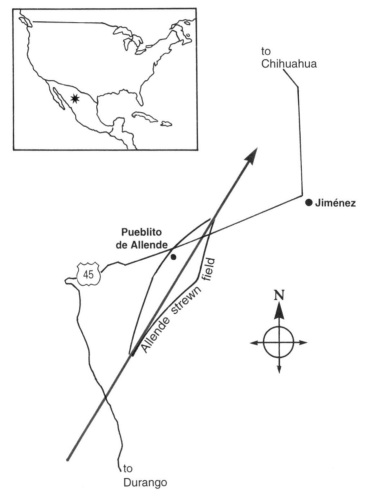

Allende meteorite strewn field and distribution ellipse. The long axis of the strewn field is more than 30 miles. −Redrawn with permission from the Smithsonian Institution

field ever investigated. Predictably, the largest masses fell at the far north-northeast end of the strewn field. A specimen weighing over 240 pounds that had shattered into many smaller pieces on impact was found at this point. More than 2 tons of meteoritic material was recovered in the first few weeks after the fall. Meteorites are still being found in and around the strewn field even after more than twenty-five years of searching. Allende takes the prize as the largest carbonaceous meteorite fall in history. It also has the largest strewn field area known.

The Allende meteorites turned out to be rare Type CV3 carbonaceous chondrites. Among other things, they contain curious, irregular white inclusions of an aluminum-calcium-titanium mineral that may represent the first silicate minerals to have condensed out of the solar nebula 5 billion years ago. More Type CV3 material was recovered from this fall than was known in all the world before 1969.

Murchison, Australia. As if Allende wasn't enough, meteoriticists were blessed again that same year by the fall of another carbonaceous chondrite, this time halfway around the world in the town of Murchison, Victoria, Australia. It was a Sunday morning, September 28, and most residents were attending church when the fireball appeared, signaling the arrival of the Murchison meteorites. Looking more like unburned charcoal briquettes than rocks, the meteorites were scattered across a

Murchison, Australia, carbonaceous chondrite. Larger, unbroken specimen measures about 2 inches across.

5-square-mile area within the town. People gathered them from their yards and neighborhood streets. Hundreds of stones were found, the largest weighing only about 15 pounds. More than 220 pounds were recovered. The Murchison meteorites are CM2 carbonaceous chondrites, with important differences from the Allende meteorites. CM2 meteorites are water-bearing. They contain about 10 percent water, resulting in a much more friable and fragile specimen. They must be collected quickly, before the weather has a chance to destroy them. Like the CI carbonaceous chondrites (Orgueil meteorite), those from Murchison contain organic compounds, namely amino acids, that are found in all life-forms on Earth. This immediately raises questions about the origin of life, a subject of much research and debate today.

This account of important meteorite showers of the past two centuries is necessarily incomplete. Many other showers occurred (see tables), and all are significant to the student of meteorites.

SELECTED OBSERVED METEORITE SHOWERS

Fall	Date	Location	No. Meteorites	Type
Allende	1969	Chihuahua, Mexico	thousands	Stone (CV3)
Beardsley	1929	Rawlens Co., Kansas	60	Stone (H5)
Bruderheim	1960	Alberta, Canada	75+	Stone (L6)
Estherville	1879	Emmet Co., Iowa	hundreds	Stony-iron (MES)
Forest City	1890	Winnebago Co., Iowa	500+	Stone (H5)
Holbrook	1912	Navajo Co., Arizona	16,000+	Stone (L6)
Homestead	1875	Iowa Co., Iowa	100+	Stone (L5)
Johnstown	1924	Weld Co., Colorado	100+	Stone (ADIO)
Knyahinya	1866	Ukraine	1,000+	Stone (L5)
L'Aigle	1803	Orne, France	2,000+	Stone (L6)
Millbillillie	1960	Wiluna Dist., Australia	hundreds	Stone (AEUC)
Mocs	1882	Transylvania, Romania	3,000	Stone (L6)
Murchison	1969	Victoria, Australia	hundreds	Stone (CM2)
Nakhla	1911	Alexandria, Egypt	40	Stone (ACANOM)
New Concord	1860	Muskingum Co., Ohio	30	Stone (L6)
Norton County	1948	Norton Co., Kansas	100+	Stone (AUB)
Nuevo Mercurio	1978	Zacatecas, Mexico	thousands	Stone (H5)
Orgueil	1864	Montauban, France	20	Stone (CI)
Pasamonte	1933	Union Co., New Mexico	75+	Stone (AEUC)
Pultusk	1868	Warsaw, Poland	200,000	Stone (H5)
Sikhote-Alin	1947	Maritime Province, Siberia	thousands	Iron (IIB)
Tenham	1879	Queensland, Australia	hundreds	Stone (L6)

SELECTED UNOBSERVED METEORITE SHOWERS

Fall	Date	Location	No. Meteorites	Type
Brenham	1882	Kiowa Co., Kansas	thousands	Stony-Iron (PAL)
Campo del Cielo	1576	Chaco, Argentina	thousands	Iron (IA)
Gibeon	1836	Namibia	thousands	Iron (IVA)
Imilac	1822	Atacama, Chile	hundreds	Stony-iron (PAL)
Mundrabilla	1911	Western Australia	hundreds	Iron (IRANOM)
Plainview	1917	Hale Co., Texas	1,000+	Stone (H5)
Toluca	1776	Mexico, Mexico	thousands	Iron (IA)
Vaca Muerta	1861	Atacama, Chile	dozens	Stony-Iron (MES)

Watch Out for Falling Objects!

Recently, interest has increased in the potential dangers large meteorites may pose, especially to areas with dense populations. There are many reports of meteorites having fallen on buildings. In the United States at least twenty-one authenticated strikes have been recorded in the twentieth century. Four of these are especially interesting. On April 8, 1971, in Wethersfield, Connecticut, a suburb of Hartford, a stone weighing 12 ounces fell through the roof of a house and lodged in the living room ceiling while the occupants slept. The homeowners discovered it the next day. Eleven years later another stone fell through the roof of a house in the same town, less than 2 miles from the first house. This meteorite was considerably larger than the first, weighing 6 pounds. It penetrated both the roof and the ceiling of the living room and bounced off the floor into the dining room, where it ricocheted off the ceiling, knocked over a chair, and came to rest under the dining room table. Wethersfield is unique in the historical record: The odds of two meteorites landing within 2 miles of each other over a lapsed time of a mere eleven years are practically zero. Though they appear to be nearly identical L6 chondrites, they have vastly different exposure times in space, and their minerals show different degrees of shock.

The third of the four falls is unique for another reason. The Sylacauga, Alabama, fall of 1954 is the only authenticated instance of a meteorite striking a human being. On November 30 at 1:00 p.m., two stone meteorites fell 2 miles apart. The larger meteorite, weighing 8.5 pounds, crashed through the roof of a house. It penetrated the ceiling of the living room, and bounced off a radio. Its final trajectory carried it to the couch where Mrs. E. Hulitt Hodge lay sleeping. She was struck on the hip by the ricocheting rock, and even though she had been covered

Stony meteorite that crashed through the roof of Mrs. Hulitt Hodge's house in Sylacauga, Alabama.
—Smithsonian Institution

by two heavy quilts, she suffered serious bruises. Fortunately, this was an indirect hit and the meteorite's energy was all but spent by the time it struck the woman. Mrs. Hodge's meteorite is in the permanent collection at the Smithsonian Institution.

The fourth meteorite fall of special interest occurred while this book was being written. Shortly before 8:00 P.M. on October 9, 1992,

Marlin Cilz holds the 26-pound Peekskill meteorite. Michelle Knapp is to the right of Cilz and her family is at the far left. Jim Schwade (far right) *and Ray Meyer* (behind) *teamed up with Cilz to purchase the meteorite.*
—Courtesy Montana Meteorite Laboratory

The famous Peekskill "meteorite car" on display at the 1993 Tucson Gem and Mineral Show. —Courtesy of Dimitri Mihalas

a brilliant fireball appeared over the eastern United States, witnessed by thousands of people (see page 30). It was heading north-northeast and was spotted by people in Kentucky, North Carolina, Maryland, and New Jersey. Sonic booms were heard as the meteorite passed overhead. The meteorite had an estimated initial mass of about 22,000 pounds and hit the top of the atmosphere at about 8.8 miles per second. It reached its retardation point 18 miles above the Earth. Ablation ceased at that time. The meteorite(s), falling vertically due to gravity at about 164 miles per hour, impacted Earth.

Eighteen-year-old Michelle Knapp was at home in Peekskill, New York, as the fireball was being seen by thousands of high school football fans at games along its path. Suddenly, she heard a loud crash outside, disturbingly close to the house. "It sounded like a three-car crash," Michelle later told reporters. When she stepped outside to investigate, she found the trunk of her car transformed into a twisted mass of metal. A 26-pound stony meteorite had struck the 1980 Chevy Malibu, penetrating the floor of the trunk, barely missing the gas tank, and coming to rest in a shallow impact pit beneath the car. She reached down and touched the meteorite and noted it was still warm and smelled of sulfur.

Within hours, news of the extraordinary fall reached museums, collectors, and meteorite dealers around the country and in Europe. They all wanted to purchase the now-famous meteorite. As it turned out, it was not so extraordinary as meteorites go. In the technical sense, it was an ordinary chondrite, which have sold for about $3 per gram (0.035 ounce), or $35,412 for the 26-pound specimen. But this ordinary meteorite had zeroed in on an automobile, which made it special in collectors' eyes. The final bid by a consortium of three bidders surpassed $69,000. If you can't afford the meteorite, how about the car? The car was worth about $300. But how many meteorite-damaged cars exist? It appeared at the 1993 Tucson Gem and Mineral Show and is rumored to have been sold to a collector for $10,000.

Analysis of the flight path of the fireball led some to believe the Peekskill stone was not alone. It may have been part of a multiple fall. Larger pieces may have landed farther down the flight path. If so, finding them will truly be a treasure hunt.

There are several documented reports of animals killed by falling meteorites. Perhaps the best known in the United States happened during the fall of the New Concord, Ohio, meteorite on May 1, 1860. About one hundred stones weighing a total of more than 100 pounds were recovered. One fell in a field, striking and killing a colt. A dog was killed by a rare meteorite that fell in Nakhla, Egypt, on June 28, 1911. Perhaps it would have been some consolation to the dog's owner to know that the rock came from Mars.

Interior of the Peekskill meteorite shows a beautifully brecciated texture attesting to the violent history of its parent body. –Allan Langheinrich collection

SIX

Two Great Siberian Meteorite Falls

The total weight of the meteorites in a typical meteorite shower is usually less than a thousand pounds; the heaviest individual specimens average about 20 percent of the total weight. Even in densely populated areas, the occasional fall of meteorites in this weight class does little damage. But there is always the possibility that Earth will be struck by a much larger body, and the damage would be incalculable. In a period of one hundred years, what are the chances of encountering something that weighs tens of tons—or more? If the twentieth century is any indication, the chances are 100 percent, because it did occur, not once but twice in the first fifty years.

The first event is still shrouded in mystery. No large meteorites were found—only the devastating effects of the encounter are evident. The second event, only thirty-nine years later, produced craters and deposited more than 23 tons of meteoritic iron. They both took place in Siberia.

Episode One: The Tunguska Event

We may never completely understand what happened on June 30, 1908, in a remote area of central Siberia. We know that a celestial body made its final plunge to Earth, leaving barely a trace of itself behind, but the nature of that celestial body still remains a question.

The story has many elements of good science fiction. At 7:17 A.M. in the Podkamennaya Tunguska River basin (longitude 101° 57' E, latitude 60° 55' N), an enormous ball of fire raced across the sky toward the north. It passed high over sparsely populated villages for hundreds of miles and many people saw it along the way. The fireball descended slowly until it seemed to almost touch the distant horizon. Then a "tongue of flame" shot up from the horizon, appearing to "cut the sky in two," according to eyewitnesses. The tongue of flame, possibly the final explosion of the body, was followed by loud bangs and tremors that shook the ground and broke windows. People close to the explosion reported blasts of hot air that reached hurricane force. The explosive sounds were heard over a radius of more than 500 miles, covering an area of 800,000 square miles. A magnetic and meteorological observatory 535 miles south of the "impact" zone registered a seismic disturbance at precisely 7:17 A.M. This was the first seismic recording ever made documenting the fall of a celestial body and the most accurate timing of any meteorite fall. The huge body hit the atmosphere with such force that barographs at meteorological stations all over Siberia recorded the resulting pressure wave as it passed. Even more extraordinary, microbarographs in England recorded the passage of the pressure wave five hours and fifteen minutes later. The same stations recorded another pressure wave more than twenty-four hours later—the waves had circled Earth and returned before dying out.

Across Siberia, northern Europe, and Asia, the night sky glowed brightly as a result of the tiny meteoritic particles blown high into the stratosphere. The particles reached altitudes of several hundred miles, clearly outside the atmosphere, and reflected sunlight above Earth's shadow line. For several nights after the event, the sky was so bright

Location of the Tunguska event in Siberia. Drawing shows the direction of the meteorite's approach and the site of the explosion, as determined by Leonid Kulik.

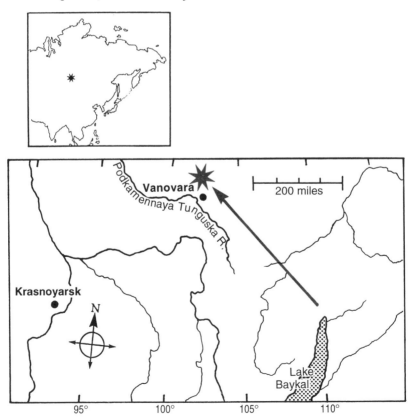

that only the brightest stars were visible. These "white nights" persisted for several days, diminishing as the particles slowly settled. Sunrises and sunsets were especially colorful, and bluish clouds (noctilucent clouds) of meteoritic particles could be seen against a red sky.

Although news of the great fireball reached the heart of the Russian state soon after the event, the political climate was not conducive to scientific investigation. Also, the presumed impact site could not have been more remote and rugged; it would take a major effort to mount an expedition. Years went by. World War I and the Bolshevik Revolution occupied the country's attention. Finally, in September 1921, more than thirteen years after the fall, the new Soviet Academy of Sciences instructed Leonid Kulik, curator of the Committee on Meteorites, to lead an expedition to Siberia to gather information from eyewitnesses. The stories he heard were incredible.

One of the nearest eyewitnesses was about 36 miles southeast of the center of the explosion. S. B. Semenov reported:

> I was sitting on the porch of the house at the trading station of Vanovara at breakfast time and looking toward the north. I had just raised my axe to hoop a cask when suddenly in the north above the Tunguska River, the sky was split in two; high above the forest the whole northern part of the sky appeared to be covered with fire. At that moment, I felt great heat as if my shirt had caught fire; this heat came from the north side. I wanted to pull off my shirt and throw it away, but at that moment there was a bang in the sky, and a mighty crash was heard. I was thrown three sagenes [about 21 feet] away from the porch and for a moment I lost consciousness. My wife ran out and carried me into the hut. The crash was followed by noise like stones falling from the sky, or guns firing. The Earth trembled, and when I lay on the ground I covered my head because I was afraid that stones might hit it. At the moment when the sky opened, a hot wind, as from a cannon, blew past the huts from the north.

Semenov apparently had felt the heat radiating off the fireball as it passed overhead. The final explosion generated shock waves strong enough to knock him off the porch. The waves were followed by more thunderous noises and a hot wind expanding away from the explosion point.

The closest eyewitnesses were the Evenki people, inhabitants of the northern part of the Krasnoyarsk district. They were camped near the mouth of the Dilyushma River, several miles north of the Tunguska River and Vanovara trading post. Akulina Potapovich told this story:

> Early in the morning when everyone was asleep in the tent it was blown up into the air, together with the occupants. When they fell back to Earth, the whole family suffered slight bruises but Akulina and Ivan actually lost consciousness. When they regained consciousness they heard a great deal of noise and saw the forest blazing around them and much of it devastated.

The Evenki nearest the explosion point told how the taiga (old growth forest of pine, cedar, and deciduous trees) was flattened and burned and some of their reindeer and dogs were killed by the explosion.

Kulik's first investigation of Tunguska did no more than gather information from eyewitnesses as far as 600 miles away from the explosion point. It did confirm, however, that something extraordinary had occurred in central Siberia. The location of the presumed impact point itself was only vaguely surmised, but the reports from the Evenki about the flattened forest were compelling. No one had ever seen anything like this. For five years, reports of the fireball's devastating effects continued to reach Kulik in Leningrad. Things moved slowly in the Soviet Union in those days. The Soviet Academy eventually decided the matter should be investigated further, and Kulik was chosen to lead an expedition. He would go to the area north of Vanovara, where he was certain a great meteorite had struck.

First Expedition

On February 12, 1927, Kulik and an assistant set out from Leningrad for the unexplored lands north of Vanovara. In those days, anyone traveling beyond the range of the Trans-Siberian Railway had to proceed by horse and sled to all points north or south. For the exploratory expedition, this meant an enormous undertaking. It involved packing enough provisions for several months and hiring a guide to take the expedition to the fall site. Traveling during the spring thaw was dangerous because rivers were flooding and the land was swampy. The party had to hack its way through the taiga for miles because the saturated ground was so unstable and the foliage so thick. Progress was slow at best. Kulik reached Vanovara on March 25, where he set up a base camp and prepared to travel the remaining 36 miles north in search of the fall site. He tried to set out the following day, but the snow was too deep for the heavily burdened horses. The men were forced to trade their horses for reindeer, and on April 8, they finally left Vanovara with an Evenki guide.

They proceeded northeast along the Chambé River until they came to the confluence of the Chambé and Makirta Rivers. It was April 13. Standing on the banks of the Makirta and looking north, Kulik saw the first of the blown-down trees. He had reached the perimeter of the devastated area, and his small band of explorers was still more than 20 miles from the estimated impact site. At this point, all the uprooted trees lay with their tops pointing southeastward.

Nothing in Kulik's experience could have prepared him for such devastation. The closest modern equivalent resulted from the eruption of Mount St. Helens on May 18, 1980. The conifer forest on the volcano's north flank was flattened, and the fallen trees pointed away from the lateral blast that exploded from the mountain's side. Even in this nuclear

Downed trees in Tunguska, about 5 miles from the point of explosion.
−L. A. Kulik photo, Smithsonian Institution

age, the power of the Mount St. Helens eruption was impressive. The energy of the Tunguska explosion was much greater.

From his vantage point, Kulik could not estimate the extent of the devastated area. He climbed several ridges and surveyed the land as far as the mountains, 7 miles to the north. The trees lay starkly outlined against the melting snow. Some were still standing but their trunks, 2 or more feet across, had snapped in two. The tops had been hurled many feet to the south. The devastation continued as far as he could see and the hills were bare. Only here and there in protected hollows were trees still standing. The destruction extended more than 10 miles to the east. He found an area where the trees had been scorched by the blast. Later he realized that the entire forest to the north had burned.

On May 30, Kulik reached a swampy basin, which he called a cauldron. It was partly surrounded by low-lying hills. Climbing these hills, he circled the rim of the basin and measured the direction in which the trees had fallen. For the first time he saw trees that were not

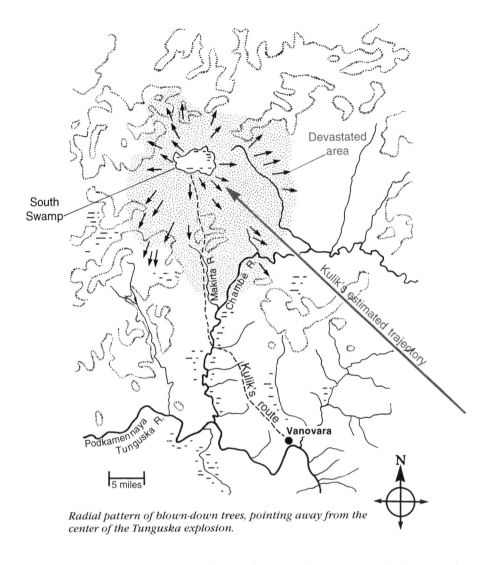

Radial pattern of blown-down trees, pointing away from the center of the Tunguska explosion.

all pointing southeast, as they had been when he was farther south. Instead, he observed a radial pattern with trunks pointing away from the basin. Later, Kulik wrote, "There could be no doubt. I had circled the center of the fall!"

Kulik concluded that the meteorite, or fragments of it, must lie somewhere in the basin. Perhaps the basin itself was part of a crater formed by the impact. In it he noticed flat, circular features that were scattered randomly. He called them "crater holes" and likened them to lunar craters as seen through a telescope. He was convinced that buried within these holes were the remnants of the great meteorite. This conclusion, reached in considerable haste and after only a cursory examination, was to obsess him for the next fifteen years in his relentless search for the Tunguska meteorite. Unfortunately, this assumption was wrong.

By a curious coincidence, while Kulik was investigating the Tunguska fall, Daniel M. Barringer was investigating an unusual crater on the northern Arizona high desert. He too suspected that the origin was a meteorite impact. No confirmed meteorite crater anywhere in the world was known at the time, and few experts were willing to say that impacts of this magnitude were even possible on Earth. Both Kulik and his counterpart in America were exploring unknown scientific territory. There is no question that Barringer's revolutionary work influenced Kulik, as he mentions it often in his reports. It was surely on his mind when he explored the Tunguska site.

Second Expedition

Kulik's report to the Soviet Academy after his return to Leningrad in the fall of 1927 was met with renewed interest. He convinced the Academy that he had found the center of the blast. The basin was ground zero. The flat holes in the basin were likely to contain meteorites. It only remained to make magnetic surveys on the "craters" to verify the exact positions of these buried meteorites.

A second expedition followed in April 1928. From the beginning it was beset by difficulties. Again, spring floods made travel difficult and slow. Illness among the expedition members forced a return to Vanovara in August. During the summer months the holes remained filled with water, making magnetic readings impossible. It wasn't until late October, when the water was frozen, that Kulik could begin his measurements. Disappointment prevailed; no magnetic readings were indicated by the instruments. Further reconnaissance of the basin and the digging of a shallow trench next to one of the main holes didn't provide any additional evidence for the existence of meteorites in the presumed fall area.

These negative results reported by Kulik when he returned to Leningrad in the fall raised some questions by Academy members. Was Kulik truly correct about the meteoritic nature of the basin and the "craters"? A few geologists saw the basin as a natural feature that had not been modified by an impact. Nonetheless, the downed trees pointing away from the site were a powerful argument. Undaunted, Kulik insisted that meteoritic material would be found there. Photographs and films taken during the second expedition kept the level of excitement at the Academy high. Despite the disappointing results, the membership elected to finance another expedition the next year.

Third and Fourth Expeditions

Drilling tools were among the equipment taken into the taiga for the third expedition. This was a much larger undertaking than the previous two. Among the members were an expert in swamp geology; E. L. Krinov, a mineralogist and meteoriticist from the Academy; and several workmen trained to operate drilling equipment. This time the team planned to stay through the winter and work during the summer and fall of 1930.

They arrived at the basin on April 6, 1929, and began the study of the "craters." A trench dug by hand through the largest hole drained the accumulated meltwater and allowed the workers to study the condition of the material within. After more than a month of difficult digging, they reached a depth of 12 feet, but still found no obvious impact features. The walls of the trench showed only undisturbed material. This did not discourage Kulik. He directed the men to bore holes into several additional "craters." This arduous task continued through the winter and well into the following spring. All this labor was done by hand. They built a log hut over the drilling site to protect the workers during the frigid winter months. Again they found no trace of meteoritic material, though they had drilled through more than 90 feet of solid frozen ground. Even Kulik was now beginning to acknowledge that his meteorite "crater" hypothesis might be in error.

During this expedition, new studies of the distribution of fallen trees around the basin placed the fall point in a swamp on its south side. "South Swamp" was permanently filled with water. Aerial photographs could not be taken because of inclement weather, and it wasn't until the summer of 1938, during Kulik's fourth and final expedition, that aerial photographs were finally made. The photographs were incomplete, but they were good enough to confirm his relocation of the fall point under the South Swamp.

During the third and fourth expeditions, special attention was given to the area surrounding the South Swamp. The swamp itself seemed normal and had it not been at the apex of the downed forest, it would have been overlooked. A few trees predating the Tunguska event remained standing. After draining one of the larger "craters," drillers found a stump that proved the hole was old and not related to the meteorite. Around the swamp, there were no signs of an energetic impact. Where was the soil and rock that would have been ejected? Where were the upturned strata? Where was the meteorite crater with its raised rim that should have been created at the moment of impact? All these signs, so evident at the Arizona crater, were missing here. Years later, after many more expeditions, the final answer would become known: The object had never reached the ground. It had disintegrated and vaporized in a final explosion several miles above the basin, leaving a leveled forest below and only the tiniest traces of its celestial origin.

This conclusion would have been devastating to Kulik. For more than twenty years, he had tenaciously battled the elements in the frigid Siberian taiga pursuing a great meteorite that didn't exist. World War II put an abrupt end to his quest. Hitler's army was on the outskirts

South Swamp, where Kulik believed the meteorite exploded. —Smithsonian Institution

of Moscow when Kulik volunteered to defend his homeland. He was wounded during the battle of Moscow, taken prisoner, and died in a Nazi concentration camp on April 14, 1942.

The Final Word?

Kulik died believing that the remains of the meteorite lay somewhere in the South Swamp. Fifteen years passed before research was resumed. In 1957, soil samples Kulik had collected from the drill holes and trenches in 1929 were finally examined microscopically. Among the soil particles were found tiny spheres of meteoritic dust that had sloughed off the incoming body as molten droplets and solidified. Some spheres were made of the iron-oxide mineral magnetite. Magnetite typically forms on the exterior of a meteoroid as it ablates in the oxygen-rich atmosphere. Other spheres were glassy, composed of silicate minerals that are typical of stony meteorites. Some spheres were composites of both types, proving they had formed together. These spheres were similar to those Donald Brownlee retrieved from the floor of the Pacific Ocean thirty years later.

Slowly, a new picture began to emerge. Perhaps early on that June morning in 1908, Earth had encountered a comet: a body composed of a mixture of ice and metallic and silicate chunks. The comet entered the atmosphere at the low angle of only 17 degrees. It traveled for 400 miles in a south-southeast to north-northwest direction, beginning above the north shore of Lake Baykal. The initial mass of the body was estimated to be about 1 million tons. It was traveling between 17 and

24 miles per second when it hit the upper atmosphere, and it retained most of its cosmic velocity throughout its trajectory. It lost nearly 95 percent of its mass in transit, so it was probably between 20,000 and 70,000 tons when it exploded about 5 miles above Tunguska. The energy of the blast was equivalent to that of a 20-megaton hydrogen bomb. Most of the energy was transferred into a shock wave that flattened more than 770 square miles of forest. The trees directly below the blast remained standing only because they presented a small cross section to the shock wave. The ice vaporized, the heat scorched the trees below, and a cloud of meteoritic particles was blasted into the upper atmosphere.

Update

The idea that a comet was the culprit responsible for flattening the forests around Tunguska nicely solved a basic problem of the missing meteorite: There was no meteorite, only ice and meteoritic dust. The shock wave preceding the comet could easily have downed the trees in a radial fashion. Mystery solved, the scientists concluded—until recently.

A great deal more is known today than in Kulik's day about the forces acting upon bodies that enter Earth's atmosphere at cosmic velocities. The force acting upon a meteoroid depends upon its velocity and mass. Its composition is an additional factor. Most asteroids have compositions of stony meteorites, which have a limited ability to withstand atmospheric forces. Stony meteorites that make it to Earth are probably no bigger than about 30 feet in diameter when they hit the atmosphere. A large percentage of their mass is vaporized, and chunks only a few feet or less in size fracture off the main mass en route, reaching Earth as meteorite fragments. Stony bodies between 30 feet and 330 feet in diameter are within a size and mass range that result in the total disruption of the bodies several miles above Earth. They explode with the force of tens of megatons, sending powerful heat and shock waves to the ground with devastating results. Stony meteoroids larger than 330 feet in diameter can resist atmospheric forces and impact Earth with a large fraction of their cosmic velocity intact. These form impact craters.

In 1983, Cal Tech astronomer Zdenek Sekanina published a paper in which he insisted that the Tunguska object was a stony meteorite, not a comet. At the shallow trajectory the object took, he believes a comet could not have remained intact to reach the lower atmosphere but would have broken up in the upper atmosphere and flashed out of existence within seconds. To reach the lower atmosphere it must have been a much denser object like a stony meteorite. Sekanina believes that the stony body was about 300 feet in diameter, placing it in a category of objects too small to survive the incredible forces imposed upon it. Enormous pressure differences existed between the forward and rear surfaces. The atmosphere rapidly compressed the forward surface, while behind the body a near-vacuum existed. This extreme pressure difference acted to flatten the body, creating even greater resistance to its forward motion.

For an object traveling at a speed of about 7 miles per second, the lower atmosphere 5 miles above Earth acts like an impenetrable wall. There, the object comes to a halt, immediately releasing its kinetic energy in an explosion equivalent to a 10- to 20-megaton blast, completely disintegrating the body. Had the object been smaller, aerodynamic forces would have been proportionally weaker, possibly breaking it up into smaller chunks, some of which could have survived to Earth's surface. The much higher cosmic velocities expected of a cometary body would have caused it to disintegrate much higher in the atmosphere with far less devastating effects than actually occurred.

Scientists seem reluctant to put Tunguska to rest. Recent observations were made that further tend to favor the hypothesis of a stony meteorite. The platinum-group metal iridium, and other rare elements in ratios consistent with a stony meteorite, were found in the spheres discovered at Tunguska; and in the mid-1980s researchers in Antarctica found ice, dating from 1909, rich in Tunguska-like meteoritic material. The material was much more widespread than thought earlier, leading to the conclusion that the meteoroid may have been in the 7-million-ton class, giving it a diameter of nearly 500 feet.

As if to confirm these ideas, the spacecraft *Magellan* photographed Venus in 1992 and revealed some curious, dark circular patterns on the surface that suggest a Tunguska-like mechanism. Some of these dark features surround impact craters, others do not. These patterns are the effects generated by the impact of asteroids into Venus's dense atmosphere.

Three impact features on Venus made by asteroids. Left: *A dark spot produced by an asteroid so small it disintegrated before reaching the planet's surface. Its atmospheric shock wave struck the surface, crushing surface materials to a fine debris, which appears dark in the radar images.* Center: *A similar circular feature but with some scattered fragments within it. A larger asteroid broke apart in the atmosphere, and its shock wave and some fragments reached the ground.* Right: *An asteroid large enough to survive passage through the atmosphere reached the surface intact, creating an impact crater and bright ejecta. Its shock wave also hit the ground, producing the dark circular area, 35 miles in diameter.* –NASA

Tunguska-like impact scars abound on Venus. This implies that such events are equally common on its sister planet, Earth, but, unlike Venus, Earth absorbs most impacts in its oceans. In this respect, Tunguska was unique. It was the first and remains the only witnessed impact by an asteroid or comet to create an area of such great devastation. Consider this: Had it entered Earth's atmosphere four hours later, it would have come down near Moscow; or, given an additional two hours, in the heart of Europe.

As if this story is not exciting enough, some people offer alternative explanations. The similarities between the Tunguska event and the effects of an atomic bomb can hardly be missed. Was Tunguska the explosion of nuclear engines in a disabled alien spacecraft? To some, the remote nature of the site suggests it was deliberately chosen by aliens to do little damage. Or perhaps a piece of antimatter explosively converted to pure energy when it contacted Earth's ordinary matter. Maybe Earth was hit by a mini–black hole. We can have fun with these esoteric ideas, but they are not really necessary—a comet or asteroid will do just fine!

Episode Two: Sikhote-Alin

Less than five years after Kulik's death, Siberia was again visited by a celestial intruder. This time there was no mystery. It was a solid iron meteoroid from space—a big one.

Mineralogical Museum in Moscow where a copy of the Medvedev painting of the Sikhote-Alin fireball is on permanent exhibit. –Courtesy of Peter Franken, University of Arizona

The morning of February 12, 1947, began normally. It was clear over the Sikhote-Alin Mountains of eastern Siberia. These mountains lie in Siberia's Maritime Province on a narrow strip of land wedged between the Sea of Japan to the east and China to the west. Farmers along the western front of the mountains went to their fields to tend livestock; loggers returned to the mountains for a day's work. Children went to school. There was not the slightest hint of what was about to occur. An artist named Medvedev in the village of Iman decided to set up his easel looking eastward at the mountains. He had just begun sketching when, at 10:38 A.M., a flaming fireball as large as the Sun suddenly appeared in the cloudless sky. It was traveling from north to south, and Medvedev was looking directly at it. The fireball was so bright that it cast moving shadows in broad daylight. Nearly every witness saw it changing color; red dominated, especially toward the end of the flight. An enormous, boiling, multicolored smoke train extended behind the fireball, marking its southerly path across the sky. Though it lasted at most only four or five seconds before it disappeared behind the hills on the western slope of the mountains, Medvedev's critical eye took in remarkable

Medvedev's original painting of the Sikhote-Alin fireball as he observed it from Iman. There are two copies of the original, one at the Mineralogical Museum in Moscow and a second held by the Krinov family.
Inset: *A Russian postage stamp commemorating the tenth anniversary of the Sikhote-Alin fall.*
—Courtesy Dr. Michael Peteav, Smithsonian Astrophysical Observatory

detail. He quickly began sketching while the image was still fresh in his mind. Soon the Sikhote-Alin fireball was recorded on canvas, the only picture in existence of the astonishing event. A copy of this painting is on permanent display in the Meteorite Exhibition Hall at the Mineralogical Museum in Moscow. A reproduction is on a Russian postage stamp issued in commemoration of the event.

Loggers closer to the descending mass saw it split into several pieces, which disappeared behind the trees; then they heard an ear-splitting explosion followed by a noise like rolling thunder. Observers close to the fall point felt the push of the shock wave as it expanded from the explosion point. It was felt 100 miles away, almost to Vladivostok, rattling and breaking windows along its path. Once the meteoroid splintered in flight, its cosmic velocity decreased to near zero, so the bodies fell as a meteorite shower, landing within a 1.2-square-mile area. The breakup of the main mass took place only 3.6 miles above Earth's surface—a remarkably low altitude, which accounts for the small area over which the fragments fell.

The impact site was discovered quite by accident the next day. Two pilots flying from the village of Ulunga to Iman passed right over the site as they cleared the crest of the Sikhote-Alin Mountains. There on the western slopes, among many splintered and severed trees, were several craters. Ejected rock and soil surrounding them contrasted with the snow that covered the ground. The pilots had witnessed the fireball the day before. They immediately associated it with the craters and reported their discovery when they reached Khabarovsk, north of Iman. The Geological Society in Khabarovsk quickly organized a search party composed of three geologists. On February 21 they flew with the same pilots over the cratered site, landed in a clearing only 7 miles south, and began their trek into the taiga. Like Tunguska, the area was overgrown with vegetation and the snow was deep. For three days they searched through the thickets without luck. Then on February 24, signs of destruction began to appear:

> As we advanced, the proximity of the place of the meteorite fall became more and more obvious, for the fragments of bedrock hurled out by the meteorite, and snapped branches of trees, began to appear in the snow. Farther on, fragments of rock weighing several tens of kilograms were found, and the snow which had been soft underfoot became dense, with a hard crust that supported the weight of a man. It was mixed with sand and clay, with large and small pieces of stone and forest debris. At last, in front of us appeared a huge crater. . . .

What a moment that must have been. It was obvious the fall had been a shower of an undetermined number of fragments. Before them, among the shattered trees, they counted thirty-three additional craters. This was just a cursory look. One hundred twenty-two craters and pits

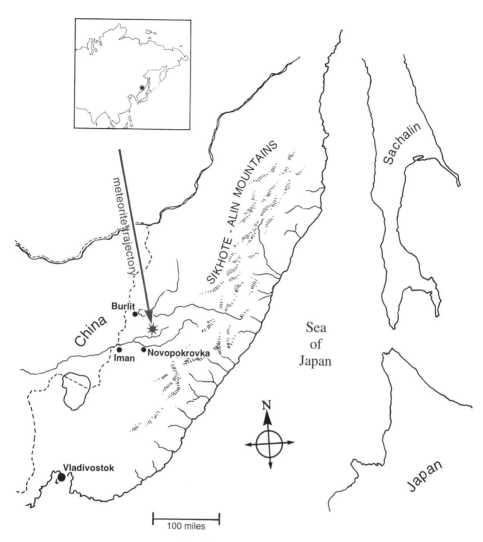

The Sikhote-Alin impact site.

were found later, hidden in the tangle of broken trees and brush. The largest was over 85 feet across and nearly 20 feet deep. It didn't take them long to partially excavate a small crater and locate the first of what would eventually be thousands of iron meteorite fragments.

This fall was truly unique in the history of meteoritics. In the past, many meteorites had been witnessed as they fell and were subsequently recovered, but there had never been an observed fall of iron meteorites of this magnitude, and certainly none producing such an extensive crater field. The meteorite crater fields where meteorites have been found are all ancient, predating human history, and were produced by iron meteorites. Time has taken its toll on these geological features, and much of the evidence of impact has been erased by erosion. This fall presented an opportunity to study the structure of freshly produced meteorite impact craters.

Serious fieldwork on the Sikhote-Alin fall began shortly after the discovery expedition communicated their findings to the Soviet Academy. Time was of the essence. It was important to examine the crater field and surrounding area thoroughly before natural processes began to modify the pristine fall. On April 27, an expedition arrived at the crater field. Among the scientists was E. L. Krinov, Kulik's assistant at the Tunguska site.

The Craters

For the next three weeks the geologists had a "field day." Moment by moment, discoveries were made. One of the first curious findings was that, contrary to expectations, the largest craters did not contain the largest meteorites, but only small fragments. Here the bedrock was shattered and scattered around the crater floors and piled up around their perimeters, forming a low rim. Around these craters, arranged in an asymmetrical pattern, was ejected soil and rock, forming an ejecta blanket. There were no signs of the intense heating that would surely have occurred had the body exploded on impact. No fused blobs of silica, often present around explosion craters, were found. The large

The largest crater at Sikhote-Alin measures 87 feet in diameter and is nearly 20 feet deep from the rim to the floor. –Smithsonian Institution

craters were excavated simply by the impact and shattering of the meteorites; they were not the result of an explosion. Such craters are referred to as penetration holes.

Another curious observation was made: The smaller the crater, the larger the meteorites in it. The dividing line between finding whole, unbroken meteorite specimens and only broken fragments seemed to be related to crater size, with 12-foot-diameter craters as the dividing point. Indeed, whole, unbroken specimens were found in almost all craters measuring 6 feet or less in diameter. This told an interesting story. The largest meteorites blasted out the largest craters, since they had the greatest energy on impact; but this greater energy also shattered the meteorites in the process, all but destroying them. Once the field workers understood this relationship between crater size and meteorite size, they could predict with some accuracy what might be found in the craters and holes.

After most of the craters were excavated, or at least sampled, it became evident that a relationship existed between the size of the crater and the mass of the meteorite that produced it. In the largest crater, 87 feet in diameter, 73 fragments were found weighing a total of only 154 pounds. By contrast, a crater only half the size yielded 232 fragments with a total weight of 391 pounds. Clearly, the largest crater lost a substantial portion of the original meteorite body. As a rule, the larger the original mass, the smaller and more numerous were the fragments produced on impact.

The Meteorites

The first geologists to reach the crater field naturally began their search for meteorites in the larger craters. What they found were iron meteorites that looked like shrapnel from an exploded bomb. (See color plate I.) A typical meteorite fragment was smooth and rounded on one side and often showed flow lines where the metal had melted and flowed along the surface. The other side was altogether different—generally flatter, with edges that were highly distorted and bent. The edges curled around from the smooth front side, forming ragged claws that could easily tear someone's skin if handled roughly. These specimens generally did not show an outer fusion crust of magnetite, indicating that they did not form in the usual way in the atmosphere. Instead, these meteorites formed when a single, large mass struck the ground and splintered into thousands of much smaller fragments.

When the second expedition arrived on the scene, the only meteorites that had been recovered were these fragments. Within only a few hours of searching, Krinov found the first whole, unfragmented specimen. Compared with the fragments they had been studying, this new find was like a meteorite from an entirely different fall. The specimen weighed about 24 pounds, was smooth on all surfaces, and bore the

A shield-shaped, iron meteorite fragment from Sikhote-Alin showing a smooth, rounded exterior with flow lines extending to a distorted, jagged edge. Specimen is 3.25 inches long.

unmistakable trademark of a meteorite—a fusion crust. The crust was bluish gray, much like an iron bar after it has been heated by a torch.

The most distinctive feature was the presence of shallow cavities called regmaglypts. The size of these cavities varied with the size of the meteorite. A meteorite weighing about 10 pounds had regmaglypts roughly thumb-sized. Even the smallest whole specimens, weighing

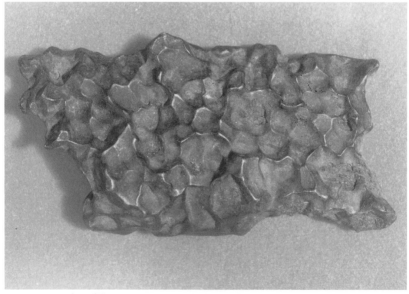

A whole individual meteorite from Sikhote-Alin. The extraordinary pitted surface structure is typical of individuals from this fall. Specimen measures 4 inches long and weighs 14 ounces.

only 0.18 gram (0.006 ounce), were covered by a full regmaglyptic structure. The size of the cavities was proportional to the total weight of the specimen, here nearly microscopic.

Separating each cavity on the meteorite was a pronounced ridge of metal formed by minerals with higher melting points, thus less subject to the ablation process. Using these whole specimens as evidence, Krinov proved that the original meteoroid had been a single body when it entered the atmosphere. It disintegrated only 3.6 miles above the ground and fell as thousands of meteoroids ranging in size from only a few grams to more than a ton. As they were heated during their brief flight, certain minerals (possibly troilite, an iron sulfide common to meteorites), composing part of the meteorite as roughly spherical nodules, ablated away faster than the other minerals. This produced the cavities on the surface, the signature of the Sikhote-Alin individual meteorites. These survived impact with Earth as whole bodies or in large, nearly whole fragments. Commonly, fragments that fit together precisely were found near each other. Because they fragmented in the atmosphere while incandescent, their ragged edges were smoothed by heating and a complete fusion crust was formed around them. The largest meteoroids impacted Earth with much higher energies and broke into thousands of jagged fragments, most of which came from the interior of the meteorite, spared from fusion crust formation. Often they broke along natural cleavage planes, planes of weakness found within the internal structure of the meteorite.

The third full expedition, 1950–1951, recovered the largest whole meteorite. It weighed nearly two tons. It was buried in the center of a crater only 11.5 feet in diameter. The mass was far larger than predicted (190 pounds for a crater 11.5 feet in diameter).

Adding the estimated total weight of the meteorites that must have formed the craters to the weight of individual meteorites scattered in the elliptical strewn field yields an approximate total weight of the mass that reached the ground—about 70 tons, of which only 23 tons have been recovered. Forty-seven tons may still lie buried in the taiga.

Another source of meteoritic material was discovered in a most unusual place. Many fir and cedar trees around the crater field showed considerable damage from direct hits by meteorites. The tops and many branches had been severed, and scarring along the trunks was evident. Sometimes the tree trunks had been split lengthwise. These observations led to a search of the trees themselves. Whole individual meteorites weighing as much as 30 pounds were found embedded in the trees. Damage to the trees was understandably most extensive around the larger craters. Jagged meteorite fragments were ejected from the craters at high speed like shrapnel from an exploding bomb. The forest is so dense in this area that it was impossible for the fragments to miss the trees. Many were found lodged in the fir and cedar trees radial to these craters, providing further evidence that the largest meteorites splintered into small fragments when they hit.

Meteoritic Dust

Running a powerful magnet through soil in the strewn field yielded meteoritic particles best seen with a hand magnifier or mineral microscope. Searchers found three types, representing three different modes of formation. Collectively, they can be considered meteoritic dust. The first type was irregular in shape, with sharp, angular edges. These obviously were torn from larger masses at impact and were tiny meteorite fragments. A second type was also produced by fragmentation, but high in the atmosphere during the breakup of the main mass. These became micrometeorites, with a typical fusion crust and a microscopically regmaglyptic surface. The third type resembled rounded globules, some almost perfect spheres (Brownlee particles). This was the melted material that sloughed off the meteoroid during the fireball stage, and probably constituted the material that formed the great smoke train.

The Distribution Ellipse

The meteoroid fragmented at a much lower altitude than the average stony meteorite. Fragmentation usually takes place 10 or more miles above the ground, but at Sikhote-Alin the altitude was only 3.6 miles. The falling meteorites could travel only a short distance before striking Earth, so there was little time or space to separate the small and large masses along the fall path. The meteorites fell along a distribution ellipse with a major axis of only 1.2 miles. The ellipse is quite narrow, with a minor axis of 0.6 mile. According to Krinov, this is the smallest strewn field known. Following the usual distribution, the largest masses fell along the forward end of the ellipse and the smallest at the rear. The largest craters are at the forward end. As a corollary, the greater devastation of the forest occurred along the forward end with little if any forest damage at the rear. There was some mixing of large and small meteorites within the ellipse, so the mass distribution is not as one would expect. This may suggest that some larger fragments broke into smaller pieces several times before finally striking Earth. This multiple fragmentation would tend to mix the masses.

Where Did the Meteorite Come From?

The Sikhote-Alin fireball was particularly well observed. Two hundred forty reliable observations were made along the Sikhote-Alin Mountains. These were sufficiently accurate to allow astronomers to calculate the orbit of the meteoroid before it entered Earth's atmosphere. The calculated elements of the orbit showed an aphelion, or farthest distance from the Sun, of 2.163 Astronomical Units (AU). The AU is the standard unit of distance in solar-system astronomy. One AU is equivalent to Earth's average distance from the Sun, about 92,900,000 miles. This places the original Sikhote-Alin meteoroid within the asteroid belt, between the orbits of Mars and Jupiter. Its perihelion point (closest point to the Sun) fell slightly inside Earth's orbit. Because the meteoroid's orbit crossed the orbit of the Earth, it was only a matter of time before the body intercepted Earth and collided with it. In 1947, its time had finally come.

The Package—A Personal Note

I have the good fortune to add a personal experience to the Sikhote-Alin story. Dr. Peter Franken, a good friend of mine on the physics faculty at the University of Arizona, frequently traveled to the Soviet Union to lecture and consult on laser physics. In 1983, he asked me if there was anything he could do for me while he was in Moscow. I took advantage of the opportunity and gave him several meteorite specimens of different types with the request that he present them to Dr. E. L. Krinov, chairman of the Committee on Meteorites at the Soviet Academy. This was years before glasnost, when one of the few avenues into the country was through scientific channels. I requested a trade for a sample from the Sikhote-Alin fall, a rare meteorite in the United States. Dr. Krinov was delighted to receive the specimens and quickly wrote me that the Sikhote-Alin meteorites were in storage but a sample would be forthcoming. Meanwhile, he sent me a book he had written on the Sikhote-Alin investigations. It was in Russian, which I couldn't read, but I was delighted to receive it.

The meteorites were a long time arriving. Nearly a year later, a wonderful wooden box appeared with a Soviet Union postmark. I carefully removed the nailed lid and reached inside to find two specimens. The first one was an ugly, ragged-edged, rusty hunk of iron. But I was not disappointed, for to my eyes it was a specimen of great value. It was obviously a fragment from one of the larger Sikhote-Alin craters. Then I unwrapped the second specimen. Just from the feel of its surface, I knew why he had given me two. There, fitting nicely in my hand, was without doubt the most beautiful meteorite I had ever seen. It was a complete individual, a typical regmaglyptic structure coated with fusion crust. Most important to me, these specimens were from the great Krinov himself. It was a wonderful moment. Two people with a common passion had met from across the world.

Thank You, Sikhote-Alin

Students of meteorites learned a great deal from the Sikhote-Alin fall. They made many discoveries about the circumstances of an iron meteorite shower, crater formation, and distribution of meteoritic material in crater fields. We now have a better understanding of what to look for in a meteorite strewn field—where the meteorites might be found and in what condition.

Aerial view of Meteor Crater looking southeast. The buildings on the north (left) rim are a small museum. The picture was taken in the 1940s before the new museum was constructed. Barringer's drill holes are evident on the crater floor. –Courtesy Barringer Crater Company

SEVEN

America's Great
Meteorite Crater

The idea of meteorites actually producing craters on Earth was hardly considered before the mid-twentieth century. It was asking enough to believe that rocks could fall from space and actually reach Earth. Rocks large enough to leave craters was going too far. After all, no one had ever witnessed the fall of a single meteorite of more than a ton in weight. At 1 ton, the depression created was several feet deep but not much bigger around than the meteorite itself; one would hardly call that a meteorite crater—it's more like a meteorite hole. Most astronomers doubted that a meteorite body large enough to form a respectable crater could survive passage through Earth's atmosphere.

The Moon is covered with craters of all sizes, some of which appear to be impact craters, so why not Earth also? This seemed to be a fair question, but the cratered surface of the Moon has no counterpart on Earth. Volcanic craters, yes; but meteorite craters, definitely not. The problem was that little was known about how meteorite craters form and what they should look like.

Meteor Crater

Picture a barren land, covered by sagebrush as far as the eye can see. Back in the 1870s, northern Arizona was land a person just passed through on the way to someplace else. It was cattle country, much as it is today. Imagine you are a cattle drover heading east out of Flagstaff. You see the landscape change rapidly as the elevation drops from 7,000 feet to 5,000 feet and you pass through pines, juniper trees, and then sage. The land becomes monotonous, an undulating plain with only a few hills to interrupt the flat horizon. After about 20 miles, you notice a low hill in the southeast. It attracts your attention because it appears whiter than the other hills. You might be curious enough to leave the trail to investigate. About 3 miles from it, you cross winding Canyon Diablo, which is usually dry. Another mile and the hill takes on a hummocky appearance. There are large boulders here and there on its flanks. When you finally reach the base, the flat summit rises 150 feet above you. As you climb the hill, you notice large, whitish limestone

West side of Meteor Crater, 2 miles from Canyon Diablo. The museum is on the north rim, to the left. The rim rises 150 feet above the surrounding plain.

boulders, which seem out of place on the otherwise featureless hill. You finally reach the summit and an astonishing scene comes into view: The ground abruptly drops away into a gigantic hole nearly a mile wide and 600 feet deep. You try to imagine what caused it, but it will not give up its secrets easily.

Today we know what it is. This is Arizona's Meteor Crater (see color plate II). It is the first crater proved to be of meteorite-impact origin. This hard-won knowledge did not come from the geologists who first studied it. Rather, it was a lawyer and mining engineer, enticed by the price of nickel-iron, who pieced together the clues that tell its history.

A Tourist's View

The highway signs invite you to "Visit Meteor Crater, a Natural Landmark." The name Meteor Crater is an obvious misnomer, since meteors are aerial phenomena and don't produce craters, but the name of this famous structure is well established. The crater is a National Natural Landmark and is privately owned and developed by the Barringer Crater Company of Decatur, Georgia. It is operated by Meteor Crater Enterprises in Flagstaff, Arizona.

About 40 miles east of Flagstaff on Interstate 40 you come to the Meteor Crater turnoff, leading south past the Country Store and an RV park. A 6-mile ride takes you to the base of the crater, where the road turns east and climbs the north flank to a parking lot. Before you is Meteor Crater's museum and souvenir shop. As in many American tourist enterprises, you must pass through the commercial area before you can see the main attraction. Steps lead down to an eating area and beyond that to the crater entrance, which is guarded by a cash register and uniformed attendant. Paying your entrance fee, you prepare to see what the famous Swedish chemist Svante Arrhenius called "the most interesting spot on Earth." After entering the museum, you are greeted by displays describing the crater's formation 20,000 to 50,000 years ago. A recorded narration describes the human history and scientific investigation of the crater and its role in the Apollo Moon program in the 1960s, when it was used as a training ground for astronauts bound for the Moon. There is even a boiler-plate shell of an Apollo command module on the grounds.

Your attention is drawn to the large picture windows and your first glimpse of the great hole. The view is stunning but as you take it in, your thoughts may wander. "The Grand Canyon is grander. . . . Yosemite is more beautiful. . . . To be sure, the hole is impressive, but open-pit copper mines in southern Arizona are nearly as big and as deep." But don't be deceived: This crater is matchless, a unique scientific treasure. The story it tells about the past and possible future—maybe as soon as

Meteor Crater from the museum's observation viewpoint. The south wall, directly opposite the museum, is 4,150 feet away. The floor of the crater is 750 feet below the rim.

tomorrow—captures your attention and dispels your indifference. "What if it had landed somewhere else," you might think. "New York or Moscow or . . . my city?" It is a sobering thought in this age of push-button warfare. Would this natural phenomenon be mistaken for a sneak attack? The science of impact cratering can be explored here. But for most visitors, the value of Arizona's Meteor Crater is its statement of our vulnerability.

Early History of Arizona's Meteor Crater

Back in the early 1870s someone, perhaps a cattleman passing through looking for rangeland, discovered the crater. The first written account of the crater dates from 1871, but it was known to Native Americans who lived on these plains nine hundred years ago. Their artifacts can still be found on the crater rim and surrounding plain. Cattlemen called the crater Coon Butte, and sometimes Crater Mountain.

It wasn't the crater itself that first attracted interest. To most people it was simply an old burned-out volcano and there were plenty of those around Flagstaff—about four hundred in the San Francisco volcanic field east of town. It was the jagged masses of iron scattered almost symmetrically around the crater that attracted attention.

In 1886, shepherds were grazing their flocks near Canyon Diablo. One of them found a piece of iron and, mistaking it for silver, tried to

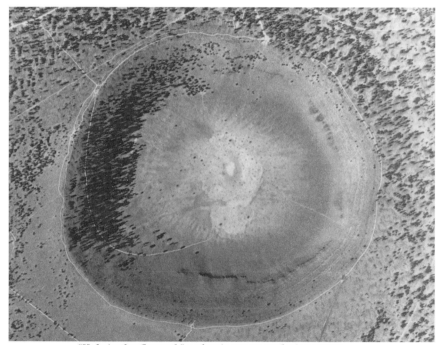

"Hole in the Ground," a classic steam-explosion crater located in central Oregon's high-desert volcanic plains. The crater is nearly the same size as Meteor Crater. –Courtesy of USDA Forest Service

Montezuma's Well, a limestone sink located about 40 miles south of Flagstaff, Arizona.

sell it. He wasn't successful. In 1891, a prospector wandered into the area and saw the large quantity of iron scattered on the ground. Thinking he had found an iron vein, he sent a sample to a mining company. They in turn sent pieces to several experts for analysis. One sample reached the hands of A. E. Foote, a well-known mineral dealer in Philadelphia. Foote knew immediately that it was a meteorite.

A few weeks later, Foote left Philadelphia to visit the crater. When he returned he published his findings, describing the iron meteorites around the crater. Incredibly, nowhere in the article did Foote try to relate the crater to the meteorites. He mentioned that the crater did not appear volcanic but made no further comment. The Zeitgeist was not yet ready for terrestrial meteorite craters.

Meanwhile, G. K. Gilbert, chief geologist of the United States Geological Survey, read Foote's paper. He was very interested because he was involved in the lunar-crater controversy—were they volcanic or meteoritic? Gilbert had been studying photographs of the Moon's craters in an attempt to prove they were formed by meteorite impacts. There might be direct evidence here applicable to his impact hypothesis.

In 1892, Gilbert sent Willard D. Johnson from the Geological Survey to investigate the site. When he returned, Johnson reported that the crater had been formed by a volcanic steam explosion. Gilbert was dissatisfied with Johnson's report (perhaps because secretly he wanted the crater to be meteoritic). Later the same year, Gilbert visited the crater himself. Incredibly, after examining the crater he too had to conclude that a volcanic steam explosion was the most plausible origin. He further suggested that if not of volcanic origin, it could simply be a limestone sink. Several were known to exist south of the crater and limestone was present in the area. He published these conclusions in March 1896.

The eye sees what the mind knows, and we can understand why Foote, Johnson, and Gilbert failed to see what seems so obvious today. Hindsight is always 20/20, and in the 1890s there was no known meteorite crater on Earth. No one knew what characteristics to look for in a meteorite crater. This crater was only 4,200 feet in diameter. Craters of that size on the Moon were barely visible through even the largest telescopes. They could not be observed in sufficient detail for us to learn much about impact craters of that size. There was nothing on which this crater could be modeled.

The obvious fact was that the meteorites were arranged around the crater. How did Gilbert account for that? He did so in a most curious way: If the crater was a limestone sink, then the meteorites fell symmetrically around the sink by chance after it formed! An even more frustrating hypothesis suggested a hybrid cause: Volcanic activity heated the deep rock beneath the present crater location; by chance, a large meteorite struck the surface and fractured the barrier of rock strata between the groundwater and the hot rock; the water came into contact with the hot rock, flashed into steam, and resulted in a violent steam explosion. Here was a hypothesis that narrowly missed the real cause. Somehow, Gilbert and other investigators could not bring themselves to suggest that meteorites could produce craters.

Fifteen years went by. Gilbert's status as a leading geologist was unquestioned and his opinion on geological matters was so respected that no geologist bothered to investigate further. Coon Butte was a volcano, period. G. K. Gilbert had said so. Meanwhile, miners hauled away tons of iron meteorites, loaded them on railroad cars, and shipped them to smelters in El Paso, Texas, where they were melted down and made into various iron products. Probably more than 20 tons were collected. There must be tools and machines still in use today that are made of celestial iron.

Daniel Moreau Barringer

In 1902, Daniel Moreau Barringer, a lawyer, geologist, and mining engineer from Philadelphia, entered the scene. In October of that year, Barringer heard reports of a crater in Arizona with iron meteorites scattered around it. He shared the scientific community's skepticism about the crater's celestial origin, but his interest was not driven by science. Barringer was also a businessman looking for mining oppor-

Daniel Moreau Barringer (1860–1929). –The Barringer Crater Company

tunities. He knew iron meteorites contain 90 percent or more iron, far more iron than is found in even the richest iron ores. In 1902, smelted iron was selling for $80 a ton. If this crater was formed by a huge meteorite, he reasoned, there would still be millions of tons of almost pure iron buried beneath it. He was well aware that the iron meteorites around the crater also contained 7 percent nickel. If the meteoritic mass weighed several million tons, then the nickel in this mass would run in the hundreds of thousands of tons. This in itself would represent a major mineral find, since no significant nickel mines existed in the United States. On the chance that his hypothesis was correct, Barringer filed claim to the land in 1903, before he even visited the site.

Typical nickel-iron meteorite fragments found on the plains around Meteor Crater. Specimen at left is 6.5 inches long.

Colorado Plateau Geology

Before looking at Barringer's work, we need to consider the geology of the region. The crater was formed in the undisturbed beds of the Colorado plateau. There are several recognizable rock layers in and around the crater. Some of these layers are also found as far away as the Grand Canyon, 100 miles to the north. Scattered here and there on the plains around the crater are deep reddish shaley outcrops, often showing ripple marks and sometimes fossil amphibian footprints. This uppermost layer is variable in thickness, being only a thin veneer on the plains, but it appears thicker in the crater walls; it is the Moenkopi formation, which was deposited after the seas retreated and the land was drying out in the early Triassic period.

The layer beneath the Moenkopi was deposited in a shallow Permian sea that once covered all of northern Arizona. Known as the Kaibab limestone, this layer is about 250 feet thick at the Grand Canyon and marks the uppermost rock layer, distinguished by its light gray-white color. The weathering of this rock in and around the crater has produced scalloped surfaces with sharp edges. Another sandy limestone called the Toroweap formation rests at the base of the Kaibab. It is only a few feet thick and is hard to distinguish at the crater. For our purposes, we can lump these two layers together.

Beneath the Kaibab-Toroweap layers is the Coconino sandstone, composed of silica grains forming a sandy, yellow-white rock. This rock shows signs of cross-bedding and represents an ancient sand-dune desert that existed long before the sea covered the area in the late Permian period. Its thickness is more than 700 feet at the crater, placing the Coconino beds both in the crater walls and beneath its floor. This rock layer is broken into a breccia hundreds of feet below the crater floor, attesting to the shock of impact.

Finally, beneath the Coconino lie undisturbed red shales of the Supai group, marking the end of the impact-modified layers. The modification and rearrangement of these layers provided Barringer with the evidence he needed to prove the crater meteoritic.

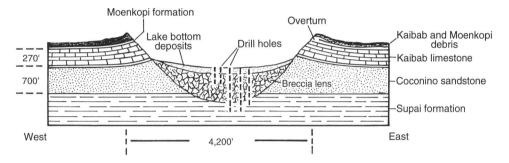

Cross section of Meteor Crater looking north shows the rock layers into which the crater intruded. Rock on the rim has been overturned, appearing in reverse order.

Drilling Operations

Barringer's first visit in 1905 was in earnest. He came prepared to determine the crater's origin. He drilled a total of twenty-eight test holes in the center. The cores showed that the rock had been badly fractured and pulverized to a fine powder, presumably by a crushing impact and explosion. He recognized that the "rock flour," as he called it, was actually Coconino sandstone. The grains of silica sand had been finely pulverized.

Bits of meteoritic iron oxide containing nickel were brought up from a depth of nearly 500 feet, from the pulverized layer. Undisturbed strata were finally encountered at about 1,300 feet.

Huge blocks of Kaibab limestone were lifted out of the crater as it was forming at the time of impact and were deposited on the rim and around the plains. Meteorites were found under the rim debris and beneath large boulders on the plain, proving that the meteorites landed at the same time the crater formed. Also, distribution of the meteorites was centered on the crater itself, leaving little doubt that the meteorites and the crater were associated. No evidence of volcanism was found. The nearest volcanism was 20 miles to the west.

In 1905, Barringer presented his first scientific paper to the American Philosophic Society, announcing the existence of a meteorite crater

Huge blocks of Kaibab limestone deposited on the east rim of Meteor Crater by explosion when the crater was formed.

nearly a mile in diameter on the Arizona plains. In 1909, Barringer further presented his findings in a scientific report to the National Academy of Sciences. The scientific community was not convinced. Nevertheless, Barringer, undaunted by skepticism, continued his pursuit. It remained only to locate the main meteorite mass.

The drilling was close to the crater center, since Barringer made the assumptions that the crater was round (it is actually squarish with rounded corners, due to perpendicular fractures in the bedrock) and that the meteorite struck from a vertical direction. Central drillings were difficult. The crater is a natural catchment basin and 100 feet of lake-bottom sediments were discovered in the cores, evidence of a lake that existed there in wetter times. The Coconino sandstone makes an excellent aquifer and groundwater was frequently encountered below this level, making the drilling of deep holes extremely difficult.

Barringer's failure to locate the meteorite in the center made him question the direction it was traveling when it hit the plains. He noticed that when a bullet was fired at an angle into mud, the bullet hole was nonetheless still round. In fact, a round hole results even at angles as low as 15 degrees from the horizontal. From this observation, Barringer reasoned that the main mass must lie somewhere under the crater rim.

The strata on the inside walls of the crater can be seen in some relief, especially where weathering has removed softer material leaving behind the more resistant outcrops. The exposed strata on the north wall are tilted about 5 degrees from the original horizontal position and away from the crater center. As the rock layers progress to the east and west sides, their tilt gradually increases until they become nearly vertical on the southeast and southwest sides. On the southern rim, a conspicuous overhanging block 2,500 feet in length is raised vertically for a distance of over 100 feet. From these observations, Barringer concluded that the meteorite came from the north and struck the ground at a low angle, blasting out millions of tons of rock before finally coming to rest under the southern rim.

The search did not begin immediately. Nearly ten years went by before Barringer secured new funding to continue. Scientists were beginning to show more interest in the crater during this period. Some agreed with his conclusions about the meteoritic nature of the crater, but few agreed with his belief that the main mass lay buried beneath the crater walls. In 1908, George P. Merrill, a noted American astronomer, visited the site. He agreed that the origin was meteoritic and even estimated the size of the meteorite as nearly 500 feet in diameter. He further estimated its cosmic velocity had been between 5 and 15 miles per second. Even at this low velocity, it became immediately evident to many scientists that a body of that size traveling at that speed possessed enough kinetic energy not only to blow a hole in the Arizona desert, but also to produce enough heat to vaporize the meteorite completely. This was not good news to Barringer's financial backers, who were becoming increasingly reluctant to finance the project.

Tilted strata on the north rim of Meteor Crater.

In 1920, Barringer again began drilling, this time in the center of the overhanging block on the south rim. As the drill reached the 1,000-foot depth, it encountered oxidized meteoritic fragments. The cores tested increasingly higher in meteoritic content. In 1923, at a depth of 1,376 feet, the drill hit a hard object, jammed, and severed the cable. This brought an end to the operation, but Barringer was elated. He was certain that he had at last located the main meteorite mass. Even with this perceived success, the drilling had overextended available funds, so Barringer returned to Philadelphia to find additional supporters.

He tenaciously resumed work in 1928 when other funds became available. This time he drilled 1,000 feet to the south of the earlier drill hole, hoping to avoid the groundwater that had been a constant problem during all drilling operations. The idea was to drill to the 1,500-foot level and slowly tunnel north until they struck the main mass. At only 750 feet, they encountered groundwater. It flooded the drill hole and eventually stopped the drilling completely.

Even while these final holes were being drilled, Barringer's hypothesis was again under attack. An American astronomer and leading authority in celestial mechanics and ballistics, F. R. Moulton, visited the crater in 1929 when the drilling operations were still in progress. Using data from Barringer's own work, Moulton concluded that the mass could have been no more than 3 million tons and possibly much less. He further showed that the body had to have been traveling between 7 and 17 miles per second. In other words, it was not retarded by the atmosphere at all, but hit Earth at cosmic velocity. A second report quickly followed. It was the coup de grâce to Barringer's hypothesis. Moulton showed that a meteorite weighing thousands of tons striking Earth at cosmic velocity would create pressures on the meteorite sufficient to "disperse the substance of the meteorite as if it had been a gas." This substantiated Merrill's earlier conclusions. Barringer's 5- to 15-million-ton mass could not possibly exist.

Predictably, such a report discouraged Barringer's backers. Soon after he returned to Philadelphia in 1929, the stock market crash put a stop to any further investigations. Later that year, Barringer died at his home in Philadelphia, ending his nearly thirty-year quest for the elusive meteorite. The parallel between the Barringer story and the concurrent search for the suspected Tunguska meteorite is ironic. Today the Arizona crater is held by Barringer's family as a public trust and is also known as the Barringer Meteorite Crater. It is a fitting tribute to the human spirit and a pioneer who dared to venture beyond the believable.

The Post-Barringer Period

Many surveys and studies have been conducted at the crater since Barringer's day. Some of these studies confirmed his conclusions, some did not. Magnetic and electrical conductivity studies between 1930 and 1951 supposedly located several large masses. Some were within the crater walls; several other masses allegedly lay under the plain about a mile from the south rim.

Most of these studies were not taken seriously by geologists, and by the mid-1950s few believed the crater had been caused by impact. Then in 1957 a young geology graduate student, Eugene Shoemaker, began investigating the crater anew. He made new studies of the folded and overturned rock strata forming the crater walls and descended into Barringer's original drill shafts to examine firsthand a cross section of the rock beneath the crater floor. He found a lens-shaped mass of broken rock breccia at the 600-foot level. Beneath this level he found shock-melted glass with meteoritic droplets embedded in the glass. It was Shoemaker's persistence that eventually convinced geologists that Barringer Crater was the youngest and best-preserved impact crater on Earth.

Other studies concentrated on the distribution of meteorites on the plains surrounding the crater. The typical elliptical strewn field expected for multiple falls was not seen here. The formation of the crater undoubtedly modified the normal distribution. Early studies showed meteorites as far as 8 miles away from the crater's center, but most were found within a radius of 2 miles from the rim, with a concentration near the northeast flank.

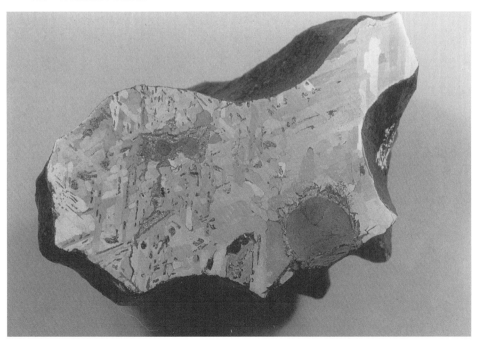

Polished and etched face of a Canyon Diablo iron meteorite, showing Widmanstätten figures. The large circular inclusion is a graphite nodule with a wedge of troilite (iron sulfide). Specimen measures 7.5 inches in its longest dimension and weighs 6 pounds. **123**

The most intensive studies were conducted by H. H. Nininger between 1939 and 1946. He collected meteorites from the plains and around the rim. These were cut, polished, and acid-etched to bring out the internal structure known as Widmanstätten figures common to many iron meteorites. Specimens taken from the plains on the west side showed the beautiful patterns typical of Canyon Diablo meteorites (the meteorites were given the name of the small canyon 2 miles west of the crater). Those from the rim, however, were twisted and deformed into odd shapes and the Widmanstätten figures were altered or absent. This alteration was due to excessive heating. The plains specimens apparently broke off the main mass before it hit and so were not subjected to the intense heating of the main mass. The rim specimens were remnants torn from the main mass at impact and had been subjected to high heating. Most specimens lay under the rocks on the northeast rim.

Nininger also discovered fused silica bombs, called impactites, on the west flank of the crater. Impactites were first found around the Henbury meteorite craters in central Australia in 1932. They are small, black masses made entirely of glass, containing minute iron fragments. The iron fragments tested positive for nickel, which is always found in combination with iron in meteorites. The fused silica bombs (they were called bombs for their similarity to small volcanic bombs typically found around volcanic vents) were produced by the melting of silicate rocks during the impact of a large meteorite. These bombs were ejected from the crater along with other rock debris that formed the raised rim. Nininger found iron particles in the silica bombs that also tested positive for nickel. The search for impactite bombs is now standard procedure in crater investigations.

Formation of Meteor Crater

Much of what we know about the crater's formation comes from the study of impact cratering on ballistic test ranges and from laboratory tests with powerful, vertically mounted gas guns directing high-speed projectiles against various soil and rock mixtures. In the late 1950s, Eugene Shoemaker studied craters produced by small nuclear detonations at the Nevada Test Site, using them as close analogues to impact cratering. Through this pioneering work, scientists established the basic mechanisms that result in impact craters. The following scenario describes the formation of explosion craters in general and Meteor Crater in particular.

Approximately fifty thousand years ago, an iron meteoroid (a small asteroid) about 80 feet in diameter and weighing 63,000 tons entered Earth's atmosphere at a low angle of about 30 degrees. (Barringer used the tilted strata as evidence for a northerly approach of the meteoroid. Evidence from meteorite distribution studies points toward a southwesterly direction. The final shape and structure of the crater may have little to do with the direction of impact.) The body, traveling about 10 miles per second, fragmented into many large pieces in flight. Thousands of small pieces broke off the larger masses just before impact, and these scattered across the plains to the west.

At the moment of impact, a shock wave was generated downward into the rock and upward through the meteorite. The crater produced was small at first, about the size of the meteorite itself. The meteorite and impacted rock were subjected to enormous pressures at the instant

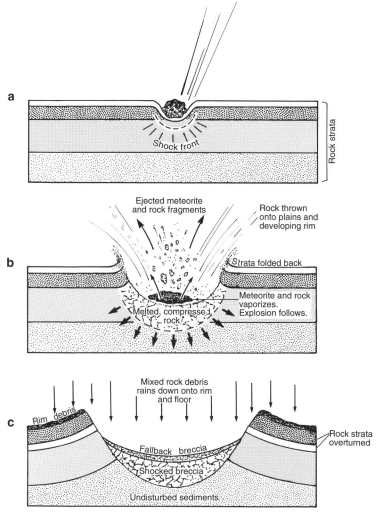

Sequential stages leading to the formation of Meteor Crater. a: Meteorite penetrates surface and sends shock wave through rock layers. b: Decompression wave displaces rock outward onto rim and plains. Shock wave traveling from front to back of meteorite expels meteorite fragments. Remaining meteorite explosively vaporizes. c: Rock and meteorite debris from dust cloud rain down on crater and surrounding plain.

of impact. These pressures compressed the meteorite and impacted rock into less than half their normal volume. The compression wave rapidly penetrated downward and outward radially, away from the impact point. A decompression wave followed, displacing rock material sideways in a conical shape along the walls of the rapidly expanding crater and out onto the developing rim. Simultaneously, the underlying rock was fractured, partially melted, and compressed into a breccia. This was the pulverized material found by Barringer beneath the crater floor in his first drillings and later studied by Shoemaker.

The shock wave passing through the meteorite traveled from the impacting front to the back. As it exited, explosive decompression took place, causing meteoritic material to be fragmented and shot out of the expanding crater like a blast from a shotgun. This resulted in the distorted specimens found on the northeast rim. The remaining meteorite was by now superheated, as its kinetic energy had been converted to heat. Both the meteorite and the compressed rock began to flow like a fluidized gas (gas under pressure). This pressure was relieved by an explosion equivalent to that of a 1.7-megaton bomb, which further widened the crater and flung rock debris out onto the lip of the crater, adding to the raised rim.

At Meteor Crater, the explosion lifted the horizontal rock strata upward and tilted them backward, away from the crater center. The first rock expelled from the growing cavity was the uppermost Moenkopi sandstone, followed in succession by the Kaibab limestone and then the Coconino sandstone, now resting on top. This inversion of rock strata is found in the raised rim and the debris blanket around the flanks. All this happened in less than a second.

Above the crater, an enormous cloud of dust, vaporized meteorite, and rock rained down, partially filling the gaping hole and mixing with the rock debris cascading off the inner walls to the crater floor. Originally, the cavity was over 1,200 feet deep, but debris from the dust cloud and fractured Coconino blocks filled the crater to half its original depth. The lighter particles in the cloud were moved by the wind—southwesterly winds in this case—and droplets of meteoritic material and silica slowly settled over the northeast side of the crater.

Today we know that meteorites weighing more than 100 tons cannot survive impact with Earth. Pieces may fragment off, but the main mass will be vaporized.

The Best Clues to Authenticity

Of the 150 or so authenticated impact craters that exist on Earth today, few show any signs of meteorites. Many of these craters may have been caused by comets, which would have left no meteorites, and others are so old that their blanket of meteorites has disappeared. Meteor Crater is a relatively young crater. But judging from the deterioration of the small meteorites still buried in great numbers beneath the plains around the crater, it will not be long before all traces of meteoritic material are gone. Much of this material has been collected through the years with sophisticated metal detectors and is now preserved in public and private collections around the world for all to see. Meteorite collecting is no longer permitted at the crater today, so the remaining material will eventually rust away. Meteorites are subject to the same rule of preservation as fossils: The only way to preserve a specimen is to collect it.

If there are no meteorites around a crater and it has weathered, so

This sign and many others like it surround Meteor Crater to remind visitors that meteorite collecting is not permitted.

that many of the telltale signs have been removed, how can scientists determine whether or not it is meteoritic? The presence of shock-melted impactite glass is a good clue. Better yet, two rare quartz minerals formed by shock were found by Shoemaker and Edward Chao in the crushed Coconino sandstone at the bottom of Meteor Crater in 1960 and 1961. These minerals, called coesite and stishovite, though chemically identical to ordinary quartz, can only form when quartz is subjected to intense pressures. Coesite requires pressures of about 150 tons per square inch, and stishovite forms at pressures five times greater and temperatures above 1,382 degrees Fahrenheit. Ordinary quartz has a density of 2.65 grams per cubic centimeter; coesite and stishovite have densities of 3.01 and 4.35, respectively. Also, the crystalline structure of coesite and stishovite differs from that of ordinary quartz. They are among the best indicators of impact phenomena.

There is one more shock-related feature that helps authenticate meteoritic craters. Curious cone-shaped rocks called shatter cones are frequently found around many old and weathered impact craters. The discovery of shatter cones dates from 1905, when they were first found in the Steinheim Basin in southern Germany. The basin is an eroded meteorite crater about a mile and a half in diameter. Shatter cones form best in limestone rocks, although they have been found in other sedimentary and igneous rocks. They are recognized by radial fractures that extend from the apex of the cones, flaring outward at an angle of from 75 degrees to about 90 degrees. They range in length from an inch or

A shatter cone formed in limestone rock. This specimen was found at the Steinheim Basin in southern Germany. Note the fine, radial fractures extending away from an apex, missing on this specimen.

so to several feet. They have been sheared along cone-shaped fractures, presumably by shock waves induced by the impact of a meteorite. The nose of the cone is always oriented toward the impact point. An American meteorite-crater expert, R. S. Dietz, first suggested in 1947 that they indicate the existence of meteorite craters.

Curiously, no well-formed shatter cones have been found at Meteor Crater. One fragment of Coconino sandstone seemed to mimic at least a partial cone shape, but few believe it authentic. Shatter cones could yet be discovered in the most likely place, beneath the crater floor. Unfortunately, the shattered floor is covered by seventy feet of lake sediments, making such a discovery unlikely.

Arizona's Meteor Crater was the first meteorite crater identified on Earth, and studying it has taught scientists much about the effects of these great impacts. While research on this crater continued during the first half of the twentieth century, other craters were being discovered around the world. By midcentury, at least fifteen craters had been found and proved to be authentic meteorite structures. Space-age photoreconnaissance satellites discovered many more. Most of them are millions of years old and many dwarf the Arizona crater. Appendix C contains a selected list of the world's authenticated and suspected meteorite craters.

First quarter Moon. The cratered hemisphere—the lunar highlands—represents the oldest terrain. Dark, circular features are huge impact basins filled with basalt, which obliterated the cratered terrain about 3.9 billion years ago.

EIGHT

Earth Is a Cratered World

There are more than 30,000 craters on the side of the Moon facing Earth, and they are not distributed uniformly; most are primarily in the Moon's southern hemisphere. The northern hemisphere is covered by vast lava flows that flooded enormous impact basins and obliterated the older cratered surface. The low crater densities on the lava plains tell us that the cratering of the Moon took place well before that time; and the ages of the rocks from the lunar highlands, where there are more craters, date the major cratering event at about a billion years earlier.

By comparison, Earth seems crater-free, though it must have been heavily cratered more than 4 billion years ago. Unlike the Moon, Earth is a geologically dynamic world in a constant state of slow change. Weathering and erosion make even the highest mountains temporary features. The movement of Earth's crustal plates through geologic history effectively destroyed the earliest of Earth's geologic structures. Thus, the earliest craters that existed on Earth have been erased. The craters we see today on the continental landmasses are young compared with the age of Earth's oldest crustal rocks.

Impact craters of the world. Asterisks indicate craters where meteorites are found. –Grieve and Robertson data. Geological Society of America Special Paper 180 (1982)

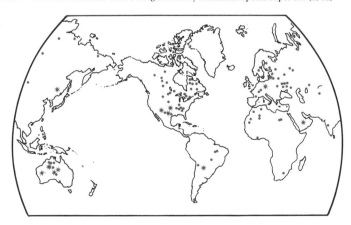

✳ Craters with meteorites
• Craters

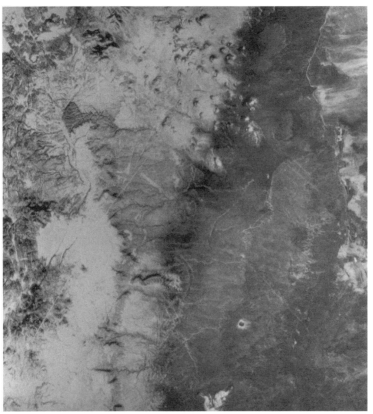

A Landsat *satellite view from 570 miles above Earth, showing the Colorado plateau east of Flagstaff, Arizona. The area measures 115 miles on a side. One-mile-diameter Meteor Crater is the small, circular feature below center, to the right. Canyon Diablo winds past on the west (left).* – NASA

The 51-mile-diameter crater Tycho is an example of a "complex" crater. The central uplift terminates in a group of hills. The rim is terraced, a collapse feature due to the inward and upward movement of rock after impact.
–*Lunar Orbiter* photo, courtesy of NASA

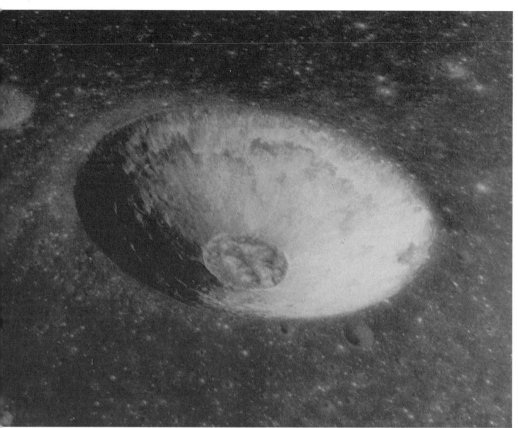

A "simple" lunar crater 4 miles in diameter. Note its sharply raised rim and steep, bowl-shaped appearance. Fallback debris covers the flat floor. —*Apollo 10* photo, courtesy of NASA

Today (1998), geologists recognize 156 impact structures around the world (a partial list appears in Appendix C). Impact geologists distinguish two basic types of impact craters, both formed by explosion. The smallest craters are called "simple" because their form is a simple bowl shape, often quite deep, with steeply raised rims and flat floors. Meteor Crater and the New Quebec Crater in Canada are examples of simple craters. "Complex" craters are much larger in diameter and usually show an uplift in the center or, if large enough, a ring of uplifted hills. The Moon has numerous classic examples of simple and complex craters. On Earth, nearly every continent has its complement of both crater types. A few are easy to visit, others require a more adventurous traveler, and such celestial scars are well worth investigating.

Craters of the United States

Odessa Meteorite Crater. Odessa Crater in western Texas was the second authenticated meteorite crater in the United States. It was discovered in 1921 on a limestone plain just outside the city limits of

Shallow Odessa Meteorite Crater, located in an oil field in west Texas.

Odessa, in an operating oil field. It is an explosion crater measuring about 530 feet in diameter. Weathering has taken its toll, for the crater is only 18 feet deep measured from the highest point on the rim. The rim averages only 3 to 4 feet above the surrounding plain. The crater was once much deeper, at least 75 feet, but as it has filled in over time, it could easily be overlooked today.

About 10 tons of iron meteorites have been found around the crater. They are similar to those at Meteor Crater, causing speculation as to whether this crater was part of the same fall. But the Odessa Crater is much older than its neighbor in Arizona. Excavations in the fill material have uncovered a fossil horse, extinct for more than 200,000 years.

Daniel M. Barringer Jr. visited Odessa Crater in 1926. He was among the first to consider it meteoritic. A 1939 magnetometer survey detected iron buried around the main crater. Subsequent excavations uncovered three additional craters, none of which were visible from the surface. All had been filled in. The largest, measuring about 75 feet in diameter, was only 75 feet from the main crater. The other two craters, 200 feet away, are impact pits measuring about 10 feet in diameter.

The Odessa crater field is similar to the Sikhote-Alin field in that many meteorite fragments were found in the smaller craters. The main crater had none, but there were thousands of small fragments on its outer walls and on the plains around it, which is typical of small explosion craters. Meteorite fragments are usually ejected with the excavated material. Shocked quartz found in core samples from a shaft drilled 165 feet into the crater's center verified its explosive origin.

The meteorites are weathered and oxidized. It is only because of the dry desert conditions that they have been preserved at all. Usually they are found with a thick layer of rusty shalelike material enclosing the specimen, and sometimes completely oxidized meteoritic shale is all that remains.

H. H. Nininger tried out the first crude metal detector on the plains around Odessa Crater in 1935. Within a few minutes, the detector located a 1-pound meteorite, just inches under the surface. He found twenty-seven meteorites in a day, which had a combined weight of 34 pounds. Years ago when I read this story in his autobiography, *Find a Falling Star,* I could easily imagine the excitement Nininger must have felt when the detector actually worked. It's something like finding a good fishing hole and catching a fish within minutes of setting the bait!

My vicarious thrill soon became reality. In 1964, my school chum Ronald Hartman and I, both graduates of Frederick C. Leonard's course in meteoritics at UCLA, decided to try our hand at finding meteorites, and Odessa was our crater of choice. Armed with a heavy, battery-laden World War II metal detector, we began combing the plains around the crater. Hours went by—no luck. About an hour before sundown, with our spirits at a low ebb, Ron hit the "fishing hole." A strong signal promised a big one. We frantically began digging. One foot down, and the signal became stronger yet. At 2 feet, our shovel hit metal. We probed the ground and determined that the object was 8 to 10 inches wide on the side facing us. Clearly, this meteorite was no hand specimen! More digging revealed a 50-pound, nearly spherical iron meteorite. We struggled to free it from its resting place. Urged on by our triumph, Ron continued to sweep the ground around the find. I had barely filled in the first hole when he had another strike. A few strong thrusts with the shovel located a second meteorite, this one weighing 25 pounds. Then a few minutes later, a 7-pounder.

Richard Norton examines a newly found, 7-pound Odessa meteorite as its discoverer, Ronald Hartman, looks on. Specimens weighing 50 and 25 pounds are in the foreground.

King of the Odessa meteorite hunters, James Williams, with Dorothy Norton, metal detector, and meteorite finds.

Eighty-two pounds of meteorites within fifteen minutes! Nininger had nothing on us!

Years later, with this experience firmly impressed in my memory, I returned to Odessa only to find the hunting much more difficult. A few local residents, with special permission from the oil company, had taken up the hunt with modern electronic metal detectors and were gradually clearing the field. I met James Williams in the oil fields as I was searching on the first day. James was the king of Odessa meteorite hunters. In his garage were more meteorites than I had ever seen in one place. Nearly a ton of meteoritic iron covered the floor, shelves, and benches, representing years of diligent searching. I arranged some larger specimens outside for a "family portrait." "There's more of 'em out there," he told me confidently, and I believe him.

ASTROBLEMES. The Meteor and Odessa Craters were obviously formed by the impact of iron meteorites, fragments of which have been found scattered around them. However, this is the exception, not the rule, for impact craters. There are at least twelve known structures in the United States lacking meteoritic material but showing strong evidence of an impact origin. Because they are ancient, they show extreme weathering. At least one is completely buried in sediments, and another is on the ocean floor off Atlantic City, New Jersey.

"I can see it. It's long and narrow!" (A true story.)

When these ancient structures were first recognized as having been formed by explosion (not necessarily by an impacting body), geologists called them cryptoexplosion structures. Later, Robert S. Dietz, an Arizona State University geologist and a recognized authority on impact structures, coined the name "astrobleme" to describe them.

Astrobleme means "star wound," which suggests their impact origin. All of them show at least some of the classic signs. Most are complex craters with raised rims and centers and brecciated, lens-shaped zones beneath their floors. Around the structures, investigators often find shocked quartz and high-pressure quartz polymorphs, fused silica, and oriented shatter cones. In all cases, either the impacting body was completely vaporized or any meteorite remnants have long since weathered away.

Sierra Madera. The Sierra Madera astrobleme is typical of the so-called cryptoexplosion structures. It is in Pecos County, Texas, about 25 miles south of Fort Stockton and can be recognized by a group of hills that cover an area 3 miles in diameter and rise 600 feet above the plains. These hills are surrounded by another low ring of hills 8 miles in diameter, hardly noticeable from the plain. Between the two is a depressed circular ring filled with alluvium. The sedimentary beds in the central hills are intensely folded, overturned, and uplifted thousands of feet. Core drillings have revealed that the floor of the depressed ring around the hills is thin, probably due to stretching as the rocks moved inward and upward to form the central uplift. The outer hills represent the rim of the crater and show folding and uplift of the rock strata. A half mile beneath the floor is a cup-shaped breccia lens about 1.5 miles wide, resembling in shape the lens at Meteor Crater.

A nearly perfect limestone shatter cone from the central uplift at Sierra Madera astrobleme. The specimen measures 2.5 inches from apex to base.

Shock-deformed rock and quartz showing linear shock features are found in the central uplift. Shatter cones were first discovered in 1959 in sedimentary beds, arranged symmetrically around the uplift. If the beds could be returned to their original horizontal state, the shatter cones would point inward and upward toward the center. Sierra Madera contains the most extensive array of shatter cones of any known impact crater.

All these features strongly indentify the Sierra Madera structure as an ancient astrobleme formed on the west Texas plains about 100 million years ago.

Australia's Meteorite Craters

Australia is the home of at least seventeen authenticated meteorite craters, and several under investigation probably will be added to the list. The dry desert climate of Australia's interior is ideal for preserving meteorites and meteorite craters.

Henbury Craters. Fortunately for crater explorers in the United States, Meteor Crater and the Odessa Craters lie in arid desert regions conveniently close to interstate highways. Most known meteorite craters are located in remote, hard-to-reach places. The Henbury Craters in central Australia are certainly remote, but they can be reached via the main

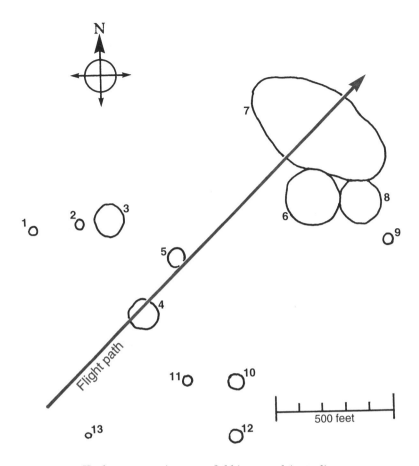

Henbury meteorite crater field in central Australia.

highway that crosses the continent from Adelaide on the South Coast to Darwin on the North. Aborigines in the area apparently knew about them; their tales of the iron stones surrounding the craters reached the South Coast, and R. A. Alderman from the University of South Australia investigated the site in May 1931. He discovered thirteen craters, about 8 miles southwest of the town of Henbury.

The largest crater is elliptical, measuring 720 feet by 360 feet, and may be two craters that merged as a result of erosion over the years. The depth measured from the slightly raised rim averages only about 50 feet. Two smaller craters lie tangent to the big crater on its south side, and still smaller craters lie several hundred yards to the southwest, ranging in size from 30 feet to more than 200 feet in diameter. The size and position of the craters illustrate the distribution ellipse of the fall. The largest craters rest at the far end of the ellipse, so the meteoroid must have been traveling from southwest to northeast and broken up in flight.

Largest crater at Henbury with two smaller, adjacent craters. —Courtesy of W. Zeitschel

An oriented 66-pound Henbury iron meteorite. Specimen measures 10 by 10 by 6 inches. —Robert A. Haag collection

Typical Henbury meteorite fragments. Specimen on lower right measures 3.25 inches in its longest dimension.

No meteorites have been found in the three largest craters, suggesting their explosive origin. Most of the meteorites were around the largest craters of the southwest group. The largest Henbury meteorites show beautiful regmaglypt structures. Some are oriented, displaying a shieldlike form produced when a meteoroid passes through the atmosphere without tumbling and ablates, primarily on the leading side.

Most of the meteorites are fragments torn from the main mass on impact. Many are thin, with bent and ragged edges attesting to the great forces to which they were subjected. Polished and etched specimens

Polished and etched Henbury meteorite reveals Widmanstätten figures. Notice the linear structure caused by mechanical stretching of the specimen.

show Widmanstätten patterns deformed into curved plates by mechanical stresses. Some meteorites show evidence of heating to over 800 degrees Fahrenheit, which destroyed their Widmanstätten patterns, especially around the edges. These were probably ejected from the two largest craters when their explosive formation caused the greatest heating. Most have rusted heavily and some have oxidized completely.

The craters may not be much more than a few thousand years old. Aborigines tend to avoid the area, and they call the meteorites "sun-trail-fire-devil-stone," which suggests that their ancestors may have witnessed the fall. Analysis of impactites (particles of fused glass) found in the craters shows the glass to be less than ten thousand years old.

Boxhole Crater. Another meteorite crater was discovered in 1937, northeast of Alice Springs near the Plenty River in the North Territory. The crater is 570 feet in diameter and about 50 feet deep. Scattered around the crater were meteoritic iron and iron oxide (iron shale). Analysis of the meteorites showed remarkable similarities to Henbury irons. They are both medium octahedrites; that is, the Widmanstätten plates show similar widths. Boxhole Crater is larger than the main crater at Henbury (actually two merged craters). It lies almost exactly on the southwest-to-northeast line of the distribution ellipse of the Henbury Craters, raising suspicions that it may be the most distant and largest single crater of the Henbury complex.

Wolfe Creek Crater. Wolfe Creek Crater, an impressive crater 2,870 feet across and 160 feet deep, was discovered from the air in 1947. It is in Western Australia, about 75 miles south of Hall's Creek. The location is remote, but adventurous explorers can reach the crater by

Wolfe Creek Crater in Western Australia. (The crater measures 2,870 feet across and 160 feet deep.) –Smithsonian Institution

car. The area is a sandy desert and some sand has invaded the east wall of the crater. The walls vary in height from 60 to 100 feet above the surrounding desert and are tilted outward, a characteristic of explosion craters. They are composed of shattered Precambrian quartzite rock. A concentration of highly oxidized meteoritic iron fragments was found on the southwest side of the crater. They have a black, crusty appearance but contain little free iron. No pure iron meteorites have been recovered.

Dalgaranga and Acraman—Australia's Smallest and Largest Craters. The smallest authenticated meteorite crater in Australia is Dalgaranga, only 82 feet across and 15 feet deep. It is on the far western side of Australia, about 60 miles due north of the town of Edah. Many small meteorite fragments have been found around the crater. Unlike other Australian crater meteorites, which are all irons, Dalgaranga meteorites are a stony-iron subclass called mesosiderites. They are highly weathered and oxidized.

Australia's largest crater is actually a nearly circular lake called Acraman Lake. Geologists first recognized it as an impact structure in 1986 after discovering large volcanic blocks embedded in 450-million-year-old sedimentary rock in the Flinders Ranges 185 miles due west of the lake. Those blocks dated a billion years older than the sedimentary rock in which they were embedded. Similar volcanic rock fragments were found in the Cawler Ranges near Lake Acraman. The rocks had blasted out of the forming Acraman crater at the moment of impact about 450 million years ago. The crater is 56 miles in diameter and located in South Australia, northwest of Adelaide.

Saudi Arabia to South America

Wabar Crater Field. In southern Saudi Arabia there is a vast desert of nothing but sand and rock. This desert, the Rub' Al-Khali, or Empty Quarter, is one of the most desolate areas on Earth—an excellent setting for a meteorite crater. The Wabar craters were discovered in the pursuit of a legend, a Bedouin story of a city called Ad-ibn Kin, or Wabar, that once existed in the Empty Quarter. According to the legend, the city was destroyed when God sent a powerful wind and fire from the sky to punish its wicked ruler. The legend also mentions a large mass of iron "as big as a camel's hump" somewhere in the desert near the Wabar site.

An English explorer, Harry St. John Philby, hoped to be the first Englishman to cross the Empty Quarter, but he was upstaged by a fellow countryman in 1931, a year before his intended crossing. Though disappointed, he found the legend of Wabar a compelling reason to carry through with his plan. Was there a ruined city in the sand? Perhaps buried treasure was there to be found. In January 1932, Philby began his trek across the forbidding desert in search of the lost city of Wabar.

Philby traveled for a month, with an entourage of fifteen tribesmen and their camels. When they reached the spot his guides called Wabar, no ancient ruined city marked the site. Instead they found two sand-filled craters Philby took to be volcanic, the largest about 300 feet in diameter (see color plate II). "I knew not whether to laugh or cry," Philby noted in his diary. He searched the area around the craters for four days, hoping to find the large mass of iron he thought might be an artifact from the ancient city. He came upon one iron fragment weighing 25 pounds and many small pieces weighing a total of about a quarter of a pound. Around the craters were countless tiny, black, silica-glass globules the tribesmen thought were valuable black pearls. Their real value was not monetary but lay in the story they were to tell scientists decades later. Philby left the site, disappointed with his failure to locate the legendary city. It must have been just that, a legend.

Philby's find was significant: two craters formed by the explosive impact of an iron meteorite. Along with the black glass were pure white stones, also glassy. The white glass was fused silica, created at the instant of impact by the intense heat that melted the pure silica desert sand. The black glass was also from fused silica. Both are impactite glass. The black glass contained tiny nickel-iron globules of meteoritic origin, formed when vaporized iron condensed into tiny droplets and mixed with the liquid glass, locking in remnants of the great iron meteorite.

Iron masses had been found long before Philby's venture into the Empty Quarter. In 1863, the British Museum acquired a 131-pound iron, said by the Bedouins to have fallen during a thunderstorm. Another mass, weighing 137 pounds, was found in 1893. Both were iron meteorites with 7 percent nickel, obviously from the same fall. Philby's find was of the same composition as the other two, but only his was associated with the craters.

The great find occurred in 1965. Bedouins insisted that al-Hadida, the "big piece of iron," did exist. The sand dunes of the Rub' Al-Khali are in constant motion, covering and uncovering everything in their path, and the dunes finally shifted to reveal the great meteorite. The Bedouin's "camel's hump" was investigated first by National Geographic correspondent Thomas Abercrombie in the fall of 1965. He found the large mass embedded in the sand near the larger crater. It was cone-shaped, a product of atmospheric shaping, and measured 3.5 feet in diameter and 2 feet thick (see color plate II). The great iron meteorite, the largest ever recovered in Saudi Arabia, weighed 4,500 pounds. An Aramco engineer, James Mandaville, visited Wabar a year later and found an additional fragment weighing 440 pounds. Both masses are on display at the King Sa'ud University in Riyadh.

Because the craters at Wabar are subject to drifting sands, it is likely the desert will eventually reclaim them. Already, the rim of the main

An aerial view of four elongate depressions possibly produced by the disruption of a 500-foot-diameter object impacting the Argentine loess at low angles.
–Courtesy of Peter H. Schultz and Ruben Lianza

crater is more than three-quarters covered and may some day disappear, just as the great meteorite did.

A New Argentina Crater Field? While flying an airplane in October 1989, Captain Ruben Lianza, an Argentine Air Force pilot, noticed a large, elongated, gougelike depression disrupting the symmetry of neatly cultivated fields in north-central Argentina. As a well-read amateur astronomer, Lianza immediately suspected an impact crater. The next morning he flew back to the area for a second look. This time he was flying 5,000 feet higher when he approached the depression, and he was surprised to find it looked much larger than he had remembered. He soon realized it was not the feature he had seen the previous day. This one was nearly identical in shape, but three times as large. Just half a mile to the west was another, nearly identical, depression. When he increased his altitude to 30,000 feet, he saw several other smaller, elongated craters, making ten in all. The craters were aligned with their long axes parallel, forming a corridor about 20 miles long and 1 mile wide. The corridor ran northeast to southwest just north of the city of Rio Cuarto, about 350 miles west of Buenos Aires.

In August 1991, a team of American and Argentine geologists and impact specialists headed by Dr. Peter Schultz of Brown University made the first ground search of the craters and found evidence supporting the idea of a grazing impact. A small Earth-crossing asteroid about

145

500 feet in diameter struck the ground at an angle of less than 7 degrees, exploded on contact, and broke into smaller fragments that produced the downrange string of gouges. The team found glassy clods called impactites, formed by the heat of impact, near the craters. The interior of the impactites showed clear signs of shocked and deformed quartz grains, evidence of the high pressures created during impact. The team found two small chondritic meteorites (class H4) on the floor of one of the smaller craters. If these are surviving fragments of the original impactor, they are the only stony meteorites ever found associated with a sizable terrestrial impact crater.

A body striking Earth at a nearly horizontal angle is extremely rare. Although scientists know of many oblong craters formed by low-angle impacts on Mars, Venus, and the Moon, this may be the first example found on Earth. Local nomadic Indians living in the area at the time could have witnessed the impact, since estimates place the age of the scars at less than ten thousand years.

Araguainha Dome, Brazil. The most impressive impact structure in South America is in the central Brazilian state of Mato Grosso, near the town of Araguainha. The *Landsat I* satellite shows a circular structure 24 miles in diameter with a 6-mile-diameter central uplift. A downdropped ring surrounds the central uplift, giving the structure an apparent 12-mile diameter viewed from the ground. Its full 24-mile extent is apparent only from satellite images.

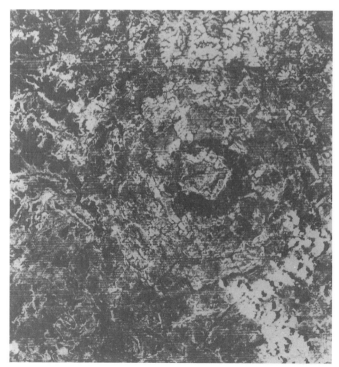

Enlargement of a Landsat *satellite photo showing the 24-mile-diameter Araguainha Dome in Brazil.* –NASA

Location of selected meteorite craters in Canada.

Robert Dietz and B. M. French investigated the Araguainha structure in 1972 and found shock-induced quartz in the igneous breccia rock of the central uplift. Shatter cones were also found. Although its age is still in question, it is probably not older than 250 million years.

Canada's Great Meteorite Craters

There are so many meteorite craters in Canada, both authenticated and suspect, that I will describe only the more spectacular ones. Most were discovered from the air, and the largest ones are best revealed by orbiting satellites. Some of these craters are so old that weathering has all but erased their once-circular shapes. All have been authenticated by the presence of shock features in the rocks. None of the craters have yielded meteoritic material, probably attesting to their great age and the enormous energies released by the meteoroids on impact.

New Quebec Crater (A.K.A. Ungava and Chubb Crater). New Quebec Crater is a simple, 2-mile-diameter crater first located on aerial photographs of the Ungava Peninsula in Quebec Province in 1950. Its extremely remote location northeast of Hudson Bay makes it difficult to study. The crater is a remarkably circular water-filled bowl resting in Precambrian granites. The walls are great blocks of shattered granite that rise an impressive 590 feet above the level of the permanent lake and about 325 feet above the surrounding plain. The crater floor plunges 1,184 feet below the rim. Although no meteoritic material is present, shocked quartz and other shocked minerals have been found

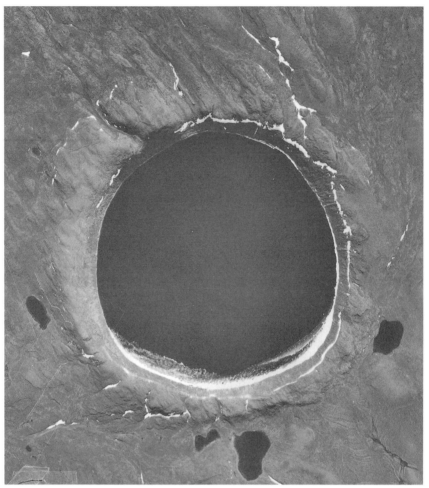

Two-mile-diameter New Quebec Crater northeast of Hudson Bay, Canada.
–Photo No. A16116-110. Copyright 1958. Her Majesty the Queen in Right of Canada. From the collection
of the National Air Photo Library with permission of Energy, Mines and Resources, Canada

around the crater. Shattered granite blocks forming the rim and scattered on the plains were obviously ejected from the crater during its
explosive formation 5 million years ago.

Brent Crater. In 1951, another simple impact crater of great antiquity
was found through aerial surveys. This crater near Brent, Ontario,
about 160 miles north-northwest of Toronto, was difficult to detect
from the ground, since it is a shallow dish 2.36 miles across and only
200 feet deep in the center. It has resisted erosion over the years
because it formed on the hard granite gneiss of the Canadian Shield,
even surviving the passage of glaciers. Any meteoritic material surviving the impact would have rusted away long ago. Holes drilled in the
crater's center have revealed a layer of glacial till from the last glacial

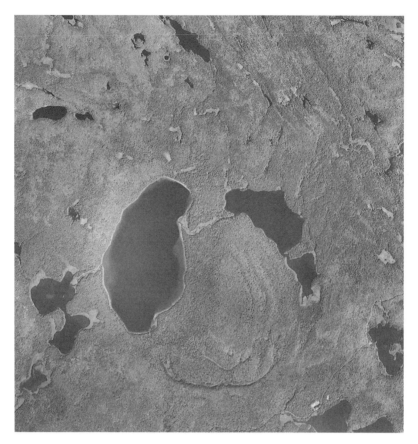

Brent Crater in Ontario, Canada. Lakes cover about one-third of the shallow 2.36-mile-diameter structure. –Photo No. A27477-96. Copyright 1989. Her Majesty the Queen in Right of Canada. From the collection of the National Air Photo Library with permission of Energy, Mines and Resources, Canada

period covering an 800-foot-thick layer of sedimentary rock that completely fills the crater. This rock dates from the Ordovician period, 450 million years ago, making the crater at least that old. Beneath the sediments nearly 1,000 feet down is an enormous lens of broken rock 2,000 feet thick in the center. This lens is similar to the breccia found beneath Meteor Crater. The shattered rock shows signs of intense heating and shock from a great explosion. Drill cores revealed a melt zone in the first 500 feet of the lens, where extensive melting and recrystallization of the original rock took place. The meteoritic mass must have penetrated to that depth. Beneath the lens is a zone permeated by fractures. All of this points to an impact origin between 450 million and 800 million years ago, the age of the granite gneiss.

Holleford Crater. This ancient crater lies about 30 miles south of Ottawa, Ontario, near the Canada–United States border. It was first discovered in 1955, during a systematic search through aerial photos

Holleford Crater near Ottawa, Ontario, Canada. This weathered crater is nearly lost in the cultivated fields surrounding and overlaying it. A two-lobed lake, now dry, partially fills the shallow depression. –Photo No. A27083-66. Copyright 1987. Her Majesty the Queen in Right of Canada. From the collection of the National Air Photo Library with permission of Energy, Mines and Resources, Canada

from the vast collection of the Canadian Air Photo Library. These photographs show an inconspicuous circular feature 1.5 miles in diameter almost lost among a complex patchwork of farmland.

The crater was originally formed in gneiss rock of the Precambrian Shield, which covers all of Ontario. The original depth was 780 feet, but it was filled in by marine sediments deposited over the crater in early Paleozoic times. Today the crater remains buried, with only the slightest trace of a depression seen from the air. The crater is hardly noticeable from the surface, since its depth is less than 100 feet. Drill cores taken along the edge of the depression revealed a buried remnant of a rim about 60 feet down, only slightly reflected on the surface by the overlying sedimentary rock. Most of the rim is missing, attesting to its ancient age of probably over a half-billion years. Additional drilling located a lens of broken basement rock, a certain sign of impact. Adding to the evidence for an ancient meteorite crater, coesite was found in the lens core-samples. Holleford Crater is far too old for any meteorites to have survived.

Deep Bay Crater. This water-filled basin was first spotted from the air in 1947. It is located on the southern end of Reindeer Lake in eastern Saskatchewan, about 275 miles north of Prince Albert. It has a remark-

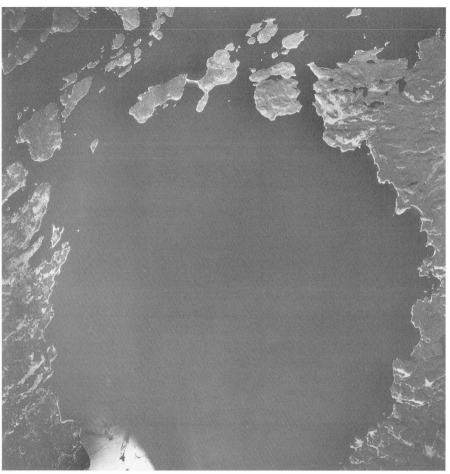

Deep Bay Crater in eastern Saskatchewan, Canada. This impressive 8-mile-diameter, water-filled structure lies on the southern edge of Reindeer Lake. –Photo No. A25906-63. Copyright 1981. Her Majesty the Queen in Right of Canada. From the collection of the National Air Photo Library with permission of Energy, Mines and Resources, Canada

ably circular shape, though serrated along the edge, and has a diameter of about 8 miles. It is connected to the lake as an outlying bay but differs from the lake in several important respects. Its depth is 660 feet, which is three times deeper than Reindeer Lake, and unlike the lake, it has no islands. Its rim, though greatly eroded, still rises an average of 270 feet above the water level. Signs of impact are evident. The crater was formed in Precambrian gneiss. This rim rock, and rocks along the shoreline, show effects of shattering and uplift. Numerous faults and fractures encircle the crater. Concentric fractures centered on the crater but several miles beyond its rim indicate that the original depressed zone was nearly twice the size of the existing crater. Undisturbed sedimentary rock several hundred feet thick fills the crater. Jurassic-age fossils found in the sediments establish an age of 140 million years. A breccia zone probably exists beneath the sediments to a depth of more than 10,000 feet. No shocked quartz, impactite, or meteoritic material have been found.

Other Great Scars?

Earth's largest astroblemes represent a much more devastating kind of impact than those that produced craters only a mile or two in diameter. They were created not by bits and pieces of asteroidal bodies a few hundred feet across, but great masses measuring miles across—far too large to be deviated in their paths by the skin of Earth's atmosphere. Only forty years after geologists began to accept the idea of Earth being struck by extraterrestrial bodies, they have begun to realize the enormous implications of this reality. Impact geology was born with the discovery of Meteor Crater's true origin. Tectonic forces acting over millions of years to shape Earth's crust was a revolutionary idea in the 1960s, and geologic structures created by cataclysmic events in an instant of real time, rather than over millions of years, seemed to most geologists nearly unthinkable.

New astroblemes are being found every year as manned and unmanned spacecraft scrutinize Earth as never before. A large circular structure, 62 miles in diameter, was discovered in 1966 in Quebec. That structure is partially filled with water, forming Manicouagan Lake. Manicouagan Crater is 214 million years old, formed at the close of the Paleozoic era, and has an obvious central uplift, evident from space. Scientists once thought it was Canada's largest impact structure, but now a huge elliptical crater near Sudbury, Ontario, heads the list as the largest and oldest—155 miles across and 1.8 billion years old.

The largest impact crater known is in South Africa, called the Vredefort Ring. From space the crater appears partially buried by younger rock, but its 186-mile-diameter rim is still discernable. Intensely shattered rock, shatter cones, and shocked quartz verify its impact origin over 2 billion years ago.

The Vredefort Ring in South Africa, photographed from the space shuttle. The southern part of the ring is nearly obliterated by new rock layers laid down after the ring formed. —NASA

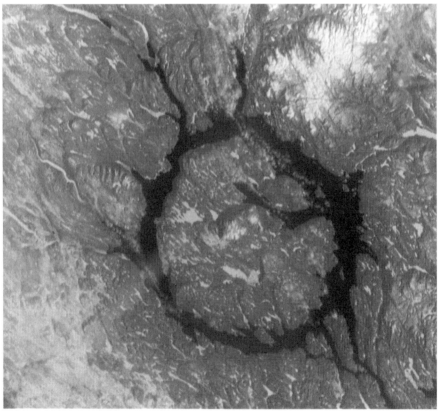

Canada's largest impact structure is 40-mile-diameter Manicouagan Crater in Quebec, seen here from Earth's orbit. —NASA

Part II

WHAT IS A METEORITE ?

The nearly perfectly oriented Adamana stony meteorite found in a landfill in 1995 near Holbrook, Arizona. —Robert A. Haag collection

Fine flow features on the black crust of a Murchison CM2 carbonaceous chondrite. —Courtesy Michael Casper, Meteorites, Inc.

NINE

How to Recognize a Meteorite

Most meteorites are found by people looking for interesting rocks, and not by scientists. Usually, the only characteristics the finders note is that their specimens are black and heavy. The great majority of meteorites are heavier than terrestrial rocks per unit volume. This is because they contain nickel-iron alloy or iron-bearing minerals. Of the three basic types of meteorites—irons, stones, and stony-irons—the irons contain about 98 percent nickel-iron, the stones as much as 23 percent nickel-iron, and stony-irons about 50 percent nickel-iron. Besides being heavy, they are all usually black. But this is not enough to determine whether a rock is a meteorite. Many people expect meteorites to be made solely of iron, but most meteorites are stony and look superficially like terrestrial rocks.

Meteorite Statistics

The 1985 edition of the *Catalogue of Meteorites* lists a total of 2,611 authenticated meteorites. (Since the *Catalogue* was published, over 700 new non-Antarctic meteorites have been recovered. Add to that the thousands found in Antarctica since 1985, and the number today is considerably higher. For the present discussion, the 1985 total can be accepted without seriously affecting the following percentages.) Of these, 1,813 (69 percent) are stony, 725 (28 percent) are irons, and 73 (3 percent) are stony-irons. Statistics can be misleading, however, as the numbers seem to suggest this is the stones-to-irons ratio of all recovered meteorites. Not true. Most meteorites found by weekend rock hunters are irons, simply because irons exhibit the characteristics people expect a meteorite to have: They are heavy (iron) and black. People usually overlook the far more common stony meteorites, which resemble ordinary rocks.

A much better idea of the true ratio of stones to irons is derived from checking the number of stones and irons recovered from witnessed falls versus those in "finds" (falls that are not witnessed). Of the 725 known irons, only 42, or 5.8 percent, are from falls. For the 1,813 known stones, 853, or 47 percent are from falls. The numbers are conclusive: Most meteorites are stones. In fact, more than 94 percent of witnessed falls are stones, suggesting that most of their parent bodies have stony compositions. Iron parent bodies must be much rarer.

Stony-iron meteorites come in last. There are 73, and only 10 are from falls. Stony-irons account for only 2.8 percent of the total known meteorites, making them the rarest of the three major types.

External Characteristics

Most meteorites in the field have a distinct appearance, and they usually contrast with surrounding local rocks. The best way to learn to recognize them is to examine as many samples as you can. Museums tend to exhibit specimens that have extraordinary shapes, enormous size, or an eye-catching internal structure revealed by laboratory processing. I believe this, in a way, is a failing of the public museum. A geology student doesn't learn about minerals by studying elegant specimens in a mineral museum, because geologists seldom find such prize specimens in the field.

So we will look at the characteristics you are most apt to find when a meteorite has been lying on Earth's surface for years, as well as the cleaned specimen, with the effects of weathering and oxidation removed. Both are important.

Besides external appearance, meteorite hunters should look for characteristics best seen with a 10X magnifier. And since most stony meteorites contain elemental iron, they are moderately attracted to a magnet. These two tools should be the meteorite hunter's constant companions.

Meteorite Shape

Meteorites are shaped by the forces encountered during atmospheric passage. If the meteoroid had an angular shape in space, its random tumble through the ablating atmosphere will round the edges, giving the meteorite a roughly spherical or elliptical shape.

Occasionally, a meteoroid will enter the atmosphere and quickly stabilize itself, so it is not rotating or it is rotating with its axis along the direction of its motion. This will selectively shape the meteoroid's leading face, ablating the corners so it takes on a blunt point. The opposite face, being protected, accumulates ablated material from the forward end. The result is the conical shape we call oriented. Roughly 5 percent of stony meteorites show some degree of orientation. Whether a meteorite becomes oriented depends on its original shape. Spherical meteoroids will tend not to orient in flight but will simply tumble, maintaining their spherical shapes as they ablate. Flattened or angular-shaped meteoroids are more likely to orient themselves during flight. Oriented meteorites are prize specimens in a museum's collection.

Stony meteorites tend to be much smoother than irons, and any pits and cavities on them are generally shallow. They usually break up in

Two Plainview, Texas, stony meteorites. The larger specimen is elliptical, weighs 7 pounds, and measures 9.5 inches in its longest dimension. The smaller specimen is more rounded, weighs 1.5 pounds, and is 3.5 inches across.

Beautiful oriented meteorite from Plantersville, Texas. This stone, which weighs 4.6 pounds, fell in 1930. –Smithsonian Institution

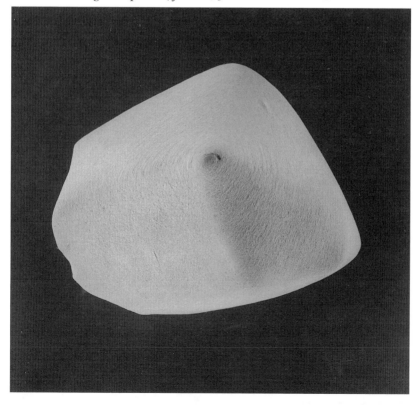

Unbroken, individual stone from Tenham Station, South Gregory, Queensland, Australia. Note the angular shape, rounded edges, and dark fusion crust. Specimen measures 3 inches across.

Fresh break in a Nuevo Mercurio stone, showing no fusion crust on the broken sections. Stone measures 2.5 inches across.

the atmosphere and fall in many smoothly ablated pieces, arriving on Earth as irregular stones with gently undulating surfaces. The original sharp-edged flat breaks may have been rounded, but the flat surface often remains. Stony meteorites often break when they slam into the ground, producing angular shapes at the fracture points. When a break is fresh, the interior is usually much lighter in color and shows no sign of heating.

The shapes of iron meteorites are much more varied. They are usually more angular, especially if they fragment on impact. Sometimes they are oriented; in fact, H. H. Nininger estimated that as many as 28 percent of all irons are oriented. The most famous example is the 15.5-ton Willamette, Oregon, meteorite. It has a nearly perfect cone shape and a broad, nearly flat rear face with large pits (see photo on page 249). Oriented irons often have a smooth forward end, with pitting and grooves becoming progressively elongated toward the rear (see photo on page 140).

Pitting is much more common in irons than in stones. The pits have the shape of shallow, elongated depressions like thumb marks made in soft putty, and are often called thumbprints. Technically, they are regmaglypts or piezoglypts. The edges of the pits are usually sharply defined and somewhat raised. They are probably produced by differential ablation, that is, lower-melting-point minerals (troilite) existing as nodules close to the surface, which tend to melt away, leaving empty

A 31-pound Canyon Diablo iron meteorite showing spherical pits (regmaglypts)
where lower-melting-point inclusions ablated out of the specimen.

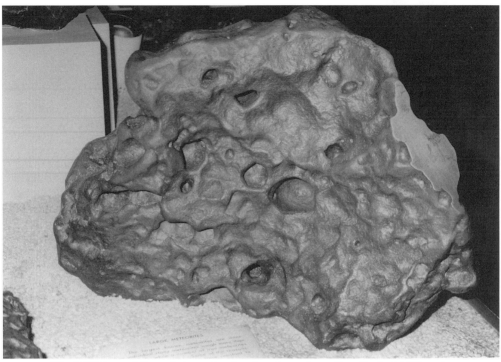

The 2,573-pound Goose Lake meteorite, showing extraordinarily deep pitting. A small slice has been sawn off the upper right side. –Smithsonian Institution

pits behind. If a low-temperature mineral is prevalent in a meteorite, the ablation process can sculpt it into extraordinary shapes. Specimens from the Mundrabilla, Australia, iron are characterized by this kind of extensive pitting (see color plate II). The famous Goose Lake meteorite found in Modoc County, California, in 1938 has remarkably deep spherical holes. The aperture of the holes is actually smaller than the diameter of the cavity. The reason these holes occur is not known. Often a hole will be ablated completely through the specimen, as in the Tucson Ring meteorite (see photo on page 260).

Iron fragments produced either by disruption in the atmosphere or impact with the ground often show twisted and distorted edges, as in the Sikhote-Alin and Henbury irons. Irons often have sharp points along the edges of the pits, accentuating the thumbprinted surface. These points were probably produced by ablation.

Stony-iron meteorites tend to have the roughest surface, especially the pallasites, which are composed of a network of iron filled with olivine crystals. Often the olivine grains drop out, leaving the iron network exposed. This is typically true of meteorites that have weathered over a period of years, producing a coarse texture.

Fusion Crust

If a recovered meteorite is fresh enough, it will have an outer crust. This is an ablation effect: As the exterior of the meteorite reaches its

162

Four-pound Odessa iron, with numerous sharp points.

melting point, the material begins to flow. Most of the liquid rock ablates away into the smoke train. What remains coats the meteorite with a fusion crust.

The crust of most meteorites is usually black, although this depends to some extent on composition. The crust of stony meteorites is a glassy material similar in composition to the silicates composing the interior. If iron is present, it usually oxidizes into magnetite, which rapidly mixes with the glass and colors the crust black. Sometimes the composition

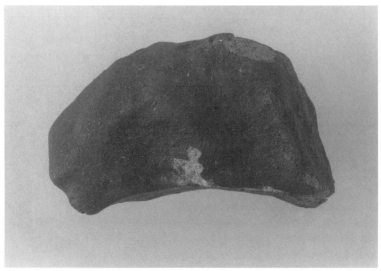

Black fusion crust on a Holbrook stone. Note the light interior. Specimen measures 2.7 inches across.

A 0.01-inch crust on a cut slab of a Bruderheim stony meteorite.

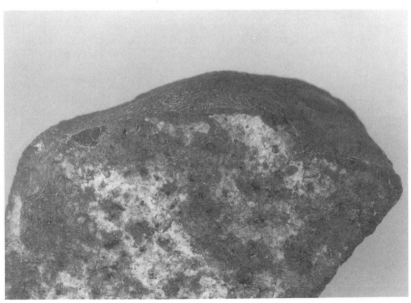

*A 0.04-inch crust revealed in broken segment
of a Nuevo Mercurio stony meteorite.*

of the meteorite can be surmised by examining the crust. A few rare meteorites have light-colored crusts. The Norton County achondrite, lacking free iron, has a light-brown crust that is still markedly different from the interior (see color plate II). Another rare type of stony meteorite, called a calcium-rich achondrite (eucrite), composed of silicates that are more readily melted, produces a shiny black crust. It is quite different from the dull black crusts of ordinary stones (see color plate III).

The crust on stony meteorites varies in thickness from an average of less than a millimeter (0.039 inch) to several millimeters, and often varies in thickness from front to rear on oriented meteorites. The forward end has the thinnest crust and the rear has the thickest, since the melted glass accumulates toward the rear.

Often a meteoroid will break up in flight. If this happens at a high altitude while the meteorite retains much of its cosmic velocity, the freshly broken face will also form a fusion crust, called a secondary crust. It can usually be recognized because it is thinner than the primary crust, having formed later in flight. Also, the broken face suffers less ablation and is usually not as smooth as the primary crust.

A thin secondary crust formed on a fragment of a Nuevo Mercurio meteorite broken during atmospheric flight. Note the thick primary crust surrounding the broken section. Specimen is 0.8 inch in diameter.

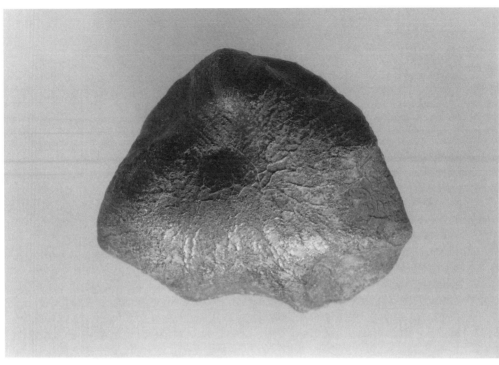

Oriented eucrite achondrite from Millbillillie, Western Australia, showing fine flow structures radiating from the forward end. Specimen measures almost an inch across.

Fine contraction cracks on a fusion crust, produced by rapid cooling. Specimen is from Nuevo Mercurio.

The surface frequently is lumpy, an expression of the interior texture of the meteorite. If it breaks up after reaching its retardation point, no fusion crust will form on the freshly broken face. It may break further on impact, producing more fresh faces without fusion crusts.

Iron meteorites have the thinnest crust of all, usually only a small fraction of a millimeter thick. A fresh crust is blue-black to black and looks like freshly welded steel. It is composed entirely of the iron oxide magnetite. This crust is fragile and easily destroyed if the meteorite weathers for even a short time.

Fusion crusts often show interesting striations that are obviously flow structures. These indicators of flight direction are most common on oriented specimens, extending from the front along the sides to the rear. Flow structures are best seen with a hand magnifier or mineral microscope. Only a relatively fresh meteorite will retain these delicate features.

Some stony meteorites have fine fractures in the crust resembling crazing in a pottery glaze, a result of rapid cooling and contraction of the fused glass.

External Features of Weathered Meteorites

Roughly two-thirds of the known meteorites are finds. They have been exposed to wind and rain for years, even centuries. Earth is an alien environment for a meteorite. Here it is subject to the same mechanical and chemical weathering as terrestrial rocks. Temperature extremes, especially in desert environments, cause expansion and contraction. Meteorites become cracked and split by frost-wedging in cold climates and may be tumbled in streams or blasted by wind-driven sand. Many old meteorites are covered by mineral stains. Chemical weathering (oxidation) is even more destructive and eventually strips away an old fusion crust. Remarkably, meteorites from the same shower differ in their resistance to weathering. Fusion crusts often vary from black to light brown, the latter being more chemically weathered (see color plate III). Old meteorites may have deep cracks, often along internal veins, which allow water to penetrate and slowly transform the interior. The minerals in meteorites become oxidized, recrystallized, and chemically altered, eventually destroying the cosmic characteristics that make meteorites so valuable.

Iron meteorites quickly rust, a process that destroys the original magnetite crust. Canyon Diablo irons are typical (see color plate III). When they are dug up, they have a thick, scaly coating of rust. To some extent, the rust protects the meteorite by slowing further oxidation; it can be removed with a stiff wire brush, revealing another black coating of magnetite, this time terrestrially deposited. Beneath that black coating is the original silvery nickel-iron. A meteorite can be tumbled for several hundred hours until the hard magnetite coating has been completely removed, leaving behind a meteorite "nugget."

A weathered stony meteorite found near Willard, New Mexico, in 1982. Nickel-iron grains have all but disappeared.

The interior of common stony meteorites usually contains nickel-iron grains. In some, as much as 25 percent of the iron in the meteorite is metal, that is, iron in its elemental state. This iron may be subject to oxidation and rusting even though it is in the interior of the meteorite. Badly weathered stones often show only occasional flakes of iron where once it was plentiful (see color plate III for another example).

A hard deposit of calcium carbonate, called caliche, is frequently seen on iron meteorites found in desert climates. This unsightly white material clings tenaciously to the meteorite, defying easy removal by mechanical means. However, it can be chemically dissolved by painting the specimen with fairly strong sulfuric acid. The acid attacks the carbonate, but this practice is harmful if the acid permeates fractures in the specimen. If the meteorite is strictly for display rather than research, chemical removal of the carbonate should be done before removal of the rust layer, to provide some protection for the meteorite. In wet climates, iron meteorites oxidize rapidly, turning into a laminated, shalelike material. At this stage of weathering, it can no longer be called a meteorite, since it has been completely terrestrialized.

With the exception of two "fossil" meteorites from the Ordovician period, 450 million years ago, the oldest meteorites, some 700,000

White caliche deposits on a Canyon Diablo iron.

years old, come from Antarctica. Terrestrial conditions are not conducive to a meteorite's survival.

Lawrencite "Disease"

Meteorites exposed to the elements are destined to crumble into powder. And once they are placed in protective display cases or drawers, their weathering troubles are not over.

The most serious type of weathering in many meteorites continues even after they have been carefully collected and preserved. A "cancer" often found in iron-bearing meteorites causes them literally to fall apart. It is the result of a ferrous chloride mineral, lawrencite. Though not a meteoritic mineral, it forms within the meteorite by diffusion of atmospheric chlorine. Lawrencite reacts with oxygen and water in the air, and oxidizes to form ferric chloride and ferric hydroxide (limonite). It forms a green-brown goo that exudes between the Widmanstätten plates in iron meteorites.

Two things may then happen to the ferric chloride. It may be reduced by contact with the iron to ferrous chloride again, beginning the cycle anew. Or it may react with water to produce more limonite and hydrochloric acid, which immediately begins to eat away at the meteorite specimen, forming a brownish rust between the Widmanstätten plates that often expands into waves on the cut surface (see color plate IV). As the reaction takes place from within, the

Widmanstätten plates are slowly lifted out of place, causing disintegration of the meteorite.

The Brenham, Kansas, pallasite is particularly susceptible to lawrencite disease. It is distressing to see a beautiful slab of the Brenham pallasite, its network of nickel-iron surrounding yellow-green olivine crystals, slowly disintegrating. When the iron is attacked, the enclosed olivine loosens and drops out of the meteorite. Eventually, the iron network rusts through, fracturing the meteorite into smaller pieces (see color plate IV).

Since the reaction takes place in air and develops inside the meteorite, there is little one can do to prevent it. Some museums keep their lawrencite-ridden specimens in sealed cases filled with dry nitrogen in an effort to control it. Fortunately, not all iron-bearing meteorites contain lawrencite and the lawrencite content seems to vary markedly in specimens from the same fall. In the field, heavily rusted meteorites are a sure sign of the presence of lawrencite. Cutting them and exposing their interiors will only exacerbate the problem.

"I'm afraid you have the Lawrencite disease. I prescribe an alcohol bath and oven treatments."

Iron Oxides

Most people think of meteorites as being heavy, pitted, and black. These superficial characteristics often do apply to meteorites, but unfortunately, they also describe many terrestrial stones. The most popular "meteor-wrong" is magnetite, a common terrestrial mineral often found lying on the surface in the deserts of the American Southwest. This metallic mineral is simply an oxide of iron and is mined as iron ore. Magnetite is black and leaves a black streak when rubbed against a harder substance, like the unglazed surface of a porcelain tile. It is dense, heavy, and attracted to a magnet. Its cousin, hematite, is also an oxide of iron but is usually more earthy looking and can take on many hues from light brown to red to black. It streaks reddish brown and usually does not have magnetic properties. Stony meteorites usually do not leave a streak like the ones metallic minerals do, since their crusts are primarily glass. Iron meteorites may leave a faint streak if the magnetite crust is fresh.

Artifacts

Terrestrial iron is rarely found in its natural state; that is, as pure iron. This is because it is easily oxidized and is usually found in one of its oxidation states or in combination with other metals making up common rock-forming minerals such as olivine or pyroxene. A specimen that is strongly attracted to a magnet and is metallic-looking when

A bomb-case fragment mistaken for a meteorite.

scratched or cut is either an iron meteorite or an artifact. Rusting iron artifacts—bits and pieces of old mining tools and machinery that are probably over a century old—are scattered throughout the desert. The western mountains and deserts are littered with iron artifacts. Many have all but lost their original identities. A relatively recent addition to the growing list of artifacts are the remnants of exploded bomb cases left behind by the military during war maneuvers in the western deserts. At first glance, I have been fooled more than once when asked to examine such specimens. And it's remarkable how a century or more of weathering can transform a broken piece of miner's pick or a railroad spike into what seems to be an iron meteorite. An examination of a freshly ground surface, however, usually reveals the impostor.

Skin Deep

Deserts are great places to hunt for meteorites, since the vegetation there is sparse. The desert abounds with rocks, both rounded and angular. The angular rocks are broken fragments of larger rocks, but the rounded rocks have an aqueous history; they have been tumbled for

Mineral stain on a rounded terrestrial rock gives it the appearance of a fusion crust, especially if the rock has a light interior.

172

countless centuries in intermittent desert streams, gradually becoming rounded. Many slowly acquire a desert varnish—a thin veneer of manganese mineral-stain that darkens through time. This rounded, stained rock mimics a stony meteorite, especially if a piece is broken, revealing a lighter interior. Again, a critical examination of the interior usually reveals the rock's terrestrial character.

Rocks with Holes

Most people envision a meteorite as a rock full of holes. To be sure, meteorites, especially irons, often do have pits and deep holes created by the ablation process. The most common terrestrial rock with holes is lava rock. In the western United States, which has a long history of volcanism, remnants of lava flows are everywhere. Most of this lava is the dark igneous rock called basalt. Usually erupted quietly as rubbly masses from fissures around the flanks of cinder cones, basalts often mix with gases that form bubbles. When the basalt cools and hardens, the bubbles remain as small cavities in the rock. The dense rock (basalt contains plenty of iron), dark and riddled with cavities, is a classic "meteor-wrong."

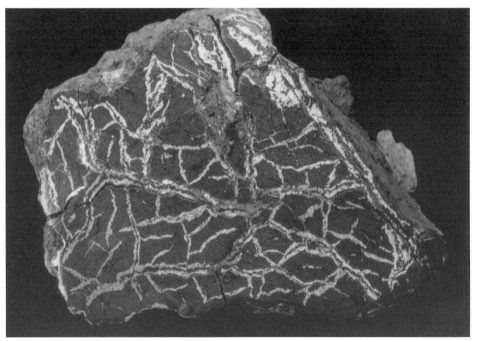

This graphite nodule from Meteor Crater, Arizona, was once within a large iron meteorite. The metal surrounding the inclusion probably weathered away, but veins of metal were injected into the nodule at the time of impact. Metal-bearing meteorites often contain veins of metal, a sign of impact shock. Specimen measures 5 inches across.

Highly textured interior of a CV3 carbonaceous chondrite from Axtell, Texas. Well-defined circular and subcircular chondrules typical of CV3 chondrites crowd the cut surface along with occasional white calcium-aluminum inclusions, distinguished from the chondrules by their irregular shapes.
—Courtesy Michael Casper, Meteorites, Inc.

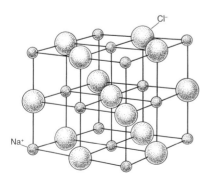

Arrangement of sodium and chlorine atoms in the unit crystal lattice of common table salt.

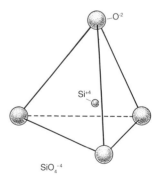

The silicon-oxygen tetradedron.

TEN

Chondrites—
The Common Stony Meteorites

Looking at the exterior of a stony meteorite reveals little about the fascinating world within. Although they contain minerals common in Earth's crust and mantle, meteorites have truly cosmic internal structures that are the source of the fascination for stony meteorites. To touch a rock from space, an "alien" rock unlike anything found on Earth, is a truly unique experience.

Scientists sort meteorites into types according to structure and mineralogy. Early classification schemes were based on easily recognizable characteristics, but, more recently, analytical tools and methods based on chemical and isotopic analysis have been used to determine meteorite types. As scientists learn more about how meteorites are related, the classification schemes change. Over the past decade, the harvest of meteorites in Antarctica has yielded meteorites that do not easily fit into the current classification system, which will surely result in further changes in the future.

Fortunately, a person doesn't have to become involved with isotopic analyses to be able to classify stony meteorites. But you do need to know something about minerals, since mineralogy is one of two important criteria in meteorite identification.

Minerals in Stony Meteorites

Rocks found on Earth are aggregates of minerals combined in various complex ways. Minerals are naturally occurring elements or compounds with a specific composition. They usually form crystals that reflect the orderly arrangement of their constituent atoms. We're all familiar with common minerals like halite, or table salt, which crystallizes as perfect cubes. Halite's crystal faces are the external expression of the internal atomic arrangement of its sodium and chlorine atoms. This atomic arrangement is called a crystal lattice.

The minerals that make up the crustal rocks of Earth are called rock-forming minerals. They are various combinations of only eight different elements: oxygen, silicon, sodium, calcium, potassium, aluminum, magnesium, and iron. Four oxygen atoms combine with a silicon atom to form a four-sided, pyramid-shaped atomic structure called a silicon-oxygen tetrahedron. In this crystal lattice, the relatively large oxygen atoms occupy all four corners of the tetrahedron, with the small silicon atom suspended between them. This tetrahedron, as the complex ion SiO_4^{-4}, is the basic building block of the silicates, which make up more than 90 percent of the rock-forming minerals in Earth's crust. The tetrahedral units combine in various ways, resulting in different silicates, each with a unique crystal structure. Along with the tetrahedra, each mineral makes room for one or more metal atoms to join the crystal structure. These are usually iron, magnesium, sodium, potassium, calcium, or aluminum, all common metals in Earth's crust.

The Three Primary Minerals in Stony Meteorites

Olivines and pyroxenes are the most abundant minerals in stony meteorites; but unlike terrestrial rocks, the common stony meteorites contain nickel-iron alloys in the matrix as unoxidized metal grains. Interestingly, quartz (SiO_2), a major constituent of common igneous rocks such as granite and rhyolite, is relatively rare in meteorites.

Feldspars are another group of common minerals in terrestrial rocks. They are composed of sodium, potassium, and calcium combined with aluminum and the silicon-oxygen tetrahedron. The white crystals in a typical granite rock are sodium feldspar. Feldspars usually are accessory minerals in stony meteorites.

Olivines. Olivine in meteorites is easily detected with a 10X hand lens. It is most often yellow or brown, depending on the amount of iron in the mineral. In Earth rocks, it often appears as tiny olive-green crystals in dark iron-bearing rocks such as gabbro and basalt.

Olivine is a term for a group of minerals with similar structure but varying composition. Its general formula is $(Mg,Fe)_2SiO_4$. Magnesium (Mg^{2+}) and iron (Fe^{2+}) ions have approximately the same atomic size and can readily change places with each other in the crystal lattice. The superscript (2+) means that the atom has lost two electrons and is positively charged. Such positively or negatively charged particles are called ions. It can readily combine with the silicon-oxygen tetrahedron, which is negatively charged. The relative amounts of magnesium and iron determine the type of olivine crystallizing out of a cooling magma. This ionic substitution is called a solid solution. Mineralogists illustrate the solid solution of minerals by a triangular composition diagram called a ternary diagram.

Pyroxenes. Like olivines, pyroxenes are subject to solid solution, and they also contain various percentages of magnesium, iron, and calcium. The difference lies in the composition of the silicon-oxygen component of the molecule. Olivine is based upon SiO_4 chains, while pyroxenes are based on chains of SiO_3, resulting in entirely different crystal structures.

Nickel-Iron Minerals. A rock with silvery metallic grains in the interior and an attraction to a magnet is probably a stony meteorite. As much as 23 percent of the iron in stony meteorites can be in the elemental state, usually alloyed with nickel and called "metal" in meteorite jargon. The remaining iron is found in combined form as oxides, sulfides, carbides, and phosphides or in olivines or pyroxenes. It is referred to as "low iron" or "high iron" content, depending upon the relative amount of iron in these minerals.

The iron as metal is always alloyed with nickel, which can vary from less than 5 percent to as much as 25 percent of the total metal. These nickel-iron alloys have distinguishing characteristics best seen in the iron meteorites.

COMPARISON OF COMMON MINERALS IN EARTH'S CRUST AND METEORITES

Mineral	Description	Common Earth Rock	Meteorite
Feldspar	An aluminum silicate with varying amounts of sodium and calcium.		5–10% in chondrites
	Plagioclase ($NaAlSi_3O_8$ to $CaAl_2Si_2O_8$); 39% in crustal rock	Granites, basalts	Ca-rich achondrites, mesosiderites
	Orthoclase ($KAlSi_3O_8$); 12% in crustal rock	Granites	Rare
Pyroxenes	A large group of silicate minerals with varying amounts of iron, magnesium, and calcium $(Fe,Mg,Ca)SiO_3$; 11% in crustal rock	Mafic rocks, basalts	All stony meteorites, mesosiderites
Olivine	A silicate with varying amounts of iron and magnesium $(Fe,Mg)_2SiO_4$; 3% in crustal rock	Mafic rocks, basalts	All chondrites, pallasites, mesosiderites
Quartz	A very common rock-forming mineral (SiO_2) on Earth; 12% in crustal rock	Silicic rock	Very rare
Amphibole	A large and complex family of dark, hydrous ferromagnesian silicate minerals; 5% in crustal rock	Most crustal rock	Rare; in enstatite chondrites
Mica	A common complex hydrous silicate mineral containing varying amounts of magnesium, iron, potassium, and aluminum; 5% in crustal rock	Silicic and mafic rocks	Absent
Nickel-iron	Alloys of iron, nickel, and cobalt	Very rare	Common in all meteorites
Troilite	An iron sulfide (FeS)	Pyrrhotite on Earth	All meteorites in varied amounts
Magnetite	A common black oxide of iron (Fe_3O_4)	Common iron ore	Carbonaceous and LL-chondrites, fusion crusts
Serpentine	A class of water-bearing silicates with varying amounts of iron and magnesium $(Fe,Mg)_6Si_4O_{10}(OH)_8$	Common metamorphic mineral derived from olivine	Carbonaceous chondrites

Ternary Composition Diagram for the Olivine Group

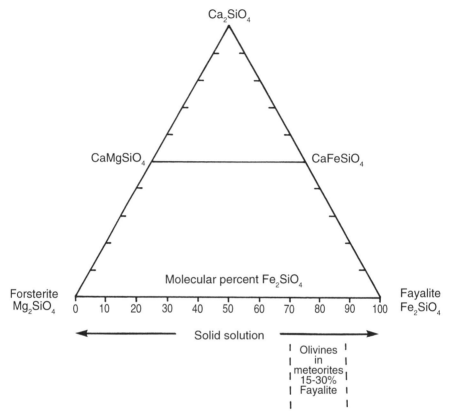

Diagram shows various compositions possible for olivine. A solid solution exists between magnesium, iron, and calcium. In meteorites, the forsterite-to-fayalite ratio is the most important.

The most common olivines are found in the series involving magnesium and iron. This ranges from forsterite, with all magnesium and no iron, to fayalite, with all iron and no magnesium. Six other minerals of the series found between the two end members have various percentages of magnesium and iron. Olivines with various ratios of calcium and magnesium or calcium and iron are relatively rare. Mineralogists usually state the composition of an olivine by determining its fayalite content; that is, the percentage of iron in the olivine. In all but the rarest stony meteorites, the olivines are magnesium-rich, usually with 15 percent to 30 percent fayalite.

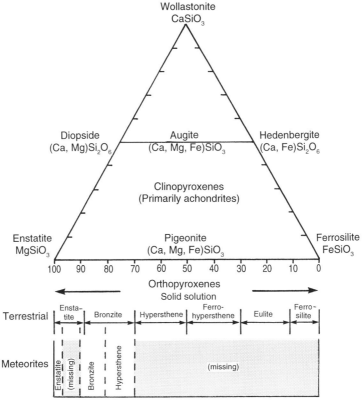

Ternary Composition Diagram for the Pyroxene Group

Wollastonite
CaSiO$_3$

Diopside
(Ca, Mg)Si$_2$O$_6$

Augite
(Ca, Mg, Fe)SiO$_3$

Hedenbergite
(Ca, Fe)Si$_2$O$_6$

Clinopyroxenes
(Primarily achondrites)

Enstatite
MgSiO$_3$

Pigeonite
(Ca, Mg, Fe)SiO$_3$

Ferrosilite
FeSiO$_3$

100   90   80   70   60   50   40   30   20   10   0

Orthopyroxenes
Solid solution

Terrestrial | Ensta-tite | Bronzite | Hypersthene | Ferro-hypersthene | Eulite | Ferro-silite

Meteorites | Enstatite | (missing) | Bronzite | Hypersthene | (missing)

Diagram shows possible compositions for pyroxenes. The orthopyroxenes, a solid solution between magnesium and iron, are the most important in meteorites, usually ranging from the enstatite end-member through hypersthene. The bronzite and hypersthene nomenclature for terrestrial pyroxenes extends to higher iron content than the nomenclature for meteorites. Also, the orthopyroxenes do not form a complete solid solution in meteorites.

Like olivines, the most important pyroxene series involves the solid solution between magnesium and iron. These are the orthopyroxenes. The end members (the pyroxenes containing either all magnesium or all iron) are enstatite (MgSiO$_3$) and orthoferrosilite (FeSiO$_3$). Stony meteorites are named for their pyroxene component; for example, meteorites that have about 20 percent FeSiO$_3$ are called olivine bronzite chondrites, indicating that they contain olivine and bronzite (the magnesium-rich pyroxene) and that their structure is chondritic (composed of tiny spherical structures). One rare type of stony meteorite containing no olivine and no iron-bearing pyroxenes is called an enstatite chondrite. A common stony meteorite that contains between 22 percent and 30 percent FeSiO$_3$ is called an olivine-hypersthene chondrite.

The remaining pyroxenes are known collectively as clinopyroxenes. They have various compositions involving calcium, with magnesium and iron. All are important accessory minerals in the classification of stony meteorites.

Accessory Minerals

Of the three minerals that make up the most common stony meteorites, olivine usually ranks first, between 40 percent and 50 percent by weight. Pyroxenes come second with 15 percent to 25 percent, and third, nickel-iron metal with from less than 3 percent to as much as 23 percent by weight. The remaining minerals are accessory minerals and usually make up only about 10 percent to 15 percent of the total weight.

Iron Sulfide. Common stony meteorites usually contain a sulfide of iron (FeS) called troilite, named after an eighteenth-century Italian meteorite investigator who was also a Jesuit priest, Father Domenico Troili. Troilite is easily distinguished from other iron minerals by its bronze color.

Oxides of Iron. The most common iron oxide is magnetite (Fe_3O_4), a component of the dark, glassy fusion crusts on stony meteorites. It is also an important mineral in carbonaceous chondrites. Some magnetite found in weathered meteorites, both stone and iron, is the result of oxidation after landing on Earth.

Plagioclase Feldspar ($NaAlSi_3O_8$–$CaAl_2Si_2O_8$). Feldspars are common silicate minerals in Earth's crustal rocks. They are also found in small quantities, 5 percent to 10 percent by weight, in most stony meteorites and mesosiderites. Sodium-calcium-aluminum silicates are called plagioclase feldspars. Like the olivines and pyroxenes, plagio-

Cross section through an H5 chondrite from Plainview, Texas, showing elemental nickel-iron metal as small, shiny, white flakes dispersed uniformly in the matrix. The round, gray inclusions are chondrules. Specimen measures 4.5 inches across.

clase forms by solid solution with exchanges of sodium and calcium ions. The most common plagioclase in stony meteorites is primarily sodium-bearing, but a few rare stony meteorites are known to contain almost pure calcium feldspar (achondrites).

Structure of Stony Meteorites

Because meteorites pass through the atmosphere so rapidly, their interiors remain unchanged by heating—they are usually pristine, revealing cosmic characteristics not seen in terrestrial stones. The two major types of stony meteorites, chondrites and achondrites, are distinguished by their internal structure.

Ordinary Chondrites

The most common type of stony meteorite is called an "ordinary chondrite" (OC). About 85 percent of the observed stony meteorite falls are in this class. The remaining chondrites are carbonaceous (C) and enstatite (E) chondrites, each with unique individual characteristics.

Chondrules. Little spherical inclusions called chondrules give the chondrite meteorite its name. The texture is cosmic because chondrules are not found in terrestrial rocks. A few rare types of stony meteorites classified with the chondrites contain no chondrules. At one time the presence of chondrules was the primary criterion for classifying chondrite meteorites. Today scientists recognize that this texture can be and usually is modified by the thermal or chemical history of a meteorite. A better criterion for classifying chondrite meteorites is the elements that compose them. With the exception of the lightest gases, hydrogen and helium, chondrites have an elemental composition very close to the Sun's. It is as though pieces of the Sun, minus the light elements, had condensed into solid matter to form chondrites.

Average chondrules are a millimeter (0.039 inch) in diameter. It is not uncommon to see much larger ones, and a few have exceeded 7 or 8 millimeters. Some chondrites have such prominent chondrules that they protrude from the surface and broken faces of the specimen. Chondrules are easily pried out of the matrix in which they are embedded, and sometimes only the cavity that once contained a chondrule remains.

No one knows how chondrules form. Meteoriticists used to think they simply formed as individual spherical droplets, condensed from the original solar nebula, which later became trapped in the matrix of an accreting parent body. This is no longer thought possible. There is no question that chondrules were once melted, but current evidence suggests they formed from preexisting solid grains that were somehow reheated to the melting point and then combined with one another to form chondrules. They remained in a liquid state for only a brief time before solidifying and accreting with material from other parts of the solar nebula, forming the matrix in which they are found today.

Typical chondrule field in a cut stony meteorite. The largest chondrules are about 0.08 inch in diameter. Chondrules are best seen after the specimen has been cut and ground flat to a dull luster.

Prominent chondrules protrude from the broken face of this L4 olivine-hypersthene chondrite from Bjurböle, Finland. Specimen is 2.3 inches across.

A cavity once occupied by a small chondrule about 0.02 inch in diameter in an Allende CV3 carbonaceous chondrite.

Photomicrographs of thin sections of chondrites showing chondrule types. (Also see color plate IV)

Barred olivine chondrule (BO) from the CV3 Allende chondrite. The bars and rim comprise a single crystal. The bars are set in clear glass. Chondrule 0.039-inch diameter. 25X magnification in crossed polarized light.

Radial pyroxene chondrule (RP) from the Gao H5 chondrite. Thin blades of pyroxene radiate eccentrically from a nucleation point. Thinner laths are unresolved. The once circular shape (lower right edge remnant) has been chemically eroded around the edge, forming a lobate, fan-shaped, 0.023-inch diameter chondrule. 98X magnification in crossed polarized light.

A porphyritic pyroxene chondrule (PP) in crossed polarized light from the Semarkona LL3 ordinary chondrite. Chondrule is composed of large subhedral clinoenstatite crystals, many showing multiple parallel twinning planes. –Dr. Rhian Jones, Institute of Meteoritics, University of New Mexico

A porphyritic olivine chondrule (PO) from the Gao H5 chondrite. Well-shaped olivine crystals are set in turbid glass typical of late petrologic types. Chondrule 0.051-inch diameter, 25X magnification in reflected light.

Porphyritic olivine pyroxene chondrule (POP) from the Faith H5 chondrite. Large angular crystals composing the interior are olivine. The surrounding rim is low-calcium orthopyroxene. The nearly circular black bleb in the interior is iron sulfide (troilite). Chondrule 0.027-inch diameter, 39X magnification in crossed polarized light.

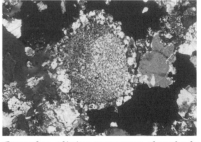

Granular olivine pyroxene chondrule (GOP) from the Nuevo Mercurio H5 chondrite. Tiny crystals composing the interior are an almost equal mix of olivine and pyroxene grains. Larger crystals of both minerals form a rim. Chondrule 0.020-inch diameter, 98X magnification in crossed polarized light.

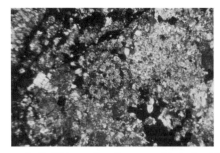

Left: *Thin section in crossed polarized light of a relict chondrule from the L6 Forrest B chondrite barely showing its original circular shape within a matrix of recrystallized olivine. This texture is typical in late petrologic type chondrites where recrystallization destroys the original chondrule texture.*

Right: *Thin section in crossed polarized light of a relict radial pyroxene* (center) *almost hidden among the large recrystallized olivine grains from the L6 Forrest B chondrite.*

Chondrules are made of olivines and pyroxenes, similar in composition to the surrounding ground mass. When chondrites are cut, polished, and examined through a microscope, they are a wonder to behold (see color plate IV). Pyroxene in chondrules often shows an eccentric, radiating structure not seen in any terrestrial rocks. Olivine sometimes shows a barred structure with individual crystal faces running parallel to each other. A common chondritic structure may have relatively large crystals of olivine or pyroxene embedded in a glassy matrix. Technically, this texture is called porphyritic. Porphyritic texture is commonly found in terrestrial igneous rocks. Chondrules can also have a fine, granular texture with embedded glassy fragments. Many chondrules are surrounded by rim material of a different composition.

Although several types of textures can be distinguished, they should not be thought of as separate entities. Commonly, textures within an individual chondrule show a bewildering mixture of types.

Petrologic Types—The Chondrule Connection

Cut sections from a few dozen ordinary chondrites show that the chondrule density and definition in the embedding groundmass vary from specimen to specimen. Some chondrites have abundant, well-defined chondrules, while others have far fewer, and these seem to merge into the groundmass. Other chondrites seem to have almost no well-defined chondrules at all. These characteristics have proved to be a useful tool in classifying chondrites from type 1 to type 6. Chondrules in the ordinary chondrites vary from well-defined and plentiful (type 3) to less-distinct and sparse (types 4–5) to nearly absent (type 6). This variation represents stages of metamorphism caused by heating, either during or after the chondrules accreted to form the original parent body. Type 3 shows little if any alteration of the chondrule texture,

PETROLOGIC TYPES

Chemical Type		1	2	3	4	5	6
		Chondrule Texture					
		Absent	Sparse	Abundant / Distinct		Increasingly / indistinct	
Ordinary Chondrites	H			░	░	░	░
	L						
	LL						
Carbonaceous Chondrites	CI	░					
	CM		░				
	CR		░				
	CO			░			
	CV			░			
	CK			░	░		
R-Chondrites[a]	R			░	░		
Enstatite Chondrites[b]	EH			░	░	░	
	EL			░	░	░	

| <150°C | <200°C | 400°C | 600°C | 700°C | 750°C | 950°C |

← Increasing aqueous alteration   Increasing thermal metamorphism →

[a] R-chondrites have brecciated textures with light clasts and dark matrix. Typically, the clasts show high petrologic types (5–6), and the matrix shows low petrologic types (3–4).

[b] The lack of a type 4 EL-chondrite is probably artificial. The existence of types 3, 5, and 6 suggests a type 4 exists but has yet to be found.

with the original structure still preserved. Beyond type 3, increasingly higher thermally altered metamorphic stages have made the chondrules less and less distinct. Recrystallization of the original chondrule minerals causes the chondrule boundaries to diffuse uniformly into the matrix of the meteorite, as new crystal growth takes place between the chondrules and matrix. Ordinary chondrites must have been heated after they formed to display these various metamorphic stages. That heating is not sufficient to melt the meteorite; recrystallization takes place in the solid state. The degree of recrystallization is related to the degree of heating and the length of heating time. Petrologic types 1 and 2 do not occur in ordinary chondrites. A few meteoriticists describe a petrologic type 7 in which there is no trace of the original chondrules. Other meteoriticists insist that such highly metamorphosed meteorites are the result of complete melting, possibly during energetic impact events, since their textures go beyond the most extreme solid state recrystallization.

These textural differences are easy to see with a magnifier. The late petrologic types (5–6) often show a brecciated texture, being composed of fragments cemented together by shock-melting, suggesting an impact history. These breccias are unmistakable.

Petrologic type 1 chondrites are a curiosity, since they do not contain chondrules. They are classified with the chondrites because their chemical compositions are related to that of type 2 chondrites.

Types 1 and 2 are not ordinary chondrites and are classified under carbonaceous chondrites. Both types show alteration by water. Of the two, type 2 chondrites are probably more primitive, with less aqueous alteration.

Another important criterion in determining the petrologic type of a chondrite is the degree, or state, of chemical equilibrium. Early petrologic type chondrites (1–3) are unequilibrated because their individual minerals show wide variations in chemical composition as though they formed and cooled before the chemical reactions were complete. The minerals are not in chemical equilibrium. If these chondrites could be heated to a metamorphic state, their individual minerals would become more homogenized as they continued to react chemically; that is, all of the individual olivine grains would have the same composition as would all of the individual pyroxene grains. Later petrologic types (4–6), in which heating has apparently continued, show much greater homogeneity among the minerals and are called equilibrated. Unequilibrated meteorites are more primitive and contain a sampling of the first minerals to form.

Chemical Types of Ordinary Chondrites

Another important distinguishing characteristic of the ordinary chondrites is their chemical and mineral compositions. Although much less obvious than chondrule texture, some mineral characteristics are easy to see with a 10X magnifier or even the naked eye.

Three categories of ordinary chondrites are recognized, distinguished by total iron and total metal content. Iron can occur in combination with small amounts of nickel or sulfur (as FeS), which appears in stony meteorites as tiny silvery grains scattered in the groundmass and causes the chondrites to be slightly magnetic. Total metal refers to this uncombined iron plus FeS. Iron also occurs in meteorites in combination with other elements, most often in the silicates olivine and pyroxene. Total iron refers to iron in both the free and combined states.

H-Chondrites. Of the observed chondrite falls, 38 percent are H-chondrites. The *H* stands for "high-iron." This group contains 25 to 31 percent total iron by weight, with between 15 and 19 percent of the iron in the uncombined metal state, making H-chondrites easily attracted to a magnet. After the high percentage of metal, these stony meteorites are primarily composed of equal amounts of the minerals olivine and pyroxene. Since the pyroxene is usually bronzite, H-chondrites were, until recently, referred to as olivine-bronzite chondrites, to distinguish them from the L-chondrites. Though meteoriticists consider this specific mineral designation obsolete, it is still useful to the collector since it reminds us of the pyroxene common in the H-type. The composition of the olivine in ordinary chondrites is given as the percentage of the iron-rich olivine, fayalite, in the total olivine content. For

H-chondrites, fayalite is between 16 and 20 percent, usually written, $Fa_{16}-Fa_{20}$. The first chemical test performed on a new ordinary chondrite is for percent fayalite, which places it among the three basic types. All ordinary chondrites vary in petrologic type from 3 to 6.

L-Chondrites. The *L* in L-chondrites stands for "low-iron." They still contain between 20 and 25 percent total iron, but the amount of free iron metal is only about 4 to 10 percent. Unlike the H-chondrites, these meteorites are not strongly attracted to a magnet, and there are fewer metal flakes visible in their interiors. Olivine is abundant but has a fayalite content of 23 to 26 percent, since more iron has been oxidized in the L-chondrites. The iron-rich orthopyroxene, hypersthene, occurs with the olivine. Like the H-chondrites, scientists formerly referred to the L-chondrites by their common pyroxene, hypersthene, thus the label olivine-hypersthene chondrite. L-chondrite falls are the most common of the three ordinary chondrites, composing 46 percent of the observed chondrite falls.

LL-Chondrites. The least common member of the triad of ordinary chondrites, with only 8.5 percent observed to fall, is the LL-chondrites. *LL* stands for "low total iron" and "low metal" content. LL-chondrites contain the least amount of iron: from 19 to 22 percent total iron and less than 1 percent to at most about 3 percent metal. The olivine in this

CLASSIFICATION OF CHONDRITES

Mineral Types	Petrologic Types	Primary Distinguishing Characteristics
Ordinary Chondrites (OC)		
(H)	H3-6	high metal, high total iron
(L)	L3-6	less metal, low total iron
(LL)	LL3-6	low metal, low total iron
Carbonaceous Chondrites (C)		
Ivuna (CI)	CI (1)	chondrule-free, aqueously altered, hydrated phyllosilicates
Mighei (CM)	CM2	sparse, small chondrules, aqueously altered, 50 percent hydrated silicates
Renazzo (CR)	CR2	primitive chondrules, metal, aqueously altered
Ornans (CO)	CO3	tiny chondrules, metal, CAIs
Vigarano (CV)	CV3	large chondrules, CAIs, slight aqueous alteration
Karoonda (CK)	CK3-6	large chondrules, dark silicates, CAIs
R-Chondrites		
Rumuruti (R)	R3-6	high iron oxidation, brecciated clasts R5–6, matrix R3–4
E-Chondrites		
(EH)	EH3-5	high metal, high total iron, small chondrules, highly reduced
(EL)	EL3-6	high iron, lower total iron, larger chondrules, highly reduced

Textures of chondrules in chondrites of petrologic types L3, H5, and L6.

Sharply defined chondrules of an L3 chondrite, showing no metamorphism (Moorabie, New South Wales, Australia).

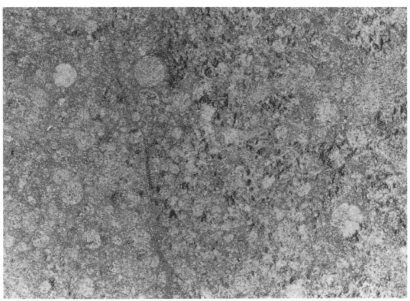

Fewer and less-well-defined chondrules of an H5 chondrite, showing some recrystallization (Faith, South Dakota).

Chondrule texture of an L6 chondrite is almost completely destroyed by heating and recrystallization. Careful scrutiny shows chondrules crystallized into the matrix (Bruderheim, Alberta, Canada).

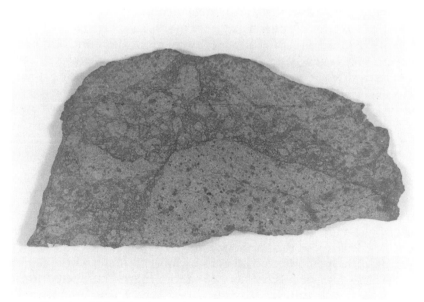

Brecciated L6 olivine-hypersthene chondrite from Nakhon Pathon, Thailand. Specimen is 4.7 inches in largest dimension.
—Robert A. Haag collection

class is especially rich in iron with fayalite between 27 and 32 percent. In the past, LL-chondrites were called amphoterites, but this name is no longer used. Like the H- and L-chondrites, the LL-chondrites vary in petrologic type between 3 and 6.

The Oxygen Connection. The ratio of free (reduced) iron to combined (oxidized) iron is the fundamental criterion for distinguishing H-, L-, and LL-chondrites. Of the three, H-chondrites contain the least oxygen and LL-chondrites the most oxygen, because oxygen occurs in the silicate minerals as part of the silicon-oxygen tetrahedron, with iron attaching itself to the tetrahedron. Current understanding of the chemical makeup of the solar system suggests oxygen was least plentiful close to the Sun and progressively more abundant with distance from it. So the H-chondrites, with their lower oxygen content, probably formed closer to the Sun than either the L- or LL-chondrites. One additional chondrite, not an ordinary chondrite, completes the oxygen connection. These are called the E-chondrites.

E-Chondrites

E-chondrites are rare, representing less than 2 percent of the stony meteorites, and only twenty-four are known. They must have formed in an oxygen-depleted environment, since most of their iron occurs either as metal or in combination with sulfur, forming the iron-sulfide mineral troilite. In ordinary chondrites, pyroxene contains both magnesium and iron in a ratio depending on the availability of oxygen. Not so the E-chondrites. Their pyroxene contains no iron, only magnesium. This almost pure magnesium silicate pyroxene is called enstatite, and accounts for about 65 percent of the mineral content of these meteorites. Thus, they are called enstatite chondrites, or E-chondrites. It's not surprising to see a fayalite content of less than 1 percent. Like the ordinary chondrites, E-chondrites are subclassified into H and L types depending on total iron. EH-chondrites have more total iron (about 30 percent) and more metal. EL-chondrites have less total iron (about 25 percent) and less metal.

Their low oxygen content suggests that they formed even closer to the Sun than the H-chondrites, possibly inside Mercury's orbit.

R-Chondrites

The R-chondrites represent the newest chondrite group. The need for a new chondrite group began with the 1977 discovery of a meteorite near Carlisle Lakes on the northern edge of the Nullarbor Plain, Western Australia. This meteorite and others subsequently found worldwide did not fit neatly into any of the other chondrite groups, but meteoriticists were reluctant to call the "Carlisle Lakes" type a new group, calling them anomalous instead. No one had seen a Carlisle Lakes type meteorite fall, and all finds had weathered during their many years on Earth, which blurred their distinguishing characteristics. Scientists needed a freshly fallen, unaltered specimen.

No one realized at the time that people had witnessed an R-chondrite's fall nearly forty years earlier in 1934 near Rumuruti, southwestern Kenya. A shower of meteorites fell over a 0.5-square-mile area, and a number of these stones were collected. In 1938, the Berlin Museum of Natural History acquired a 67-gram specimen, which remained there unstudied until 1993. That meteorite had the characteristics of the Carlisle Lakes meteorite but fortunately lacked the weathering. It became the type specimen used to define the characteristics of the new group, called R-chondrites, or Rumurutiites. Only nine are known: Carlisle Lakes, Rumuruti, six from Antarctica, and one from the Sahara Desert.

R-chondrites are brecciated with light clasts embedded in a dark, fine-grained matrix. The light clasts show a petrologic type 5–6 while the dark matrix shows type 3–4 metamorphism. There are few chondrules compared to H- or L-chondrites and iron-nickel metal is nearly absent. Most of the metal is in the sulfides pyrrhotite (FeS) and pentlandite $[(Fe,Ni_9)S_8]$ or combined with olivine. The pyrrhotite is magnetic, giving the meteorites a weak magnetic attraction. The fayalite content averages about Fa_{39}, which is the highest iron-bearing olivine content of any of the chondrites and is chemically the most distinctive characteristic. Rumurutiites have the highest iron oxidation of the chondrite class.

Carbonaceous Chondrites

Carbonaceous chondrites are rare, primitive, and organic—the least aesthetically attractive of the meteorites, and also the most profoundly interesting, for they contain tantalizing clues to the origin of life.

Carbonaceous chondrites are some of the most complex and heterogeneous of all meteorites. In chemical and mineral composition they differ widely from ordinary chondrites and from each other. Their internal textures vary from well-formed chondrules to no chondrules at all. They show practically no thermal metamorphism, and compared to other meteorites, their composition is closest to that of the Sun. For these reasons, carbonaceous chondrites are considered the most primitive meteorites. They are so different that uninformed geologists usually cast them aside as curious terrestrial stones. (I once presented one to a college laboratory class in mineralogy and asked the instructor and students to identify it. After considerable argument, they concluded it was some sort of chemically altered sedimentary rock. They were on the right track but they had the wrong planet!)

Carbonaceous chondrites must have formed in an oxygen-rich environment. Most of their metal is either locked in the silicon tetrahedron or combined with oxygen to form oxides—magnetite for example—or combined with sulfur to form sulfides.

The single most important characteristic of these chondrites is the presence of water-bearing minerals. In many specimens, there is evidence of water having slowly percolated through the interiors not long after formation. This water, at temperatures from above freezing to 280 degrees Fahrenheit, reacted chemically with the original minerals, resulting in a kind of chemical metamorphism. This chemical alteration created hydrated silicate minerals similar to those in terrestrial clays. Serpentines are the hydrated minerals forming the bulk of many of these meteorites. The presence of hydrated minerals (though not absolutely diagnostic, since one type does not contain them) strongly suggests a carbonaceous chondrite.

Hydrated minerals make these meteorites very fragile. Easily broken or crushed, they crumble to powder if left out in the weather. It is remarkable that they make it through the atmosphere intact. More than half of all recovered carbonaceous chondrites were seen to fall and were recovered soon thereafter.

Types of Carbonaceous Chondrites

Like normal chondrites, carbonaceous chondrites are distinguished by petrologic types (1 to 6) and by chemical and mineral content. They are designated by the letter C and are further distinguished by their similarity to one of six different carbonaceous meteorites, which are prototypes for each subgroup: CI, where the I stands for the Ivuna meteorite, which fell in Tanzania in 1938; CM, with the M standing for the Mighei meteorite, which fell in Ukraine in 1889; CV, the V standing for the Vigarano meteorite, which fell in Emilia, Italy, in 1910; CO, the O standing for the Ornans meteorite, which fell in Doubs, France, in 1868; CR, the R stands for Renazzo, a meteorite that fell in Ferrara, Emilia, Italy, in 1824; CK, the K stands for Karoonda, a town in South Australia, where a CK chondrite fell in 1930.

CI Carbonaceous Chondrites. Petrologic type 1 carbonaceous chondrites are the most friable and delicate of the five petrologic types. They

The Vigarano CV3 carbonaceous chondrite, the prototype meteorite from which the V-type carbonaceous chondrites are named. Specimen is 3.5 inches across.
—Robert A. Haag collection

CM2 carbonaceous chondrite from Murchison, Australia, showing a field of olivine chondrules and light-colored, high-temperature silicates. Specimen is about 1.5 inches long.

are also the rarest; only five specimens are known. All were observed falls, since this type must be found quickly or they will fall apart in Earth's aqueous environment. Recognized by a lack of chondrules and abundance of hydrous minerals, they contain the highest percentage of water—20 percent—of any carbonaceous meteorite. When heated in a closed container, the water is easily driven off and condenses on the side of the enclosing vessel. The primary minerals are serpentines and epsomite (epsom salt, or calcium sulfate), both containing water molecules. There is some evidence for deposition of magnesium and calcium sulfate from an aqueous solution. Iron is represented by the presence of magnetite, making these meteorites slightly magnetic even though no metal is present. Crystals of olivine are sparsely distributed in the matrix.

This description applies only to the CI specimens of the five-member type 1 group; it is designated type 1 (CI). The famous Orgueil meteorite is in this subgroup.

CM Carbonaceous Chondrites. CM carbonaceous chondrites are all type 2, designated CM2. They contain less water than the CI group, only about 10 percent, but their mineral compositions are essentially the same. Unlike the CI meteorites, CM2 chondrites contain high-temperature silicates, which can be seen as light-colored inclusions in the matrix. The most obvious difference is the presence of chondrules, though they are somewhat sparse compared to type 3. The chondrules are olivine (forsterite) in composition. Both CI and CM2 contain carbon, some of it in the form of complex organic compounds. CI meteorites are richest in carbon, containing 3 percent to 4 percent compared to 2 percent to 3 percent for CM2 meteorites.

The CM2 group is by far the most abundant of the carbonaceous chondrites. Of the fifty-seven known, eighteen were observed falls. The Murchison meteorite is a well-known example of a CM2.

CV Carbonaceous Chondrites. The CV carbonaceous chondrites are closer to the structure and composition of ordinary chondrites. Their densities are relatively high, around 3.5, and their matrixes are primarily

A beautiful example of an oriented CV3 meteorite from Allende. Chondrules and CAIs are obvious even without cutting the specimen. Specimen weighs just over a pound and is 3.3 inches wide.

iron-rich olivine. Less water is locked in their mineral structures, so they are sturdier and less prone to disintegration during exposure to the elements. Only sixteen are known, but in terms of total weight, there is probably more CV than any other carbonaceous chondrite group. The prolific fall of the Allende CV3 in 1969 accounts for this. Over 2 tons were recovered and additional specimens are still being found. Therefore, a private collector is more likely to have an Allende as a sample of a carbonaceous chondrite than any other type. In general, the CV meteorites contain much less carbon than those in the CI and CM groups.

CV meteorites exhibit petrologic type 3. A striking characteristic of the CV3 meteorites is the presence of large, sharply defined chondrules 1 millimeter or more in diameter made of magnesium-rich olivine. In the Allende meteorites, chondrules occupy about 30 to 40 percent of the field. These are often surrounded by iron sulfide (see color plate V).

The most distinctive characteristic of CV meteorites is the presence of large, white, irregular inclusions in the dark-gray matrix. They are easily seen on the face of a broken piece and generally are several times larger than the chondrules, often making up 5 percent to 10 percent of the meteorite. Analysis shows them to be a mixture of high-temperature oxide and silicate minerals of calcium-aluminum-titanium composition. They are called CAIs (calcium-aluminum inclusions) or, simply, white inclusions. CAIs have been under intense study since their importance was discovered in the late 1960s. It now appears they have a great deal to tell us about our own origins. Isotopic anomalies in the form of unusual isotopic proportions in the elements found in the crystal structure of these inclusions suggest that some material forming the CAIs came from preexisting interstellar grains that mixed with the forming solar nebula. Nearby supernovae may have further provided unusual isotopes. CAIs must be among the earliest formed minerals, most certainly predating Earth.

CO Carbonaceous Chondrites. Like the CV chondrites, CO chondrites are petrologic type 3. Though similar to the CV chondrites, they differ in a number of ways, easily seen with a hand magnifier. First, the

An unusually large chondrule, 0.4 inch in diameter, protrudes through the surface of an Allende specimen.

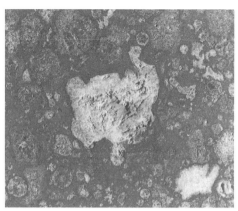
Photomacrograph of a calcium-aluminum inclusion from an Allende CV3 specimen.

chondrule size, averaging only 0.008 inch in diameter, is noticeably smaller than in CV3 chondrites. The chondrules are so closely packed together that the matrix is sometimes hard to see. CV chondrites have a chondrule-to-matrix ratio of about 1 to 3. In CO chondrites this ratio is reversed with only 20 to 30 percent matrix. An even more visible characteristic is the presence of iron-nickel inclusions scattered as flakes through the meteorite. The average CO chondrite contains about 6 percent by weight of metal. CV3 chondrites usually contain considerably less metal, as little as 0.5 percent by weight in the most oxidized specimens. CO chondrites are even more rare than CV chondrites with an observed chondrite fall rate of only 1 percent.

CR Carbonaceous Chondrites. Scientists recognized this chondrite group because of new discoveries in Antarctica. The 1985 *Catalogue of Meteorites* lists Renazzo, a meteorite that fell in Italy in 1824, as a "type II" CM2, but recently scientists named Renazzo the type specimen for this new group of carbonaceous chondrites. In general, the important defining characteristics of carbonaceous chondrites are hydrated silicate minerals, magnetite, well-formed chondrules, and a lack of iron as metal. CR chondrites contain hydrated minerals formed by hydrothermal alteration, but like the CO chondrites, CR chondrites are sufficiently reduced so they contain about 10 percent by weight iron-nickel metal and iron sulfide. The metal content is the most easily distinguished characteristic. The metal is found in the fine-grained matrix and as inclusions in the chondrules. Roughly 50 percent of the meteorite is relatively large chondrules (0.027-inch average diameter) and chondrule fragments.

CK Carbonaceous Chondrites. Meteorites that were earlier labeled CV4–5 are now a new class, the CK carbonaceous chondrites. There are only thirteen known, most found in Antarctica. The only observed fall of a CK chondrite occurred in 1930 in Karoonda, South Australia. It is the type specimen for all CK meteorites. A large mass, the only CK chondrite available to the collector, was found in Maralinga, Australia, in 1974.

All CK chondrite meteorites show various degrees of metamorphism, displaying petrologic types 3–6 and some showing shock veins, suggesting an impact history. The cut surfaces of these meteorites appear blackened with a sooty substance so that the structure is difficult to distinguish. The blackening agent is fine particles of magnetite and dark sulfides dispersed in the silicates of the groundmass and chondrules. Like most other carbonaceous chondrites, CK chondrites are highly oxidized, show no metal grains, and have iron-rich olivine and pyroxene. This group is similar to CV and CO chondrites but differs in bulk chemistry.

A Question of Carbon—Life in Meteorites?

The interest in carbonaceous chondrites stems in part from their relatively high carbon content; thus, the name "carbonaceous." Carbon compounds give these meteorites a dark-gray color. They are not the only meteorites containing carbon, however. Many contain carbon in the form of carbides, graphite (amorphous carbon), or even diamonds. The richest carbonaceous chondrites contain about 5 percent carbon. Some of it is in the form of carbonates, which were probably formed by the chemical action of water. The remainder are organic compounds.

The presence of organic compounds immediately raises the question of a biological origin. Organic matter was first noted by the Swedish chemist Jons Jacob Berzelius in 1834; but the crude analytical methods of the nineteenth century could not identify the compounds. This would have to wait until the mid-twentieth century.

Analysis of several carbonaceous chondrites, with special emphasis on the Orgueil meteorite, which fell in France in 1864, brought on a controversy still reverberating in meteorite laboratories. Many organic compounds in these meteorites turned out to be long-chain saturated hydrocarbons, similar to waxes and paraffins. Others were fatty acids of large molecular weight. The discovery of amino acids was especially interesting, suggesting a possible biological origin. But the balloon burst soon afterward, when subsequent tests showed the specimens were simply contaminated through handling and storage.

Though discredited, the idea of amino acids in carbonaceous meteorites remained very appealing. With the original discovery of organic compounds in these meteorites early in the nineteenth century, the idea that life could have been transported to Earth on a meteorite began to gain support. There were many attempts to find evidence of "fossil" life in several meteorites, which, interestingly, did not include carbonaceous chondrites. The stakes were high. The discovery of life-forms in meteorites would provide convenient answers to two outstanding questions in science: How did life originate on Earth, and is there life beyond Earth? Finding evidence of life in meteorites would suggest meteorites had seeded Earth with the first life and that life must therefore exist elsewhere in the universe. By the 1870s, a controversy raged in the European scientific community, especially after several renowned scientists lent their support to the idea. Then provocative reports appeared, claiming discovery of fossil remains of life-forms similar to sponges and corals in some of the meteorites. What began as an intellectual argument turned into a scientific brawl. Insults and accusations flew back and forth among Europe's best scientists, and the press had a field day. Fortunately, the furor quieted down by century's end. The consensus was that the "fossil" forms were not fossils at all, but only normal mineral crystals and pieces of chondrules that appeared to take on fossil-like shapes under the microscope.

But the idea did not go away. In the 1930s it resurfaced when Charles B. Lipman, a bacteriologist from the University of California at Berkeley, claimed to have found and cultured live microorganisms from eight different meteorites—including one iron specimen. Again the press—this time the *New York Times*—took the initiative and published a provocative article stating that Lipman had found life in meteorites. And once again, controversy flared. Lipman's experiment was repeated and the same microorganisms were found. This time they were identified as common Earth bacteria. The meteorites had been contaminated.

The idea of life in meteorites, or at least organic compounds we associate with life on Earth, is not easily dismissed. The controversy faded but the search quietly continued. In 1961, Fordham University scientists Bartholomew Nagy and Douglas J. Hennessy, and Warren Meinschein of Esso Research Corporation, addressed the question again in a report on their analysis of the Orgueil carbonaceous chondrite. This time the sophisticated equipment used could accurately determine the composition of minute quantities of elements and compounds in the meteorite. Their conclusions were sensational. Hydrocarbons resembling the by-products of living organisms were found. They were well aware of the historical controversy surrounding this study and were understandably cautious about their conclusions. Still, they were certain enough to leave readers of the *New York Times* with this stunning inference: "We believe that wherever this meteorite originated something lived!"

It did not end there. Only months after Nagy presented his analysis, he reported finding "organized elements" in the same meteorite. He described several types with distinctive characteristics, concluding that they were "possible remains of organisms." No one seemed to dispute Nagy's claim of complex hydrocarbons in the Orgueil meteorite, but few scientists accepted the premise that they were products of biological activity. They correctly pointed out that such compounds could and probably did form in the solar nebula during planetary formation, nearly

5 billion years ago. The "organized elements" turned out to be inorganic minerals, normal silicates and limonite with unusual structures.

In the early 1960s the United States committed itself to landing a man on the Moon. Laboratories were set up with special equipment and techniques to avoid contamination of the anticipated lunar samples. Within months of the first two lunar landings, two carbonaceous chondrites fell: Pueblito de Allende (type CV3) on February 8, and Murchison, Australia (type CM2), on September 28, 1969. It was a rare opportunity to study two carbonaceous chondrites, using analytical procedures developed for the study of lunar rocks and the search for organic matter. The results were astonishing. In the Murchison meteorite, amino acids and long-chain hydrocarbons were found together. Though nonbiological, some of the amino acids are present in Earth life-forms. At last, conclusive evidence for the existence of organic compounds essential to life on Earth was finally found. By 1985, seventy-four amino acids had been positively identified. Eight of these are associated with protein synthesis in living systems and eleven other less-common amino acids are also biologically significant. The remaining fifty-five amino acids are not found on Earth.

Startling discoveries like these lead to some fundamental questions about the synthesis of organic compounds in the solar system. Scientists generally agree that the amino acids found in meteorites were formed by nonbiological processes. But the fact that such complex molecules could form and survive under the harsh conditions of the early solar system is remarkable. We must consider the conditions leading to early organic evolution on Earth. Specifically, did amino acids form on Earth or were they seeded by carbonaceous chondrites or comets, impacting after Earth's crust formed? An even more compelling question logically follows. If biological amino acids can form so readily on smaller bodies in the solar system, could life have evolved elsewhere at the same time it began here?

The search for life on other planets has already begun. The *Viking* landers that set down on Mars in July and August 1976 examined soil samples for the by-products of living microorganisms; none were found. A further test for organic compounds in the soils of Mars also yielded negative results. The question remains unanswered. The experiments were limited: They could not test for all possible life-forms, and they could only test the soil within a few feet of the landers. The absence of evidence is not necessarily evidence of absence. What is so prevalent on Earth seems at first glance to be elusive elsewhere in the solar system.

Pathfinder on Mars

The search continued on July 4, 1997. After a hiatus of twenty-one years, a Mars lander called *Pathfinder* safely set down with Sojourner, a toylike roving vehicle, aboard. The *Viking* landers were sentinels in perpetual watch, as they had no ability to wander around and could only reach out a few feet to dig into the Martian soil. But Sojourner had the run of the place. Once engineers back on Earth deployed ramps and released the shackles restraining it during the bouncy landing, Sojourner drove out onto the Martian landscape.

A steady stream of pictures taken by *Pathfinder* and Sojourner cameras revealed a vast floodplain with huge rocks slanted away from the direction of water flow that once inundated the landscape. Scientists selected this spot, Ares Valles, as a place where water probably once existed in abundance. Here, within the outflow channels of the Ares Valles, life may have thrived, if only for a relatively short time. Sojourner could not directly hunt for signs of life, but it traveled to selected rocks and performed whole-rock analyses.

The first rock Sojourner examined, nicknamed "Barnacle Bill," was an andesite, a common volcanic rock on Earth. Its presence on Mars tells a story of successive melting and remelting within the Martian crust. This exciting discovery suggested that Mars experienced geologic processes similar to those on Earth. The second rock, called "Yogi," was even more appealing. This large boulder near the lander was familiar but not common to Earth. It resembled a shergottite, a basaltic rock erupted by a Martian volcano more than 1 billion years ago—a rock similar to others that had been propelled to Earth by the force of an ancient impact on Mars. About a dozen meteorites possess characteristics that indicate they are from Mars. Yogi confirmed that Mars rocks have come to Earth.

Researchers found a rock of similar composition lying on blue ice in Antarctica in 1984. This meteorite, ALH 84001, is the most famous rock in the world today—its interior contains evidence for possible past life on Mars. Formed through igneous processes within the crust of Mars, ALH 84001 is an achondrite, another important class of stony meteorites.

Mars roving vehicle rests against the Mars rock "Yogi" and performs a whole-rock analysis. Yogi is a familiar volcanic rock, a shergottite, samples of which we have on Earth. This analysis confirmed that Mars rocks have made their way to Earth in the geologic past.
—NASA/JPL photo

A slice of a famous achondrite that fell in the Brazilian village of Ibitira on June 30, 1957. It is a calcium-rich eucrite showing no brecciation but a highly vesicular texture. This texture is similar to the vesicles or holes in terrestrial lavas. –Courtesy Michael Casper, Meteorites, Inc.

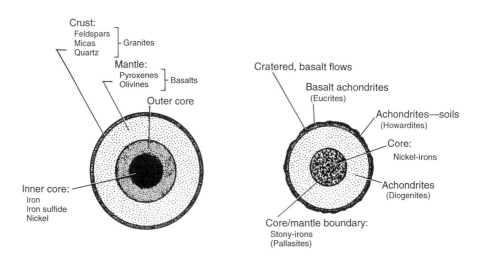

A differentiated planet with core, mantle, and crust. Achondrites probably represent the crust of a differentiated parent body.

ELEVEN

Achondrites—The Crusts of Planets

Chondrites account for about 85 percent of all known meteorites. The rest are achondrites, stony-irons, and irons. These three types of meteorites have one thing in common: They were formed by melting, either on or deep within a planetary parent body. They once were probably chondritic, but that structure was destroyed during the planet-building process, early in the solar system's history. The parent bodies were somehow broken apart, resulting in three classes of meteorites that have little in common with the chondrites.

Differentiation

Originally, Earth and the inner planets had somewhat uniform compositions throughout, similar to that of chondritic meteorites. Their masses grew by accumulation of rocky bodies (planetesimals), and the forming planets gradually heated up. The nearly constant impact of meteoritic bodies released heat, melting surface and near-surface rocks. The additional mass compressed the planets, releasing gravitational energy and converting it to heat deep in the interiors. The decay of radioisotopes trapped in the rocks added to the other heat sources, providing sufficient heat to melt the planets entirely. Now in a semi-liquid-to-liquid state, they became giant gravitational separators. Heavy elements like iron, nickel, and some of the rare, iron-loving noble metals such as gold, platinum, and iridium separated from the viscous bodies and gradually sank to the centers, forming heavy cores. Lighter elements and minerals accumulated around the cores, forming dense basaltic rock and creating the planets' mantles. The lightest minerals, like feldspars, quartz, and micas, floated to the top and formed a thin, outer crust. Thus the rocky planets differentiated into mineral zones.

The Igneous Process

Achondrite meteorites are exciting because they are samples of planetary bodies similar to Earth. Their parent bodies must have undergone differentiation, and it seems the same geological processes that built the crustal rocks of Earth must have taken place as well. To understand the origin of these important meteorite types, it is worth-while to take a brief look at the primary rock-building process on Earth.

Rock that forms from a "melt," or magma, deep within a planetary body is called igneous rock. It is the most common kind of rock. Derived from the mantle, it has almost equal amounts of plagioclase and iron-bearing pyroxene. One example is basalt, a dark rock that is ex-truded onto the surface during volcanic eruptions. Usually, the feldspar and pyroxene crystals are very small, giving the rock a smooth, fine-grained texture, because the cooling process was rapid when the liquid rock was erupted. If left to cool slowly deep in the crust, basaltic rock will grow crystals easily visible to the naked eye.

Minerals in the rock crystallize out of the original basaltic magma in a definite sequence as it cools. These newly formed crystals can either react with the magma, changing to new minerals, or settle to the bottom of the magma chamber, accumulating as a layer of a single mineral. For example, if calcium plagioclase (a feldspar rich in calcium) crystallizes out of the magma and accumulates in a layer on the bottom of the chamber, the resulting rock is a plagioclase cumulate. Pyroxene or olivine crystals could also form the cumulate, since they are present in the mantle as well.

A classic example of this process is the basalt sill. A sill is made up of igneous rock intruded horizontally into overlying rock layers. Olivine

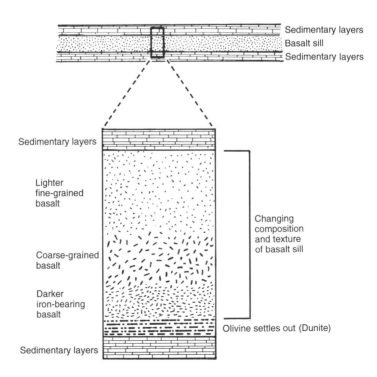

Cross section of a basalt sill. The composition of the sill changes as the minerals crystallize out of the magma in a specific sequence. Note the olivine layer at the bottom.

CLASSIFICATION OF ACHONDRITES

Achondrite Class	Primary Mineral Composition
Asteroidal Achondrites	
Eucrites	Pigeonite and calcic feldspar (basaltic)
Diogenites	Orthopyroxene-hypersthene, bronzite
Howardites	Brecciated eucrite/diogenite mix
Aubrites	Orthopyroxene-enstatite
Ureilites	Olivine-pigeonite-graphite
Angrites[a]	Pyroxene-augite (basaltic)
Brachinites	Olivine, orthopyroxene, and clinopyroxene
Mars Rocks (SNC)	
Shergottites	Pyroxene and maskelynite glass (basaltic)
Nakhlites	Pyroxene-augite and olivine
Chassignites	Olivine (dunite)

[a] Until recently, only one angrite was known, a single stone that fell in 1869 in Angra dos Reis, Brazil. In the 1980s, three angrites were found in Antarctica, two on Lewis Cliff and one at Azuka, bringing the total number to four.

is the first mineral to crystallize out of the magma, forming a nearly pure olivine rock called dunite. Then basalt of plagioclase-pyroxene composition forms above the olivine layer. As minerals crystallize out of the magma, the magma's composition changes.

Our basic knowledge of the igneous process comes from the study of Earth, and it can be applied to rocks from space. If some meteorites are the result of similar igneous processes, they should show textural and mineral properties equivalent to those of igneous rocks found on Earth.

The Achondrites

Achondrites are vastly different from chondrites. The name achondrite means "without chondrules." The lack of chondrules used to be the primary criterion for distinguishing them. But we have already seen that at least one chondrite has no chondrules and there is at least one achondrite that has chondrules. Meteoriticists have placed these rare meteorites into seemingly contradictory positions because their mineralogy and chemistry are closely allied to their respective meteorite classes. The lack of chondrules suggests that achondrites have a different history from chondrites. Chondrites formed by the accretion of chondrules and matrix material into a parent body, apparently without sufficient heat to completely destroy the chondrule structure by thermal or shock metamorphism. Achondrites formed by a melting and recrystallization process on a chondritic parent body. Achondrites are more coarsely crystalline than chondrites, and are more like terrestrial igneous rocks than any other meteorites. In essence, achondrites are the igneous rocks of other worlds.

Achondrites are among the rarest of stony meteorites, though they are twice as plentiful as carbonaceous chondrites. Of the 1,813 known stony meteorites, there are only 132 achondrites, about 7 percent of the total. Chondrites tell us about the origin of parent bodies. Achondrites tell us about the igneous processes that occurred on these bodies.

Types of Achondrites

Achondrites are divided into two broad groups based on mineral composition: calcium-rich and calcium-poor achondrites. Calcium-poor achondrites contain 3 percent calcium or less, and calcium-rich achondrites have between 5 and 25 percent or more. This chemical difference is difficult to observe directly in a specimen. Fortunately, the two types do show visible characteristics that usually enable a knowledgeable investigator to tell them apart. For simplicity's sake, I will ignore the calcium-based subdivisions and lump all achondrites together, concentrating on the more obvious distinguishing characteristics.

Aubrites—Enstatite Achondrites. Named for a fall near Aubres, France, in 1836, aubrites are also known by the name of their most abundant mineral, enstatite. Aubrites are almost pure $MgSiO_3$, but unlike most enstatite chondrites, they have little FeNi metal or iron-bearing pyroxene. The lack of metal in either combined or reduced states results in a light-colored meteorite. A fresh specimen has a gray-white or light tan crust, a distinguishing characteristic of enstatite achondrites.

Of the eleven aubrites known, ten have a brecciated texture. The term breccia describes rock made of a mixture of angular, broken pieces of the same rock (a monomict breccia) or of different rocks cemented together to form new rock (a polymict breccia). The best known of the brecciated enstatite achondrites is the 1-ton mass that fell in Norton County, Kansas, in 1948 (see color plate II). Until that fall, aubrites were one of the rarest meteorites.

Another aubrite fell near Cumberland Falls, Kentucky, on April 9, 1919, creating a fireball and smoke train.

Diogenites—Plutonic Meteorites. Diogenites are named for a Greek philosopher of the fifth century B.C., Diogenes of Apollonia. Some historians credit him with having been the first person to suggest meteorites come from beyond Earth. Diogenites are about as rare as aubrites, with only fifteen known. These curious meteorites are almost pure orthopyroxene: They are made up of large crystals of iron-rich hypersthene and bronzite, with about 12 percent iron by weight. About 25 percent of the pyroxene is actually the iron end-member ferrosilite, a mineral rarely found on Earth. Their higher iron content makes these meteorites darker than aubrites. The large crystals must have formed from cooling magma and accumulated on the bottom of the magma chamber. Like aubrites, diogenites are cumulates of mostly a single pure mineral. The large crystals mean that diogenites are plutonic; that is,

Cut and polished section of a brecciated aubrite that fell near Cumberland Falls, Kentucky. The angular fragments are pure enstatite; the rusted iron flakes are a rarity in aubrites. Specimen is 4 inches across.

they must have formed within their parent body under slowly cooling conditions, which allowed large crystals to form. They are the parent body's intrusive igneous rocks. (See color plate V.)

Eucrites—Basalt Achondrites. Eucrites are the most common achondrites. Fifty-five grace meteorite collections today. They are closely related to terrestrial and lunar basalt lavas and tend to look like them. Many even have the gas holes typically found in terrestrial basalts.

The main physical characteristic distinguishing eucrites and diogenites is their interior textures. Eucrites mimic lavas from runny basalt on Earth, with textures that are almost microscopic. They are composed of very small interlocking crystals, showing evidence of flow on the surface of the parent body. When magma cools quickly, it produces small crystals, contrasting with the large pyroxene crystals found in diogenites. Eucrites are calcium-rich achondrites, since they are made of calcium-rich plagioclase feldspar crystals (anorthite). These are often encased in the iron-magnesium-calcium pyroxene called pigeonite. The pigeonite matrix usually contains tiny grains of iron metal, making these meteorites slightly magnetic.

A thin section of a terrestrial olivine basalt showing thin needles of plagioclase feldspar and small anhedral crystals of olivine and calcium-rich pyroxene.

A thin section of the eucrite achondrite from Millbillillie, Australia, showing plagioclase needles in the matrix and large clustered plagioclase crystals giving a texture called glomeroporphyritic. Tiny olivine grains are embedded in the plagioclase. Dark matrix material is pigeonite.

205

Interior of a howardite showing large dark clasts of diogenite material in a fine-grained eucritic matrix. This meteorite, called Kapoeta, fell in Sudan in 1942. It is unusual because it contains dark inclusions of carbonaceous chondrite material. –Specimen courtesy of Dr. Elbert King, University of Houston

The most obvious characteristic of a fresh eucrite is its shiny, black fusion crust (see color plate III). The relatively high calcium and iron content produces the wet-looking fusion crust, which is distinctly different from the dull, grayish-black fusion crusts of the calcium-poor diogenites.

Howardites—A Planetary Soil. Diogenites and eucrites confirm that igneous processes were at work on a parent body. Their crystal textures show both slow and fast cooling, implying both intrusive (granitelike) and extrusive (basaltlike) igneous activity. Soils on Earth are the product of the weathering and erosion of rocks, primarily caused by the action of water, ice, and oxygen gradually breaking them down into their constituent minerals. On an airless and therefore weatherless parent body, erosion is the result of collision with other meteoritic bodies. Frequent collisions break the surface rocks of the parent body into fragments, with a resultant gradual mixing of fragments of surface rock (eucrites) with shallow-depth rock (diogenites). The impacting body, probably chondritic in composition, would also mix with the fragments.

The formation of soils on the Moon provides a close analogy to soil formation on parent bodies. Lunar rock fragments are gradually broken down into grains, which, after billions of years' exposure to high-energy cosmic rays and micrometeorites, weld together to form a loose soil. At the same time, in an airless environment, particles from the solar wind damage the mineral crystals exposed on the surface. Frequent collisions with meteorites during the first half-billion years of solar system history stirred the forming soil, exposing new crystals to the incoming solar wind.

Howardites are brecciated achondrites that appear to be the bridge between diogenites and eucrites, exhibiting all the characteristics of the mixing process described above. They contain fragments and crystals of both diogenites and eucrites, as well as chondritic material; they are polymict breccias and their crystals show damage from the solar wind. Lunar soil samples reveal nearly identical amounts of damage from exposure to solar wind. The parent body responsible for howardites

Photomacrograph of the interior of a ureilite from Kenna, New Mexico. The angular grains are olivine and the low-calcium pyroxene pigeonite. The dark material along the grain boundaries is carbon.
–Robert A. Haag collection

must have existed fairly close to the Sun, possibly as close as Earth and certainly no farther away than Jupiter, because the solar wind drops off dramatically as distance increases. The evidence points to their origin being the asteroid belt between Mars and Jupiter.

An asteroid suspected of being the parent body for howardites is Vesta, one of the largest and brightest of the known asteroids. Spectroscopic studies of light that reflects to Earth from the crystalline surface rocks of Vesta show it has a composition similar to a eucrite-diogenite mixture.

Ureilites—The Orphans. Ureilites are as rare as diogenites, with only seventeen known specimens. They are a curious type of meteorite that does not seem related to any of the other achondrites. They are composed primarily of olivine and the orthopyroxene pigeonite, but the distinguishing characteristic is the presence of carbonaceous material found between the olivine and pryoxene crystals. Inasmuch as carbonaceous chondrites contain abundant carbon, perhaps the ureilites are somehow related to these chondrites. But unlike carbonaceous chondrites, in ureilites the carbon is in the high-pressure form—diamonds. The diamonds are microscopic and were formed instantly by the intense shock of impacting bodies, a testament to the violent history of these cosmic rocks.

Brachinites. Scientists have recently recognized brachinites as a new class of achondrite. The name comes from the type specimen found near Brachina, South Australia, in 1974. Brachinites, composed almost entirely of small equigranular olivine grains, resemble a type of Mars meteorite called a chassignite and were originally mistaken for it. Scattered among the olivine are small amounts of clino- and orthopyroxene. Metal is either very rare or absent; total iron is near 20 percent but is tied to the olivine. Brachinites are the rarest achondrite.

Rocks from Mars—The SNC Subgroup

Three types of achondrites are grouped together under the classification SNC. Each letter signifies the type of meteorite. The *S* stands for shergottite, the name given the type of meteorite that fell in 1865 in the Indian state of Bihar, near the town of Shergotty. There are five

A shergottite achondrite, one of the SNC meteorites suspected to be from Mars. The shiny black fusion crust is indicative of its high calcium content. The fine-grained interior is composed of tiny pyroxene crystals. Specimen is from Zagami, Nigeria, weighs just over 6 pounds, and is 6.4 inches in its longest dimension.
—Robert A. Haag collection

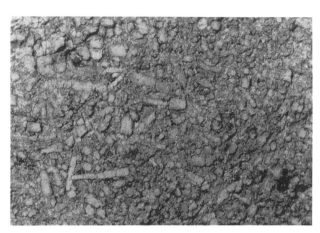

Photomicrograph of a shergottite, showing elongated pyroxene crystals (augite) stacked in random directions but primarily on their sides. Shergottites are cumulates that formed as the pyroxene crystallized out of a melt and accumulated on the magma chamber floor.
—Robert A. Haag collection

known shergottites. The *N* stands for nakhlites, named for an achondrite that fell in Nakhla (near Alexandria), Egypt, in 1911. There are only three nakhlites known. *C* stands for a meteorite that fell in Chassigny, France, in 1815, called a chassignite.

These three achondrite types have so much in common that many meteoriticists believe they came from the same parent body. They seem similar to eucrites, being igneous rocks with basaltic compositions. They are all cumulate rocks with large crystals of plagioclase, pyroxenes, and olivine. Some crystals in shergottites appear roughly oriented, as though they had lain horizontally on the floor of a magma chamber. Nakhlites are cumulates, composed of crystals of the calcium-rich pyroxene, diopside, and olivine. These minerals give the interiors of nakhlites an olive-green appearance. Chassignites (there are now two, one recently being found in Antarctica) are mostly olivine, with small amounts of pyroxene and chromite. A rock of terrestrial origin with this composition would be called a dunite.

These three SNC achondrites possess some unique characteristics. All three contain metallic oxides and water bonded within the crystal lattice of a pyroxene mineral. They are the only achondrites containing water, suggesting that they may have formed through volcanic activity involving water.

Isotopic methods are used to date the rocks. Ordinary chondrites show either no melting or partial melting in their histories, and unlike achondrites, these meteorites are relatively unaltered; isotopic studies show them to be as old as the solar system. The achondrites are the product of melting and recrystallization on a parent body. With few exceptions, their original compositions were destroyed. Thus, most are younger than chondritic bodies, but not by much. This is not true of the SNC achondrites.

As a rock melts and recrystallizes, its isotopic clock is reset. For example, the vast lava flows on the Moon occurred in huge basins that formed about a half-billion years after the crust solidified. These basalts were forced quietly from fissures and flooded the basins. The isotopic clock in the basalts shows they solidified between 3.3 and 3.7 billion years ago, making them the youngest rock yet found on the Moon.

Isotopic studies of the SNC group of meteorites date the crystallization at about 1.3 billion years ago. These rocks formed 2 billion years after igneous activity ceased on the Moon. Where could such rocks have formed 1.3 billion years ago? It is unlikely that small asteroids could have retained the necessary internal heat to erupt magma more than 3 billion years after their formation. It must have been a much larger parent body, which retained sufficient interior heat 1.3 billion years ago. We don't have to look far: The planet Mars is a prime candidate.

Photo reconnaissance missions to Mars by orbiting *Mariner* and *Viking* spacecraft have revealed towering shield volcanoes and extensive volcanic fields in the northern and western hemispheres. Vast lava flows in the region have erased the earlier craters in these hemispheres of the planet. Estimates place the most recent volcanic flows at a billion or more years ago. Could the SNC meteorites have come from Mars? The only mechanism energetic enough to transfer rocks from Mars to Earth is an impact from a small asteroid, which would have shocked the Mars rocks. Examination of the shergottites shows some evidence of shock melting and glass formation, but less than expected.

If these rocks came from Mars, they would have carried traces of the Martian atmosphere. The noble gases argon, krypton, xenon, and neon have been detected in the shergottites at levels that closely match the relative abundances in the Martian atmosphere, as found by the *Viking* landers.

Life in Mars Rock ALH 84001?

The search for meteorites among the ice fields of Antarctica had been under way for six years when Roberta Score, a member of an American team, picked up a strange-looking greenish rock on December 27, 1984, near Allan Hills. Researchers believed they were looking at the typical green color of the pyroxene hypersthene, and classified the specimen as a diogenite—just another achondrite from the asteroid belt. For nearly ten years, ALH 84001 (the first meteorite listed in the 1984 cache of meteorites from Allan Hills) remained at the Johnson Space Center in Houston. Then, in 1993, scientists took another, closer look and found trace minerals in ALH 84001 that didn't belong in a diogenite. A new analysis of this 4-pound rock showed that it closely resembled the rare SNC meteorites that come from Mars.

A new Mars rock is not taken casually, and the announcement brought requests for samples from scientists worldwide who hoped to study the mysterious green specimen. The rock did not fit into any of the known SNC groups; it was a new species of Mars rock. ALH 84001 was also ancient. The other eleven SNC meteorites are under 1.3 billion years old, but isotopic studies determined that ALH 84001 is 4.5 billion years old, as old as the original crust of Mars and older than the oldest Earth rocks by nearly 1 billion years. Cosmic ray isotopic analysis showed that ALH 84001, blasted off the surface of Mars by an impacting asteroid, had spent 16 million years wandering the solar system before arriving on the ice fields of Antarctica.

All this was interesting, but not earthshaking. Then, on August 7, 1996, a public announcement by NASA scientists at the Johnson Space Center in Houston rocked the scientific world. Researchers had found evidence of microbial life in ALH 84001. The announcement was met with reactions ranging from elation to skepticism to outright disbelief. Perhaps the most thought-provoking response came from the world's astronomical spokesman and scientific skeptic, Carl Sagan: "Extraordinary claims require extraordinary proof."

For two years prior to the announcement, a team led by David S. McKay and Everett K. Gibson Jr. had quietly studied the strange rock. ALH 84001 had crystallized out of a magma in the first few million years of Mars's existence. Sometime between 3.8 and 4 billion years ago, an impact shocked the rock, fracturing it throughout. Then, immersed in water containing dissolved carbon dioxide (the Martian atmosphere is primarily carbon dioxide), carbonates precipitated out of the water and lined the fractures. Carbonates are no strangers to meteorites, but the ALH 84001 carbonates—tiny, orange brown globules only a quarter of a millimeter across—are unique. A black mineral, composed of remarkably pure tiny crystals of magnetite and iron sulfide, rims the globules. Normally, oxidized iron and reduced iron, especially in the presence of carbonate, do not chemically coexist, but certain anaerobic bacteria on Earth can easily manufacture these iron minerals. Probing deeper with the electron microscope, the scientists found bacteria-like "fossil" forms within and around the carbonates. Remarkably, they resemble rod-shaped bacteria known to exist on Earth nearly 4 billion years ago, only these are one hundred times smaller.

The Tharsis region on Mars, where extensive rift systems spawned volcanic processes more than 1 billion years ago. Several large shield volcanoes can be seen in this view from the Viking *orbiter. The huge canyon Valles Marineris is central on the disk.* –NASA

The *Viking* landers of the mid-1970s did not find any trace of organic compounds in the Martian soil, but when scientists looked closely at the carbonate blobs, they found polycyclic aromatic hydrocarbons (PAHs) scattered within them. PAHs are large organic molecules often associated with life processes on Earth. They form on decaying organic matter. PAHs, thought to be formed by inorganic processes, are present in other meteorites.

McKay and the other team members admitted that, taken separately, none of these findings are definitive proof that early life-forms existed in the rock. But taken collectively, the carbonates, the "fossil bacteria," and the PAHs provide strong and compelling evidence. Still, many scientists point out that the carbonates could have formed under temperatures as high as 1,200 degrees Fahrenheit, too hot for life to survive; that PAHs exist in interstellar clouds, in antarctic ice, and in meteorites resulting from completely inorganic processes; and that the so-called fossil bacteria are too small to contain the genetic material necessary for life to form and evolve. There is even debate on the age of the carbonates. Age estimates range from 3.6 billion years to as little as 1.3 billion years. Most scientists believe that water has not flooded the Martian plains for over 3 billion years, constraining the time of formation of the carbonates.

This debate will continue. The possibility of discovering life beyond Earth is the compelling force driving the scientists. With ALH 84001, a renewed interest in Mars as a strong candidate for past life has emerged, marking a new age of Mars exploration.

Four-pound Mars meteorite ALH 84001 as it appeared before slicing. −NASA and Johnson Space Center

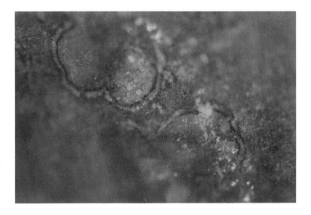

Compounds rich in magnesium and iron encircle 50-micron-diameter carbonate globules in ALH 84001. These mineral assemblages strongly resemble those produced by primitive bacteria on Earth. These carbonates may have harbored early Martian life. −NASA and Johnson Space Center

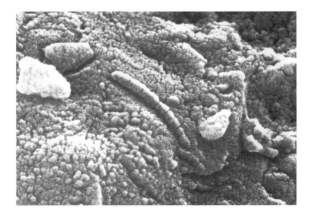

This high-resolution, scanning electron microscope image within a carbonate globule reveals a tiny, 100-nanometer-long bacteria-like structure that may be the remnant of early Martian bacteria. These structures are one hundred times smaller than 3.5-billion-year-old fossil bacteria found on Earth. −NASA and Johnson Space Center

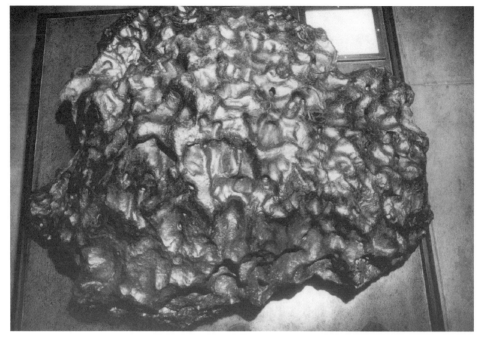

Iron meteorite from Quinn Canyon, Nye County, Nevada. This 3,190-pound medium octahedrite was found 90 miles east of Tonopah and may be from the meteor of February 1, 1894. Specimen is in the Chicago Field Museum of Natural History.

Dorothy Norton sits beside the 3-ton Old Woman meteorite at the BLM office in Barstow, California. The great iron rests on a flat-sawn section where a 900-pound slab was removed at the Smithsonian Institution.

TWELVE

Iron Meteorites—
The Cores of Planets

If achondrites come from the crust of another world, iron meteorites must be from its core. Achondrites contain little iron, since iron would have accumulated toward the center of the planet when it was in a melted state. To chip a piece of crustal rock off an asteroid by impact is one thing, but to sample the mantle and core is something else again. The core of an asteroid is inaccessible unless it is completely shattered by another impacting body. Even if the crust and mantle rock were somehow stripped away, leaving only the iron core, it would be difficult to break a solid iron body. Yet museums contain ample evidence that this has happened. The largest meteorites ever recovered are irons, many times larger and more massive than any individual stony meteorite. They are very durable and resist fracturing and weathering better than stones.

Types of Irons

When I was studying meteoritics in the 1960s, the classification of irons was fairly simple. Internal structure was the most important criterion, based on the amount of nickel alloyed with iron. Iron meteorites could be classified by simple inspection, without any special tools.

But now the situation has changed. As chemical analysis has become increasingly more precise, it is now possible to detect minute amounts of elements such as germanium, gallium, osmium, and iridium in meteorites. These elements are called siderophile (metal-loving) elements because they have an affinity for metals such as iron. There are only trace amounts in Earth's crust, since most of them have accumulated with iron in Earth's core. Iron meteorites contain relatively high concentrations of siderophile elements (though still in trace quantities) compared with crustal rocks. It is even possible that the trace amounts of siderophile elements found in Earth's crust were renewed after core formation by the impact of chondrites, which, since they are not differentiated, still have their complement of siderophile elements.

Today, meteoriticists use a new chemical classification system that divides the irons into twelve groups based on concentrations of trace

STRUCTURAL CLASSIFICATION OF IRON METEORITES

Structural Class	Texture	Chemical Class[a]
Hexahedrites (H)	Neumann Lines	IIA
Octahedrites (O)	Widmanstätten bands	
	Coarsest (Ogg)	IIB
	Coarse (Og)	IAB, IIIE
	Medium (Om)	IID, IIIAB
	Fine (Of)	IIIC, IVA
	Finest (Off)	IIID
	Plessitic (Opl)	IIC
Ataxites (D)	Structureless	IVB

[a] Appendix F contains an explanation of the chemical classification of irons.

elements (see Appendix F). This scheme obviously is not useful to the collector or field-worker, so I will use the older, structural classification. It is still valid and is closely allied to the trace chemical scheme.

The structural scheme identifies three types of iron meteorites: hexahedrites, octahedrites, and ataxites. Each has a distinctive structure based on the amount of nickel it contains. Nickel is the most abundant siderophile element in the cosmos after iron, hundreds of times more plentiful than the other siderophile elements. Nickel is alloyed with iron in most iron meteorites, ranging from a little less than 5 percent to as high as 50 percent. The structure of an iron meteorite depends on the ratio of nickel to iron and the cooling rate of the planetary core where it originated.

Alloys of Nickel-Iron

The two most important nickel-iron alloys found in iron meteorites are kamacite and taenite, also known as alpha-iron and gamma-iron. Both minerals belong to the isometric crystal system and show cubic (hexahedral, or six-sided) crystals. These alloys begin to form as the liquid metal cools below about 2,500 degrees Fahrenheit (1,370 degrees C). Just what alloy forms depends on the nickel content, the temperature at the time of recrystallization, and the rate of cooling. Even though the melt solidifies as it drops below 1,832 degrees Fahrenheit (1,000 degrees C), iron and nickel atoms continue to migrate by diffusion through the forming crystals until it cools to below 932 degrees Fahrenheit (500 degrees C).

As the taenite and kamacite alloys form, they divide into three fields as cooling progresses. For a melt containing more than about 30 percent nickel, only the taenite structure exists, since diffusion of nickel stops before the kamacite structure can form. On the low-nickel end, if the original melt contains only about 5 percent nickel, all the taenite changes to kamacite before diffusion stops. Between these two nickel percentages, kamacite and taenite can coexist and the solidified

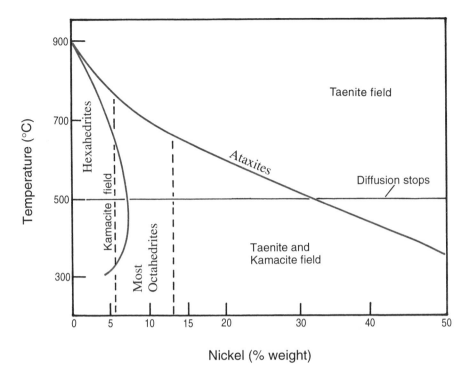

Phase diagram in figure has the following labels:

Temperature (°C) — vertical axis with values 300, 500, 700, 900

Nickel (% weight) — horizontal axis with values 0, 5, 10, 15, 20, 30, 40, 50

Phase diagram predicts the stability of the nickel-iron alloys kamacite and taenite at various temperatures and various abundances of nickel.

nickel-iron becomes a mixture of both. In this case, their crystals will be intergrown. Because most iron meteorites contain between 7 percent and 13 percent nickel, their mineral compositions contain both alloys.

In summary, the three main types of iron meteorites will be composed of kamacite only, taenite only, or both kamacite and taenite in varying amounts. This difference in composition leads to diagnostic differences in their internal structures.

Hexahedrites

The name hexahedrite refers to the crystal system in which its main nickel-iron alloy, kamacite, is found: the isometric system. Kamacite is a nickel-iron mineral with a low nickel content of between about 4.5 percent and 6.5 percent. When crystallized, kamacite forms a cubic crystal with six equal sides at right angles to each other—a hexahedron. Hexahedrite meteorites are large, cubic crystals of kamacite.

Sometimes these meteorites manage to make it through the atmosphere retaining something of their hexahedral appearance. If they break, it is usually along crystal faces, so even the fractured pieces tend to retain their crystalline shape. Such was the case with the Calico Rock, Arkansas, hexahedrite. Found in 1938, it was not identified as a meteorite until 1964. I was at the Griffith Observatory in Los Angeles the day my college buddy Ronald Oriti, then the observatory's curator of meteorites, showed me a curious, heavy, brick-shaped rock made of iron. He had gotten it in trade from the curator at Meteor Crater, who had received it by mail from the original owner. Oriti suspected it might be a meteorite even though its shape was not typical. He cut off a small piece and we tested it for nickel. It passed the test. From its shape I never would have guessed it was an authentic meteorite.

The next step in classifying the meteorite was to polish a face and etch it with acid to reveal any structure. Within minutes, a network of thin, parallel lines appeared, establishing beyond a doubt that this was a rare hexahedrite. Its shape suddenly made sense. Hexahedrites are pure kamacite, which has a hexahedral shape: The meteorite was a single large crystal.

Internally, hexahedrites are featureless except for many fine, parallel lines running along the plane of the crystal. These are called Neumann lines after Franz Ernst Neumann, who first studied them in

Neumann lines in a hexahedrite from Calico Rock, Arkansas. This 2-inch-long specimen contains 5.4 percent nickel.

1848. They are cross sections through very thin plates within the hexahedral crystal, and are the sites of twinning planes where other crystals of kamacite could intergrow. Although twinning planes are common among minerals on Earth (clusters of intergrown quartz crystals are common examples), these twinning sites were probably produced by shock. They would normally be invisible to the naked eye, but the kamacite crystal has apparently suffered an intense shock, causing the crystal lattice to slip along planes parallel to the face of the hexahedron. This slippage makes the twinning planes become visible as parallel lines and demonstrates that hexahedrites have suffered impact shock in space and were probably broken along natural cleavage planes of the original large kamacite crystals.

Photomacrograph of Neumann lines in hexahedrite from Calico Rock, Arkansas, showing two sets of parallel-running twinning planes.

Octahedrites

Most iron meteorites contain between 7 percent and 13 percent nickel. The phase diagram (see page 217) illustrates how this nickel content results in mixtures of low-nickel kamacite and high-nickel taenite. The intergrowth of these two alloys leads to a striking structure that is diagnostic for octahedrites.

a.

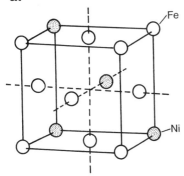

Face-centered
crystal lattice, taenite

b.

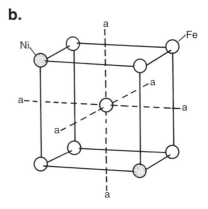

Body-centered
crystal lattice, kamacite

a: *Face-centered crystal lattice of taenite.* b: *Body-centered crystal lattice of kamacite. Note that the taenite structure allows for a tighter packing of atoms than the kamacite structure.*

Octahedrites

Both kamacite and taenite belong to the same crystal system, which includes a cubic crystal lattice similar to that of table salt. As the mixture cools, six-sided (hexahedral) taenite crystals form first. Atoms occupy all corners and centers of the cube's faces. Thi is called a face-centered cube. Fourteen atoms of iron and nickel pack each cube. Taen is much more densely packed with atoms than kamacite. The kamacite lattice is a bod' centered arrangement of atoms in which an atom occupies the center and corners of t lattice but not the sides. Only nine atoms pack the cube of the kamacite space lattice. both cases, nickel atoms can substitute for iron in the lattice, and because of the more tightly packed structure of taenite, it can contain more nickel atoms.

Because taenite is a face-centered crystal and kamacite is a body-centered crystal, they grow in different crystal forms. When intergrown, the crystal that results has elements

Widmanstätten Figures

The intergrowth of kamacite and taenite manifests itself in the most remarkable and beautiful structure to be found in any class of meteorites. To reveal this structure, one must cut an iron meteorite and grind it to a fine, mirrorlike finish. Then a dilute solution of nitric acid is brushed onto the polished face. Kamacite is more easily dissolved by the acid than taenite, causing the taenite to stand out in relief on the etched face. A series of crisscrossed plates called lamellae gradually

220

both forms. As the now-solid mixture continues to cool to a critical 932 degrees Fahrenheit, where diffusion stops, taenite cannot retain all its iron, and kamacite begins to grow within the structure of the taenite. Kamacite gains these iron atoms at the expense of the taenite. Both alloys increase in nickel content as the temperature drops. For example, at 1,292 degrees Fahrenheit, kamacite has only 4.5 percent nickel and taenite about 10 percent. When the temperature reaches 1,112 degrees Fahrenheit, the kamacite's nickel content has increased to very near its maximum of 6.5 percent and the taenite has reached 20 percent.

Kamacite grows at specific sites on the taenite cube as it slowly changes the face-centered cubic lattice to a body-centered lattice. The progressive growth of kamacite is shown below. On the left is taenite in its hexahedral form with all three axes equal. Kamacite plates begin to grow on the taenite crystal by truncating the corners of the cube at 45-degree angles (center). As growth of the kamacite plates continues, the truncated corners finally meet at all three axes on opposing sides of the cube. The new form, on the right, is an eight-sided dipyramidal shape made of eight equilateral triangles. Note that this figure is still based on the cube, since the orientation of the axes remains the same. This new crystal shape is called an octahedron, giving the name octahedrite to this meteorite class. The kamacite plates lie parallel to the faces of the regular octahedron. Octahedrites are, in essence, large, octahedral crystals.

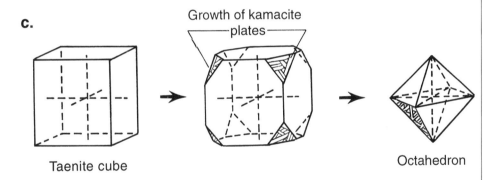

c.

Growth of kamacite
plates

Taenite cube

Octahedron

c: *Kamacite growth takes place on the corners of the taenite crystal lattice, leading to a new crystal shape with eight sides, called an octahedron, still based upon the cubic structure of taenite.*

appear on the surface (see color plate V). These plates lie on the faces of the octahedron, so they are often called octahedral figures. Since Widmanstätten figures do not appear in terrestrial iron ores and since about 85 percent of iron meteorites are octahedrites, the presence of Widmanstätten figures is diagnostic for iron meteorites in general. The etching process is described in detail in Appendix D for those who wish to try it.

Octahedral figures were discovered independently by Count Alois von Widmanstätten in Vienna and William Thompson in Naples. Thompson serendipitously discovered the figures first in 1804 while treating the Krasnojarsk (Pallas) stony-iron meteorite with nitric acid in an attempt to keep it from rusting. The procedure had little effect on the rust, but it did bring out the pattern. Widmanstätten, director of the Imperial Porcelain Works in Vienna, Austria, was investigating the structure of iron meteorites in 1808 when he too stumbled across the structure. He first produced the figures by heating a thin octahedrite slab in the flame of a Bunsen burner. Since kamacite and taenite oxidize at different rates during heating, they take on different hues, causing the figures to appear. In apparent ignorance of Thompson's earlier discovery, the structure was named for the second discoverer.

The acid etching process proved to be preferable to heating, which is risky business because the structure could be easily destroyed by reheating to the diffusion point.

When octahedrite meteorites are cut, polished, and etched, the Widmanstätten figures can appear in any orientation, depending on how the meteorites are cut with respect to the faces of the octahedron

Cuts at various angles through an octahedrite meteorite result in lamellae at various angles to each other.

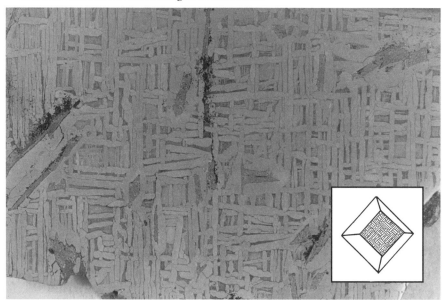

a: *90 degrees when cut perpendicular to an octahedral axis (specimen from Sam's Valley, Oregon).*

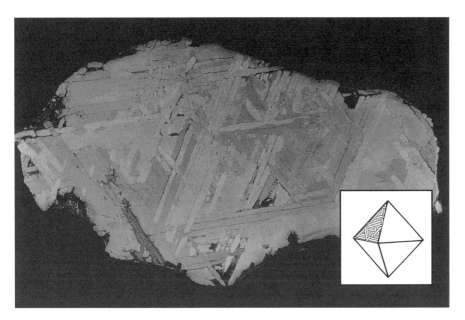

b: *60 and 120 degrees when cut parallel to an octahedral face (Toluca, Mexico).*

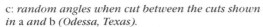

c: *random angles when cut between the cuts shown*
in a *and* b *(Odessa, Texas).*

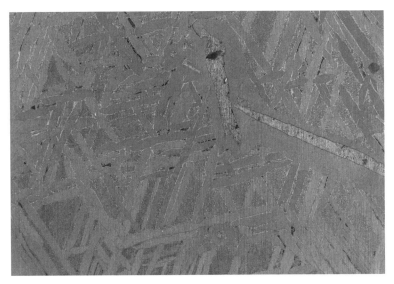

Photomacrograph showing kamacite plates (dark gray) bordered by taenite (white). Needlelike inclusion on right is schreibersite, and the dark, granular, triangular areas enclosed by the kamacite plates are plessite. Specimen is from Bear Creek, Colorado.

on which they lie. When the plates run in different directions and at various unequal angles to each other, spaces result between the kamacite plates. These spaces are usually filled with a fine-grained kamacite-taenite mixture known as plessite, which probably formed late in the diffusion process at the lowest possible temperatures. Plessite can be recognized after etching, as it usually takes on a shade of gray much darker than that of kamacite.

It's Mostly a Matter of Nickel

The amount of nickel in a meteorite determines the structure of the meteorite. This is true up to a point. Unfortunately, nature is never as simple as we might like.

Widmanstätten figures consist of both kamacite and taenite, therefore the final kamacite-to-taenite ratio becomes a vital factor in determining the structure of a meteorite. More kamacite means wider Widmanstätten figures. The ratio depends not only on the amount of nickel present but also on the rate of cooling. The amount of kamacite that forms depends on how quickly the temperature reaches the equilibrium point, the temperature at which diffusion stops and taenite ceases to turn into kamacite. The slower the cooling rate, the lower the final temperature of equilibrium. The phase diagram shows that for a given nickel concentration, the lower the temperature of equilibrium, the more kamacite forms. With a slower cooling rate and corresponding lower equilibrium temperature, more time is available for the kamacite to form.

How long does it take for a molten mass of nickel-iron to cool? This depends on two factors: the size of the mass and its rocky insulation. If the cooling mass is from the core of a parent body, it had a thick blanket of mantle rock overlain by a thin layer of crust, providing insulation. This slows the cooling rate substantially. By studying the Widmanstätten structure, a fairly accurate idea of the cooling rate at the core can be determined. This in turn yields an idea not only of the size of the core but also of the entire parent body. Those beautiful octahedral figures now take on enormous importance. They are a vital clue in deciphering the early history of the solar system, which is locked within these rocks.

Octahedrite Structural Classification

An iron meteorite with a total nickel content of between 4.5 percent and 6.5 percent is pure kamacite. No Widmanstätten structure is seen and the meteorite is classified a hexahedrite. As the nickel content increases beyond 6.5 percent, Widmanstätten patterns begin to appear as wide bands. This is the beginning of the octahedrite class. With continuing increase in nickel content, the bands become narrower. Beyond about 13 percent nickel, the bands disappear and the meteorite becomes structureless.

The width of the kamacite bands is the basis for the structural classification of octahedrites. It is a scheme anyone with a millimeter ruler and a magnifier can use to classify an octahedrite meteorite. The specimens are ranked from coarsest (widest bands) to coarse, medium, fine, and finest (narrowest bands), all based on the percentage of nickel. We can now relate the nickel content to the width of the Widmanstätten figures.

CLASSIFICATION OF IRON METEORITES BY KAMACITE AND NICKEL CONTENT

Type	Symbol	Band Width (mm)	Ni (percent)
Hexahedrite	H	>50	4.5–6.5
Octahedrites	O		
Coarsest	Ogg	3.3–50	6.5–7.2
Coarse	Og	1.3–3.3	6.5–7.2
Medium	Om	0.5–1.3	7.4–10.3
Fine	Of	0.2–0.5	7.8–12.7
Finest	Off	<0.2	7.8–12.7
Plessitic	Opl	<0.2	Kamacite spindles
Ataxite	D	(no structure)	>16.0

Three octahedrites showing (a) *coarse,* (b) *medium, and* (c) *fine*
Widmanstätten structures.

a: *Arispe, Mexico, iron, a coarse octahedrite (Og) measuring 2.9 mm (0.11 inch).*

b: *Henbury, Australia, iron,*
a medium octahedrite
(Om) measuring 0.9 mm
(0.04 inch).

Ataxites

A small class of meteorites fail to show any obvious internal structure. They are called ataxites, which means "without structure." The nickel content for octahedrites ranges from about 6.5 percent to 13 percent. At nickel concentrations higher than 13 percent or 14 percent, the fine octahedrite structure grades into a microscopic Widmanstätten

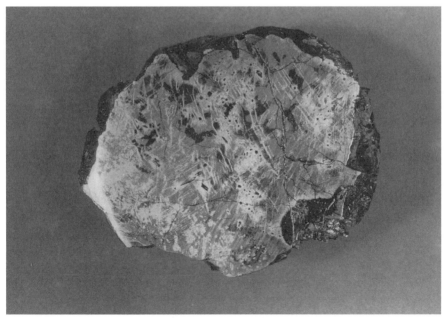

Santa Catharina, Brazil, nickel-rich ataxite containing 33.5 percent nickel. The composition is primarily taenite, with minute kamacite inclusions. Some inclusions of accessory minerals (black) are evident.

227

structure that is not visible to the unaided eye. As the nickel concentration rises above more than about 15 percent, the taenite-kamacite boundary is close to the temperature where diffusion slows considerably, so kamacite crystals have little time to grow. For nickel-rich ataxites—from 13 percent to about 25 percent—the meteorite is composed essentially of a fine-grained mixture of taenite and kamacite, i.e., plessite. If the nickel content exceeds about 25 percent, the mineral content is almost entirely taenite, with small inclusions of kamacite.

Since ataxites have no visible structure when etched, they can easily be mistaken for terrestrial iron slag. A nickel test is essential for their identification. A microscopic examination showing minute kamacite plates and taenite crystals will confirm their identity.

All known ataxites have been found long after they reached Earth; none was a witnessed fall. The largest iron meteorite in the world, the 60-ton Hoba meteorite in Namibia, Africa, is a nickel-rich ataxite.

A Word of Caution

Even with all these visible features, there still can be some confusion when trying to classify iron meteorites through structural criteria alone. I recall receiving a Sikhote-Alin meteorite from the Soviet Union in 1965. It was a jagged fragment with considerable mechanical distortion. I cut the specimen across its narrowest dimension and prepared a face for etching. After a minute of etching, nothing showed up. I continued etching for another minute—nothing. I etched the specimen for an additional three minutes and found no structure. It appeared that I had an ataxite, but I knew Sikhote-Alin was a coarsest octahedrite. I noticed that there were some fine, parallel scratches on the face, so, thinking I had not thoroughly ground out the saw marks, I reground the face and etched the specimen again. Still no structure, but again the scratches showed up—in the same positions. Then it struck me—a coarsest octahedrite has very large kamacite bands, which could measure an inch or more across. The specimen was generally flat and about an inch thick. I had one complete kamacite plate 1 inch or more across. The scratches on its face were not there through careless preparation. They were Neumann lines. Sikhote-Alin was a transitional specimen between a hexahedrite and an octahedrite.

Josephinite—A Terrestrial Nickel-Iron

Native iron oxidizes easily, therefore deposits of nonoxidized iron are rare on Earth. Josephinite is an iron mineral alloyed with minute amounts of nickel. It has been found scattered through other rock, such as basalt, and as nuggets in placer deposits. In the United States, it is found in Josephine County in southwestern Oregon. This mineral is sometimes confused with iron meteorites, since it passes the nickel test and does show an internal crystalline structure something like that of

Parallel Neumann lines cross kamacite plates in this coarse octahedrite. The irregular granular area in the upper left is plessite.

iron meteorites. Finding a nugget of josephinite is about as rare as finding an iron meteorite.

Inclusions in Iron Meteorites

Iron meteorites are more than just aesthetically pleasing specimens. The octahedral structure can display a variety of inclusions produced by accessory minerals usually present in irons. Every time I cut an iron meteorite to reveal its internal structure, I feel an air of anticipation. The inclusions are always a surprise. They make each specimen unique.

Fortunately, each accessory mineral has its own obvious characteristics, making identification easy. A triad of common iron minerals tops the list. Next comes carbon, exhibiting both crystalline and amorphous forms, then silicates common to the stony meteorites. These minerals need to be seen in the specimen to appreciate their structure and beauty.

Troilite (FeS)—A Sulfide of Iron. Troilite occurs in varying amounts in all meteorites. It is similar to terrestrial pyrrhotite but contains more iron in its lattice structure. Pyrrhotite is usually attracted to a magnet, troilite is not. Troilite is easily recognized in iron meteorites because it forms large, nearly spherical, nodules, usually surrounded by other accessory minerals (see color plate V). In stony meteorites it occurs as tiny flakes. Its bronze color is unmistakable, and if a specimen is etched in the presence of water, troilite leaves an ugly brownish stain that bleeds across its face. (Mixing nitric acid with ethanol rather than water as an etching solution avoids the stain.)

One extraordinary meteorite called Mundrabilla from the Nullarbor Plain, Western Australia, has many bronze-colored troilite inclusions in both nodules and ribbons. Troilite weathers away more rapidly than the nickel-iron alloys, leaving behind a lumpy exterior full of pits and holes.

Schreibersite (Fe,Ni,Co)$_3$P—A Phosphide of Nickel-Iron. Schreibersite is essentially a cosmic mineral, occurring only rarely in terrestrial native iron. Like troilite, it occurs in all meteorites. Schreibersite is so brittle that it looks pitted when viewed through a magnifier, even after the face of the specimen has been ground smooth. It has an unmistakable silvery luster that makes it stand out, especially after fresh etching (see color plate VI). Often it is found as a shell surrounding troilite nodules, emphasizing the troilite even further. Sometimes it forms long, needlelike inclusions, called rhabdite crystals, next to the kamacite plates.

As schreibersite is almost insoluble in acid, it stands out after a specimen is etched and can be felt with the finger. If it were teased out of the specimen, it would be attracted to a magnet.

Cohenite (Ni,Fe,Co)$_3$C—A Carbide of Iron. Of the triad, cohenite is the rarest accessory mineral and the most difficult to distinguish by eye alone. It occurs only in low-nickel meteorites such as coarse octahedrites and hexahedrites and in chondrites. In hexahedrites and nickel-poor ataxites, it often appears as tiny, needlelike crystals embedded in the iron mass. Cohenite and schreibersite are often described together, since they are strikingly similar and often occur together. Both have a silvery luster, are brittle, and are somewhat insoluble in acids. The real difference between the two lies in their composition and reaction to various chemicals. There is a simple test to distinguish cohenite. After polishing a meteorite flat, prepare a solution of copper sulfate in water, made slightly acidic with nitric acid, and brush it over the specimen. Immediately, a thin coating of copper will appear, deposited on the cohenite and kamacite but not on the schreibersite. The meteorite should be scrubbed to remove the copper, which cannot be removed without repolishing if allowed to dry. As a rule, since cohenite occurs only in nickel-poor irons, any iron with more than 7 percent nickel usually contains only schreibersite.

Carbon in Meteorites

There are three forms of carbon in meteorites. Most commonly it appears as the soft mineral graphite. Graphite usually forms nodules much like troilite and often is associated with it, intergrowing with the troilite nodule. Enormous nodules of graphite have been found in the Toluca, Mexico, irons. The large pits and occasional holes characteristic

Large graphite inclusion (black) surrounding a core of irregularly shaped troilite. Entire nodule is surrounded by schreibersite. Nodule is 0.8 inch in diameter. —Robert A. Haag collection

A 2.7-inch-diameter cavity in a Canyon Diablo iron was once the resting place for a graphite nodule. Terrestrial weathering destroyed the nodule, resulting in a thin bed of nickel-iron that rusted through, leaving a dime-sized hole.

of iron meteorites may be the result of graphite nodules ablating from the meteorite during its passage through the atmosphere. If ablation doesn't do them in, weathering often does.

Graphite is the low-pressure form of carbon. It appears either as hexagonal crystals, rarely found in meteorites, or is amorphous (without structure). The high-pressure form of carbon, diamond, is usually discovered while cutting the meteorite with a diamond saw. When the diamond saw hits a small diamond, its cutting action slows dramatically—there is no mistaking a run-in with a meteoritic diamond. The first minute diamond was found in a ureilite by Russian investigators in 1888. Three years later, small, black diamonds were found in the iron meteorites surrounding Meteor Crater. To this day, Canyon Diablo meteorites hold the record for the most diamonds recovered. These are not the perfect diamonds seen in jewelry stores. They are minute, often seen only by means of magnification, and are usually embedded in the graphite in which they must have formed. Unlike diamonds that form under high pressure deep within Earth's crust, meteoritic diamonds, most researchers believe, are a shock-related product formed instantly when a meteorite impacts Earth. Meteoritic diamonds could have formed on the original parent body during a collision with another body in space. That collision would have scattered shocked meteorites, with their cargo of diamonds, into space.

Silicates

Iron meteorites occasionally have silicate inclusions. They are especially common in octahedrites such as Odessa and Toluca (see color plate VI). Sometimes silicate inclusions are so common they make up nearly 50 percent of an iron meteorite. Silicated iron meteorites represent a transitional class between irons and stones.

A meteoritic diamond was encountered when cutting this Canyon Diablo iron. The diamond is in the upper right part of the black graphite nodule. Lines across the specimen are saw blade marks where the specimen was gouged as the diamond turned the blade aside. –Robert A. Haag collection

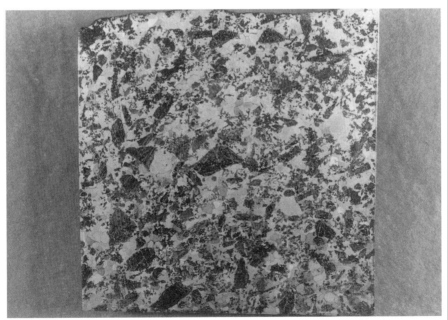

An extraordinary silicated octahedrite from Landes, West Virginia. The large, dark inclusions are pyroxenes, so plentiful that this meteorite is classified as a silicated iron. Specimen measures about 5 inches across.

A single pallasite weighing 83 pounds was found in 1915 about 5 miles north of the town of Albin, Wyoming, but was not recognized as a meteorite until 1935. The Albin pallasite is noted for its large, clear olivine crystals. –Courtesy of Michael Casper, Meteorites, Inc.

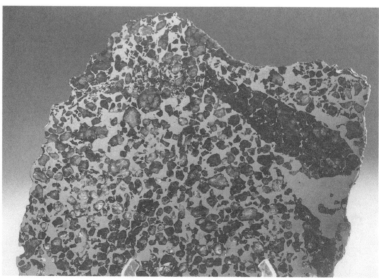

This slab was cut from a large mass of the Esquel pallasite found in 1951 near Chubut, Argentina. Noted for its high stability and clear, bright yellow olivine grains, Esquel is a prize pallasite for collectors and museums and is available through meteorite dealers. –Courtesy of Michael Casper, Meteorites, Inc.

THIRTEEN

Stony-Iron Meteorites— The Mantles of Planets

In the same way that the cores and crusts of planetary parent bodies have been sampled through their meteorite counterparts, so broken pieces of mantle material have been sampled through the third major kind of meteorite, the stony-irons. Stony-iron meteorites are mixtures of nickel-iron and silicate minerals that seem to match the mantle in our model of a differentiated parent body. They must be very rare in space; of the more than 2,600 meteorites known, only 73 are stony-irons.

Like achondrites and irons, stony-irons have suffered melting and recrystallization. They are some of the most-sought-after meteorites because of their striking appearance, especially the pallasites. Scientists recognize three possible classes. Only two of them, pallasites and mesosiderites, are represented by a respectable number of specimens. These two classes are distinguished primarily by internal structure.

The remaining class, lodranites, is most closely related to mesosiderites. These stony-irons contain equal amounts of the orthopyroxene bronzite and olivine in a discontinuous aggregate of nickel-iron. There are only two known specimens, the most recent having being found in 1979 at the Yamato Mountains site in Antarctica.

CLASSIFICATION OF STONY-IRON METEORITES

Stony-Iron	Primary Minerals
Pallasites	Iron-nickel, olivine
Mesosiderites	Iron-nickel, eucrite/diogenite minerals
Lodranites[a]	Iron-nickel, troilite, bronzite, olivine, feldspar

[a] Lodranites have also been classified as "primitive" achondrites.

Pallasites

You need to see a well-prepared pallasite only once to have its image indelibly fixed in your mind. Pallasites are mixtures of crystalline olivine and a network of nickel-iron. They are the most common of the stony-irons, with thirty-nine different specimens known. The name pallasite comes from the 1,600-pound meteorite found in 1749 near Krasnojarsk, Siberia. In 1772, the German naturalist Peter Simon Pallas took samples of the unusual stone (whose meteoritic origin was not yet recognized) and described it in his journals. The Krasnojarsk iron turned out to be an olivine nickel-iron mixture. As other stones of this composition and structure became known, they were given Pallas's name. The Brenham, Kansas, meteorite is the most plentiful pallasite known (see color plate VI). More than a ton of it has been retrieved from a shallow crater in Kiowa County, Kansas, since 1882.

Generally, pallasites contain about equal amounts of olivine and nickel-iron by volume. The olivine grains are completely enclosed by a continuous network of nickel-iron. The olivine–nickel-iron ratio varies considerably from pallasite to pallasite and even within a single specimen. The Brenham, Kansas, pallasite varies from almost pure nickel-iron to a ratio of nickel-iron to olivine of 1:2. The nickel-iron has octahedrite structure and composition, since it shows classic Widmanstätten figures when etched. Troilite and schreibersite inclusions are also usually present.

The olivine in pallasites is magnesium-rich and usually shows evidence of shock, having been fractured and fragmented. Occasionally, unshocked, gem-quality crystals are found. Many of the trapped olivine grains in pallasites appear rounded, suggesting an origin as liquid

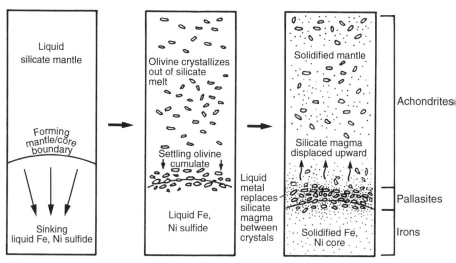

Mixing of the settling olivine and liquid nickel-iron at the core–mantle boundary within a differentiated parent body.

spheres frozen in metal. This is definitely a mixture of immiscible materials: It is, in essence, an emulsion. The olivine is probably a cumulate formed at the core–mantle boundary. As gravitational separation took place during the liquid phase of the parent body, the liquid silicate minerals separated from the heavier liquid nickel-iron. The olivine crystallized out of the silicate melt and settled toward the core–mantle boundary, which may have forced some still-liquid nickel-iron up into the olivine crystal layer, trapping the olivine in a nickel-iron network. Alternatively, if the core shrank as it cooled, it could have forced residual core liquid into the olivine layer.

Some pallasites oxidize rapidly in the open air. This is especially true of the Brenham pallasite, which tends to be highly unstable. Other pallasites exhibit remarkable stability. Unlike the Brenham specimen, which was buried in moist soil for centuries, the Imilac pallasite from Chile rested in one of the driest deserts on Earth, and oxidation has taken its toll there, too. Given enough time, meteorites rust away, even in the Atacama Desert. (See color plates VI & VII.) Earth is a hostile environment for space rocks.

Mesosiderites: Broken Rocks from a Broken Mantle

The stony-irons called mesosiderites have little in common with pallasites except that they both come from a parent body's mantle; they are formed in very different ways. Pallasites are relatively unaltered rock from the lower mantle, and mesosiderites are broken mantle rock that has been reassembled with lighter silicate rock. They are polymict breccias, that is, composed of angular rock fragments of different mineral compositions. They are troublesome to meteoriticists, since their component parts seem unrelated. The silicate-to-metal ratio is about 1:1. The metal has the mineral composition of octahedrites, with between 7 percent and 10 percent nickel, but it is granular and uniform in composition. Occasionally there are large nodules of metal that show Widmanstätten figures, probably representing remelting of the original metal with a subsequent, very slow cooling period.

Thin section of the Bondoc mesosiderite in crossed Polaroids reveals large anhedral grains of plagioclase and pyroxene. The plagioclase shows parallel bands (twinning planes) running across the crystals. 39X magnification. Bondoc Peninsula, Luzon, Philippines.

Cut slab from the Estherville, Iowa, mesosiderite. The light-colored masses are nickel-iron. Specimen is almost 4 inches across.
—Robert A. Haag collection

Cut slab from the Vaca Muerta, Chile, mesosiderite, showing a mixture of nickel-iron (light) and pyroxene. The fragments of dark stone have a composition similar to that of eucrites and diogenites. Specimen is almost 3 inches across.

The silicates are primarily plagioclase feldspar and calcium pyroxene with some hypersthene and/or bronzite. The fragments appear to be pieces of eucrite or diogenite that have no apparent association with the metal. Rounded masses of olivine are also found that may be associated with the iron, as in pallasites.

Pallasites versus Mesosiderites

The most distinctive difference between pallasites and mesosiderites is their internal structure. The nickel-iron in pallasites forms a continuous network, with cavities holding olivine. In mesosiderites, the nickel-iron is often present as jagged masses set in a stony matrix. The Estherville, Iowa, mesosiderite has this appearance, as does the Vaca Muerta, Chile, mesosiderite. Other mesosiderites show a bright, uniformly granular nickel-iron matrix, in which broken pieces of silicate minerals are embedded. The Clover Springs mesosiderite is a typical example.

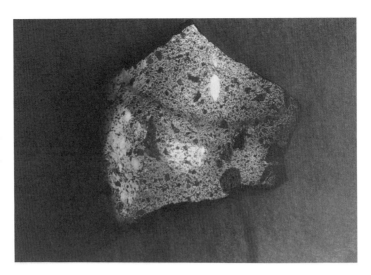

Slab of Clover Springs, Arizona, mesosiderite. Note the high nickel-iron content uniformly distributed through the specimen. Angular clasts of dark stony material are scattered through the mix.

Mesosiderites were formed on the surface of the parent body as a result of impact with another body. All the mineral components are angular fragments, many showing evidence of shock. In addition, a high-temperature quartz called tridymite is found in mesosiderites. It is a well-known shock-produced silica. The impacts must have mixed the surface eucrite material with metal and olivine from deep within the parent body or its impactor.

A Human Endeavor

The history of the solar system is hidden within the great diversity of meteorites, and each one has its own story to tell. The history of our own origins is still being written, and there is much to be learned. Evidence has come to us in a haphazard way, and until recently humans have played a passive role: We had to wait for the meteorites to come to us.

We humans could play a more active role if we could sample the rocks of planets and asteroids firsthand. We have made a beginning—our laboratories and museums are repositories for lunar rocks, gathered on the courageous first journeys beyond the confines of Earth. Robots have sampled the soil of Mars, and plans have been laid to retrieve rocks in the near future. In 1992, the Jupiter-bound spacecraft *Galileo* gave us our first close-up look at an asteroid called Gaspra. It is irregular and pitted, covered with impact craters—more craters than we had antici-pated. Grooves across its surface are fractures, attesting to its violent history. Gaspra's irregular shape suggests that it was broken off a much larger parent body. This confirms what had already been surmised: The birth of the solar system was a violent affair. Undoubtedly, chunks of Gaspra are presently orbiting among the terrestrial planets, perhaps one day to be added to meteorite collections in future museums.

The study of meteorites is a human endeavor, with all the elements found in any exploration. For most of us, meteorites are a curiosity. For a few—the meteorite hunters—they are a passion.

A slab of a severely weathered pallasite from Huckitta, Northern Territory, Australia, found in 1924. The main mass, weighing 3,100 pounds, remains the largest intact pallasite known. Nearly 1 ton of oxidized iron shale surrounded this ancient meteorite when it was found. This specimen lacks the typical bright luster of the iron network, the iron being completely oxidized to hematite. The olivine has been darkly stained.

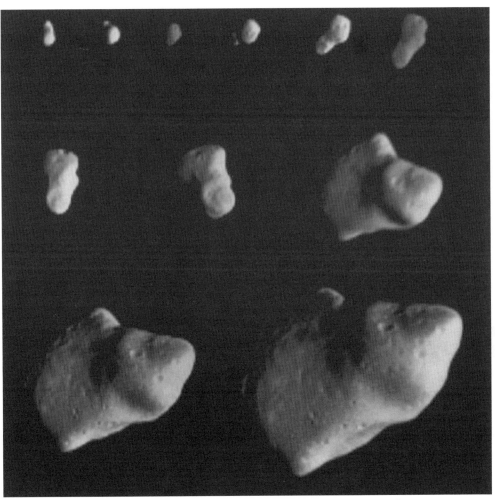

Multiple images of Asteroid 951 Gaspra taken by the Galileo *spacecraft as it approached to within 3,180 miles. Its rotation is evident during the photographic session. The long axis measures about 10 miles. More than 600 impact craters dot the asteroid's surface.* —Courtesy of Michael Belton, NOAO and NASA

Part III

METEORITE HUNTERS

Moving the 34-ton "Ahnighito," largest of the Cape York meteorites, to new quarters in the American Museum of Natural History in New York City. —Peter Goldberg, photo neg. No. M25. Courtesy Dept. of Library Services, AMNH

FOURTEEN

The Great Meteorites – Their Discovery and Recovery

All the world's great meteorites are irons. None were witnessed falls. All are presumably prehistoric. Many of these great bodies occupy places of honor in major museums in the United States.

The Great Greenland Meteorites

In 1818, the British explorer Captain John Ross sailed north along the west side of Greenland, making landfall in what he called Melville Bay. On the north side of the bay, he met Eskimos who showed him crude knives and harpoons made of iron. This was unusual, since there was no natural source of iron in the region. The Eskimos told him about a mountain with a huge, protruding, iron rock, located on the north shore of Melville Bay. They had gotten the iron for their tools by laboriously chipping pieces from this rock. Bad weather prevented Ross from searching for it, but he suspected it was an iron meteorite. He returned to England with knives made from the meteoritic iron.

Seventy-five years passed. Explorers from Europe and the United States searched in vain for the suspected meteorite. The native Greenlanders refused to divulge its exact location, and all attempts to find it failed until the famous arctic explorer Robert E. Peary joined the search in 1894. The Eskimos revered him because of his crossing of the Greenland ice sheet by dog team. Peary presented two natives with a rifle and other gifts of iron, cajoling them into leading him to "Saviksue," the great iron rock.

After eleven days of travel and the departure of one guide, his remaining guide, Tallakoteah, led him to a peninsula on the north shore of Melville Bay called Cape York. There, in a snowbank, Tallakoteah uncovered the 3-ton iron meteorite the Eskimos called the Woman. It had been severely chipped and pounded by the natives over the years, reducing its mass by nearly half. To Peary's surprise, Tallakoteah told him this was one of three great meteorites. Searching through the snow, they located the smallest of the three, called the Dog. It weighed 1,100 pounds and, like the Woman, showed evidence of having been hammered. The third and largest iron was on an island 4 miles north of the two smaller irons.

Peary delayed visiting the site of the largest iron, which natives called the Tent, until 1895, when he returned to Cape York to retrieve the Dog and Woman. Only then did he realize the formidable task before him. After partially excavating the Tent, he measured it: 12 feet long, nearly 7 feet high, and more than 5 feet thick. From these measurements he estimated the weight at 100 tons. Fortunately, he was off by a factor of three—it actually weighs 34 tons; it is unlikely Peary could have transported a 100-ton mass to his ship.

Peary and his men built timber sleds and transferred the two smaller meteorites to the coast, where his ship was anchored offshore against the permanent ice. Using ice blocks, he floated the meteorites across open water to the ship, then transferred the sled onto the permanent ice using the ship's winch. At this critical point, the Woman almost became a resident of Melville Bay. Just when they were about to raise the meteorite, the ice beneath began to cleave and give way. But the cables attached to the ship's hoist held as the meteorite was briefly submerged, then slowly lifted aboard.

Retrieving the main meteorite was another matter. Peary needed a larger ship, capable of ferrying the great mass. He returned in August of the following year with the 370-ton icebreaker *Hope,* which was equipped with four hydraulic jacks capable of lifting 30 tons each. With these he hoisted the meteorite from its resting place. A team constructed a log sled with rollers under it and placed it beneath the great weight. From a nearby village, Peary hired dozens of men to drag the ponderous weight to the crest of a hill, then roll it down the slope to the shore. It was by that time September, and ice was already threatening to lock the *Hope* in its grip for the winter. Peary left the Tent on the shore, abandoned for the moment but not forgotten, and sailed back to Boston.

In August 1897, Peary returned to Cape York to again do battle with the "demoniac iron from heaven," as he called it. He moored his ship next to a natural pier on the island and constructed a massive bridge of huge timbers to span the gap between the pier and the deck of the *Hope.* On this bridge he mounted double rows of steel railroad track. Before moving the great Greenland iron meteorite, Peary draped an American flag over it and let his daughter break a bottle of wine against it, naming it "Ahnighito." Then, arduously, slowly, with hydraulic jacks, pushing against the natural pier and the meteorite sled, the great mass inched its way toward the ship.

The *Hope* had been ballasted on the opposite side to stabilize it against the weight of the 34-ton meteorite. The Eskimos abandoned ship, fearing that the great weight of the "Saviksoah demon" would surely push the ship to a watery grave. Ahnighito finally reached the deck of the *Hope* and was lowered into the hold. A perilous journey through iceberg-filled waters and violent storms added to Peary's troubles.

The 34-ton Cape York meteorite on a timber bridge as Admiral Peary's crew prepares to slide it aboard the ship Hope. –Robert Peary photo. Neg. No. 329147.
Courtesy Dept. Library Services, American Museum of Natural History

Storms tossed the *Hope* so violently that the crew feared the massive cargo would break its restraints and smash through the ship's side. Finally, on October 2, 1897, the *Hope* sailed safely into New York harbor with its celestial cargo.

At the Brooklyn Naval Shipyard, a 100-ton floating crane lifted the Tent from the cargo hold and placed it on the pier. The world's greatest meteorite had made it to the United States.

It remained on the pier for seven years. Morris K. Jesup, president of the American Museum of Natural History in New York, finally bought it for the paltry sum of $40,000. A holiday atmosphere prevailed as a team of eighty horses pulled the meteorite through the streets of New York to the American Museum of Natural History. It remains there today in the museum's meteorite hall, along with the Woman and Dog. Peary later used the $40,000 to finance his historic trek to the North Pole.

The story of the Cape York meteorites does not end with Ahnighito's recovery. A thorough search of an iron-meteorite strewn field, even an ancient one, is often rewarding. In 1913, inhabitants of the Savik Peninsula near Prince Regent Bay found a fourth mass, weighing 3.5 tons, which was taken by a Danish expedition to Copenhagen in 1926. Then in 1963 another great mass weighing more than 15 tons was found near Agpalilik, inland from Cape York. Some believe this meteorite to be the "Man," a legendary fourth member of the Cape York family. Perhaps there are others still to be found.

Oregon's Willamette Meteorite

While the Ahnighito iron rested on the pier in New York, another meteorite adventure was unfolding on the opposite coast of the United States. In the autumn of 1902, Ellis Hughes was cutting wood in the forest near the town of Willamette, Oregon, now part of the city of West Linn, a few miles west of Oregon City. He came across a large, bell-shaped rock slightly sunken in the ground, and he guessed that this odd-shaped rock was a meteorite. He guessed correctly. It was (and still is) the largest meteorite found in the United States. Its conical shape is the result of its orientation as it passed through the atmosphere. The meteorite measures 4 feet 6 inches high, 10 feet 3 inches at its longest base dimension, and 6 feet 6 inches at its shortest base dimension. It is somewhat smooth along the sides of the cone but riddled with gigantic cavities along the oblong base that were produced by oxidation and weathering of large nodules of troilite.

The huge meteorite rested on the ground in only a slight hollow. This was not its original resting place, because it would have made a sizable penetration hole on impact. Some scientists believe it originally landed in Canada and was carried southward to Oregon by advancing glaciers during the last ice age, perhaps 100,000 years ago.

Unfortunately for Ellis Hughes, the glacier happened to drop the meteorite on property next to his own, property that belonged to the Oregon Iron and Steel Company. Legally, the meteorite belonged to them. Mrs. Hughes, perhaps seeing a potential profit in ownership, insisted that Ellis move the great iron onto their property, a distance of three-quarters of a mile through rugged and forested terrain. The great meteorite weighed 15.5 tons, and Hughes had only his fifteen-year-old son and an old horse to help him. During the winter and spring of 1903, Hughes devised an elaborate plan to move the meteorite. In preparation for the move, he cleared a narrow roadway leading to his property. Then he cleared an additional 800 feet in the opposite direction to confuse any curious neighbors.

By early summer he had built a sturdy wagon of heavy, 10-foot-long logs with wheels cut from the trunk of a large fir. Then he constructed a capstan and staked it to the ground between two large trees. For additional strength he chained the capstan to the trees. The capstan was in a direct line to his house. His trusty horse was hitched to the capstan, and a 100-foot cable was attached to capstan and wagon.

A visitor casually leans against the giant 15.5-ton Willamette meteorite in 1904, a year before it was shipped to the American Museum in New York City, where it remains on permanent display. Note the enormous cavities, some of which pass clear through the meteorite.

The meteorite was resting on its side when it was found and had to be righted, blunt end down. He managed this by jacking it up until it rolled base-down onto his wagon. As the horse labored around the capstan, the wagon slowly made its way toward Hughes's property. Inch by inch, day by day, he and his son and horse labored. Sometimes the wagon became mired in mud and had to be jacked up onto a firm board surface; these days saw no forward progress. On his best days, he moved the great mass only about 150 feet. When the cable became completely wrapped around the capstan, the capstan had to be moved 100 feet farther along the path, and again secured to the ground between two new trees.

After three months of toil, the wagon finally crossed the property line, and the meteorite came to rest in Hughes's front yard. Now he was ready to make a profit on the venture. He constructed a shed over the meteorite and charged admission to sightseers. Twenty-five cents to see the largest meteorite ever found in the United States wasn't too much to ask, Hughes thought, but it turned out to be too much for the owners of the Oregon Iron and Steel Company.

For the rest of the summer and into the fall, Hughes collected admission fees, much to his wife's delight. He gained valuable publicity from the Portland *Oregonian.* The newspaper announced the discovery of the meteorite on October 24, 1903. An article on November 3 included a photograph of the superb specimen. This drew greater crowds–and meteorite experts. F. W. Crosby, a collector for the Smithsonian Institution, visited the Hughes farm and authenticated the meteorite. The Oregon City *Enterprise* made the announcement on November 6, but the article cast a cloud over the exciting news when it mentioned a rumor that the meteorite was actually discovered on adjacent property and "clandestinely removed onto the land of those now having the meteor in possession before the discovery of the heavenly monster was announced."

Days later an attorney for the Oregon Iron and Steel Company was among the sightseers. He noticed the road cut through the forest leading directly to company property. The attorney reportedly offered Hughes $50 for the meteorite, which was quickly refused. On November 27, Hughes received another offer, which did not come with the option of saying no: to act as defendant against the Oregon Iron and Steel Company in the Circuit Court in Oregon City, for wrongful possession of the meteorite.

The company owners claimed the meteorite because Hughes had found it on their land and stolen it. Hughes argued it was the property of the Clackamas Indians, who had used it thirty years before as a sacred stone, and therefore was private property, an abandoned Indian relic, free for the taking. The jury didn't buy this argument and decided in favor of the Oregon Iron and Steel Company. When the verdict was announced on April 28, 1904, the court valued the meteorite at $150!

Before Hughes could appeal to the Oregon Supreme Court, two of his neighbors made claim to the meteorite, saying it had been moved from their property to the property of the Oregon Iron and Steel Company. As evidence, they presented photographs of a crater on their property supposedly made by the meteorite. This second trial took place on January 17, 1905, in the Circuit Court of Oregon City. Now Hughes and the Iron Company were codefendants. The trial lasted only two days and again the jury awarded custody of the meteorite to the Company. This time the jury valued the meteorite at $10,000. No meteorite in history had gone up in value so rapidly!

Hughes appealed to the Oregon Supreme Court. Meanwhile, the meteorite was placed in the custody of the Clackamas County sheriff. It was moved to the front yard of Harold Johnson, a special deputy assigned to guard the iron. There it sat for several months. During this time many souvenir hunters tried sneaking into the yard at night to break off a piece of the iron, which would ring like a bell when struck, raising Johnson from his bed—with his shotgun.

On July 17, 1905, the state supreme court sustained the circuit court's verdict, ending the battle for the country's largest meteorite.

The Willamette
meteorite on
display in the foyer
of Hayden
Planetarium at the
American Museum
of Natural History,
New York City. Note
the large ablation
cavities. —Rota-Boltin
photo. Neg. No. 323711.
Courtesy Dept. of Library
Services, American
Museum of Natural History

The Oregon Iron and Steel Company floated it by barge down the Willamette River to Portland for exhibition at the Lewis and Clark World's Fair. A year before, Dr. Henry A. Ward, president of Ward's Natural Science Establishment in Rochester, New York, and a world-class collector of meteorites, had come to study the famous specimen. He named it the Willamette meteorite, after the town nearest to the discovery site, and gave the world the first scientific description of the extraordinary meteorite, complete with photographs.

Thousands of people went to the World's Fair to see the famous meteorite. There was great interest in its final destination. Where should it reside? In Portland? Oregon City? No one doubted it belonged in Oregon. No one, that is, except the owner. While the public discussed the matter, Mrs. William E. Dodge of New York quietly offered the Oregon Iron and Steel Company $26,000 for the meteorite, thus ending the debate: The meteorite was destined for the East Coast. Mrs. Dodge gave it to the American Museum of Natural History with the stipulation that it was to remain intact as a single body. Today, two of the world's largest meteorites sit side by side in one of the world's largest meteorite collections, at the American Museum of Natural History in New York.

The Old Woman Meteorite

In the United States a meteorite is legally the property of the landowner on whose property the find was made, but sometimes ownership can become complicated. The Old Woman meteorite, the second largest found in the United States, is one example.

In February 1976, two prospectors were searching for a legendary Spanish gold mine in the Old Woman Mountains of the Mohave Desert, San Bernardino County, California. Mike Jendruczak and David Friburg

Dr. Roy Clarke Jr. stands behind the Old Woman meteorite where it was found among boulders in the Old Woman Mountains of California's Mohave Desert.
—California Desert Information Center, BLM, Dept. of the Interior

were just heading up the last canyon in the area when they noticed a strange rock among other boulders at the end of the canyon. Close inspection revealed that the rock was apparently solid iron. Jendruczak thought it might be a meteorite. The rock measured 38 inches long, 30 inches wide, and 34 inches high and they couldn't move it—it weighed more than 3 tons.

They broke off a small piece and headed home to Twentynine Palms. In March, they sent part of their sample to the Griffith Observatory in Los Angeles for identification. There, Ronald Oriti examined the specimen and tested it for nickel. The test was negative, so Oriti concluded it was not a meteorite. Had he seen the entire specimen, his opinion would have quickly changed. This "meteor-wrong" turned out to be a "meteor-right."

Friburg was not satisfied with the result and decided to send a sample and photographs to the Smithsonian Institution for a second opinion. He also sent a sample to Dr. John Wasson at the University of California at Los Angeles, who identified the specimen as a rare iron meteorite and subsequently published his final analysis. (The miners became aware of this analysis only after the Smithsonian was involved.) The Smithsonian acknowledged Friburg's letter, but examination of the specimen awaited Dr. Roy Clarke Jr., curator of meteorites, who was out of the country at the time. In September, after examining the specimen, Clarke arranged to meet Friburg in Twentynine Palms to examine the meteorite.

Friburg and Jendruczak hoped funds from the sale of the meteorite would allow them to make repairs to equipment in a milling operation they had recently rented. At the insistence of a third partner, Jack Harwood, they filed a mining claim, assuming this would protect their find. Friburg and Harwood reluctantly took Clarke to the site in a remote canyon, 1,200 feet up in the Old Woman Mountains. When Clarke saw the weathered, rust-colored iron lying in a wash near the top of a mountain, he knew his suspicions had been correct: Here was a magnificent meteorite that probably had lain in this remote canyon for thousands of years.

Consider the situation. Here was the government's top meteoriticist on the site of one of the great meteorite finds. Its scientific value was incalculable. On the other side, here were two tough miners who had searched for years in hopes of finding a gold mine. In their eyes, they had struck gold, and they hoped to sell the specimen to the government or to the highest bidder. They were convinced the rock was worth a million dollars, and they expected a sizable reward for their find.

Clarke was in an awkward position. On the return trip to Twentynine Palms, Friburg and Jendruczak pressed Clarke for a serious appraisal of the situation. Clarke explained that the question was not who held title to the specimen, which the miners claimed was theirs according to the Mining Law of 1872, but who held title to the land on which they found it. Furthermore, Clarke explained, meteorites were not a "locatable mineral" as defined by the Mining Law of 1872. And if they had found it on public land, the United States Government was the legal owner, regardless of who found it.

According to an account published by the California Mining Journal in August 1977, Clarke said that the government could claim the meteorite under the provisions of the Antiquities Act, in which meteorites found on public lands were considered objects of scientific interest and therefore belong to the government. There is no question that the meteorite was an object of scientific interest, but meteorites are not specifically mentioned in the body of the act. The law was written to apply primarily to archaeological artifacts. However, the government has interpreted the act to include *anything* of scientific interest.

The prospectors felt they had been tricked into showing Clarke the meteorite's location, and now they were being threatened with government confiscation of their prize. After Clarke returned to Washington, the prospectors tried to think of a way to remove the 3-ton iron, but the weight of the specimen and ruggedness of the terrain made ground transportation impossible. Renting a helicopter large enough to lift the rock out of the mountains was prohibitively expensive.

A 900-pound mass is cut off the Old Woman meteorite at the Smithsonian. The Smithsonian originally planned to grind, polish, and etch the huge cut face to show the Widmanstätten figures but never completed the project. The meteorite is now displayed with its cut face down.

For over a year the great iron lay on the western slopes of the Old Woman Mountains as the prospectors fought a futile legal battle with the Bureau of Land Management and the Smithsonian Institution.

On June 16, 1977, a team from the U.S. Marine Corps climbed to the site and rigged a double thickness of cargo netting around the iron. Then a large Marine helicopter lifted the iron from its resting place and set it down on the desert floor so it could be hauled away. The meteorite was displayed temporarily at BLM buildings in several southern California cities until March 1978, when it was trucked to Washington, D.C.

Smithsonian researchers removed a piece of the meteorite (about 15 percent of its bulk) for study. Chemical analysis revealed that the Old Woman meteorite, while mostly iron, contained 5.71 percent nickel, making it a relatively rare meteorite classified as a coarsest octahedrite, very similar to the Sikhote-Alin meteorite.

Meanwhile, the legal battle continued. The attorney general for the State of California twice filed to obtain a restraining order and twice was denied. At the same time, the County of San Bernardino filed and succeeded in obtaining a restraining order preventing the meteorite from leaving the state until a final judgment was issued by the United States Department of the Interior. At one point, in September 1977, Senator Alan Cranston of California released a statement that California would be allowed to keep the meteorite. The Interior Department decided to grant California custody of the meteorite in defiance of its own agency, the BLM. But in 1978 this decision was reversed, based on the legal precedent of land ownership. So the Smithsonian Institution became the final custodian of the nation's second-largest meteorite.

The government could not legally offer compensation for items it already owned, but the Smithsonian offered to compensate the miners, from nongovernment funds, with what it felt was a reasonable fee. This amount was far less than the expected $1 million dollars the finders insisted upon. Friburg and Jendruczak refused to discuss any monetary settlement that fell so far short of their expectations. It was a million or nothing—so they got nothing.

In September 1980, the Smithsonian returned the meteorite, less the piece removed for study, to California. The Old Woman meteorite can now be seen on permanent display at the U.S. Bureau of Land Management's California Desert Information Center in Barstow.

The Lost Port Orford Meteorite

Oregon is famous for its great meteorites, both recovered and lost. The story of the Port Orford meteorite begins in 1851 when the U.S. Department of the Interior contracted with Dr. John Evans to conduct a geological survey of the Oregon Territory, concentrating on the mineral and agricultural resources west of the Cascades. Evans was a physician by training, but he was a man of many talents, with experience as a land surveyor, field geologist, and naturalist. From 1851 to 1856 he explored the vast territory from the Rocky Mountains to the Pacific, from the southern Oregon line to Canada. During his survey he forwarded rock and soil samples, fossils, and plant specimens to scientists in the East. The rock and mineral specimens were sent for analysis to A. Litton, a chemist at St. Louis University in Missouri. One final shipment of specimens acquired in 1858 was sent to Charles T. Jackson, a prominent chemist in Boston.

Jackson received the shipment from Evans in the fall of 1859. Among the mineral samples was a small, jagged rock about the size of a walnut. It weighed approximately an ounce. The interior of the rock was silvery, it was attracted to a magnet, and the iron content was high. Embedded in the rock were yellow olivine crystals.

Jackson suspected the rock was a meteorite. He sent a tiny piece to Dr. Wilhelm Haidinger, a prominent meteorite expert in Vienna. Haidinger agreed the rock was a rare stony-iron meteorite—a pallasite. Jackson reported this discovery to the Boston Society of Natural History on October 5, 1859.

Evans seemed unaware he had sampled a pallasite during his explorations of the Rogue River Mountains, and he did not mention the unusual rock in his field notes. But when Jackson contacted Evans with the exciting news, Evans had no trouble remembering the specific rock and its location.

Evans and Jackson corresponded for a year and a half. In a letter dated May 1, 1860, Evans presented the following tantalizing description of the meteorite:

> My recollection is that four or five feet projected from the surface of the mountain, that it was about the same number of feet in width, and perhaps three or four feet in thickness; but it is no doubt deeply buried in the earth as the country is very mountainous, generally heavily timbered, and subject to washings from rains and melting of snow in the spring, so that in a few years these causes might cover up a large portion of it. The mass exposed was quite irregular in shape.

A pallasite of that size would be by far the largest ever recovered. Estimates placed its weight at more than 10 tons. This was indeed an extraordinary find; the largest intact pallasite known at the time was the Pallas Iron (Krasnojarsk) discovered in 1749, weighing 1,540 pounds. The news sent great excitement through scientific communities in the United States and abroad. The discovery of this amazing pallasite was on everyone's mind.

Evans said it was located in the mountains about 40 miles east of the coastal community of Port Orford in southwestern Oregon:

> There cannot be the least difficulty in finding the meteorite. The western face of Bald Mountain, where it is situated, is as its name indicates, bare of timber, a grassy slope, without projecting rocks in the immediate vicinity of the meteorite. This mountain is a prominent landmark, seen for a long distance on the ocean, as it is higher than any of the surrounding mountains.

The meteorite might be purchased from the local Indians, Evans said, with only "ten or a dozen blankets, tobacco, etc., as presents."

> There should not be the slightest difficulty in making arrangements with the Indians, for I am personally acquainted with their Chief. The principal Chief, Old John, spent several days in my camp of two men, during the height of the Indian War, when it would have been dangerous for less than three or four companies of U.S. soldiers to have passed the same mountains.

Like bait on a hook, Evans's correspondence induced several scientific societies in the United States to take action. In the spring of 1860, the Boston Society of Natural History and the Academy of Natural Sciences of Philadelphia petitioned the United States government to finance an expedition led by Evans to recover the meteorite. The timing was terrible. The country was on the brink of civil war, and Congress showed little interest in financing an expedition to recover an obscure rock somewhere in the Oregon Territory. Moreover, the government was reluctant to provide additional funds, several thousand dollars, to publish Evans's extensive report of his surveys, and the report remained stalled in the Government Printing Office for lack of funding.

Any further talk about an expedition was silenced when Evans, only forty-nine, died suddenly on April 13, 1861, the day after the beginning of the Civil War. He took the exact location of the Port Orford meteorite to his grave, and his report mysteriously disappeared from the printing office. Only sections of his field notes remain, along with his correspondence with Jackson and a catalog of the geological specimens he collected from the Oregon and Washington territories.

Nearly 150 years have passed since Evans's discovery, and the meteorite remains elusive. Since the early 1900s, interest in the lost meteorite has been revived. The idea of a huge, rare meteorite worth perhaps thousands of dollars beckoned hardy souls into the rugged coastal mountains. Thousands of people have combed the mountains in southwestern Oregon over the past half century, searching for the lost pallasite. Evans's notes have been read and reread, analyzed and reanalyzed. Some of the country's top meteoriticists have searched government archives for further clues and walked the very trails Evans described, trails that supposedly led to the meteorite. No meteorite has ever been found.

To add to the confusion, there are several mountains that could be mistaken for Evans's Bald Mountain. Today, it may be called Iron Mountain, or it may be Bray Mountain or Johnson Mountain, all of which are in the general vicinity. And the mountain may no longer be "bald." It may be covered with pine and cedar and a thick undergrowth, which could easily hide the meteorite; or mud slides—a frequent occurrence during the rainy season—could have buried the great rock.

In 1937, as a result of a rumor that the United States Government, through the Smithsonian Institution, was offering a $2 million reward for the recovery of the lost meteorite, a group of people in Oregon formed an organization for the purpose of finding the meteorite. Calling themselves the Society for the Recovery of the Lost Port Orford Meteorite, members began making annual treks into the rugged Siskiyou Mountains in Curry County, seeking the wayward meteorite. They are a high-spirited group, not easily discouraged by their lack of success over the years.

In 1929, and again a decade later, the Smithsonian sent their curators of mineralogy and petrology to the Rogue River Mountains to search for the meteorite. Neither search uncovered any traces. By 1964, Edward P. Henderson, then Smithsonian's curator of the Division of Mineralogy and Petrology, expressed considerable doubt that Evans's account was trustworthy. The fact that Evans made no mention of the spectacular find in his journals and that the specimen in the Smithsonian collection, showing no apparent distress, could not have been extracted forcibly from a large mass without damage, led him to conclude that the meteorite did not exist. Henderson jointly published a paper with Oregon's state geologist, Hollis M. Dole, which appeared in the July 1964 *Ore Bin,* a monthly magazine of the Oregon Department of Geology and Mineral Industries, in which they expressed these doubts strongly. The article did nothing to dampen enthusiasm for the search.

There the dilemma remained for more than twenty years. In 1986, a historian with a different approach to the problem made a strong case for having finally solved the mystery of the lost meteorite. Science historian Howard Plotkin, of the University of Western Ontario, Canada, uncovered strong circumstantial evidence of a deliberate hoax by Evans. Plotkin became interested in the lost meteorite in the mid-1980s, and in 1986, like so many searchers before him, attempted to walk the route Evans took in the Rogue River Mountains east of Port Orford. He decided upon Johnson Mountain as the "bald mountain" referenced in Evans's letters and conducted a magnetic survey of the mountain using a sensitive magnetometer in an attempt to locate a buried mass. This was the first time a survey using powerful electronic equipment had been attempted. Even so, no mass was found. Failing to find the meteorite raised doubts in Plotkin's mind that the meteorite ever existed in the first place, and raised suspicions that Evans had perpetrated a cruel hoax. If true, it remained for him to discover Evans's motives.

Plotkin began his search anew, but this time in the National Archives and other repositories, where he could find written material on Evans's personal and professional life. Historical accounts of Evans's work written after his death painted a picture of a competent, highly respected scientist of impeccable character, whose work in the Oregon survey was beyond reproach. But Plotkin soon uncovered an astonishing array of incriminating evidence that gave a quite different picture of Evans, the scientist. Evans lacked any formal training as a geologist. His appointment by the Department of the Interior to explore the geology of the Oregon Territory is amazing considering his lack of qualifications; but the appointment apparently was supported by recommendations from credible scientists. So he initiated the survey, ill-prepared to carry out the scientific duties of a field geologist. His exploration was useful but not scientific.

Plotkin found Evans's private life to be equally chaotic and interwoven into his professional life. Evans was a poor money manager and constantly in debt, especially during this period in his life. When he contracted with the Department of the Interior to survey the Oregon and Washington territories, he consistently overextended his budget and continually requested additional funds to complete the survey. At first, Congress was sympathetic and voted new funds to pay off the debts incurred. But as his debts continued to mount, Congress refused to cover any further expenses. Evans's mismanagement of government funds placed the remaining debt squarely on his shoulders. In addition, Evans had borrowed large sums of money to consummate land deals in Oregon, which fell through when land values dropped abruptly during the discovery of gold in Oregon and California. His combined personal and professional debt was now enormous.

It was at this point that Evans conceived an elaborate plan to coax Congress into appropriating additional funds. While passing through the Isthmus of Panama on his return trip from Oregon to Washington, D.C., in 1858, he apparently obtained the infamous meteorite, perhaps from a mineral dealer in Central America. He placed it among the rock and mineral specimens from Oregon. His hope was that considerable excitement among the scientists over the great meteorite find would encourage members of Congress to appropriate the necessary funds not only to send him back to Oregon to retrieve the specimen but to pay for the publication of his survey. The amount of money Evans needed very nearly equaled his total personal debt. His plans may have been to use the funds to pay off his debt and then claim, on his return, that the meteorite was lost, perhaps buried by a landslide.

To augment Plotkin's brilliant piece of detective work, Roy S. Clarke Jr. of the Smithsonian Institution and Vagn F. Buchwald of the Department of Metallurgy, Technical University of Denmark, simultaneously analyzed the Port Orford specimen and compared it with a well-known pallasite found in the Atacama Desert near Imilac, Chile, in 1822. The Port Orford specimen turned out to be almost indistinguishable from the Imilac pallasite (see color plate VI). Thousands of specimens from the Imilac strewn field had been circulated by the mid-nineteenth century. The two scientists concurred with Plotkin that Evans had acquired an Imilac meteorite to serve as bait in his elaborate hoax to defraud the United States government.

In 1980, when I visited the Smithsonian Institution's Museum of Natural History, I was fascinated by the small Port Orford meteorite on display in the Hall of Meteorites. When I returned in 1990, the Port Orford meteorite had vanished.

The Tucson Ring Meteorite

In 1776, a Spanish expedition established a presidio called Tucson in the Sonoran Desert. It was occupied by soldiers (first Spanish, later Mexican) charged with defending the frontier from Apaches and keeping the overland routes to California open and secure. It was a safe staging area in a hostile country, where military weapons and supplies were stored, horses stabled, and equipment repaired. An important

Tucson Ring meteorite on display at the Smithsonian Institution. The Carleton meteorite rests in the bowl beneath. –Smithsonian Institution

figure in this world was the blacksmith, who kept weapons in repair and horses in shoes. The blacksmith's shop, a simple affair, housed his single most important tool—the anvil. Typically large and massive, anvils were difficult to transport in those days, especially to such remote locations.

No one knows who discovered the great Tucson iron. Perhaps it was a military scouting party, or settlers out hunting. Whoever it was, they were making their way through the canyons of the Santa Rita Mountains about 30 miles southeast of Tucson when they came across "enormous masses of virgin iron many of which have rolled to the foot of the said range." This earliest known account of the discovery was written about 1850 by a Sonoran government official, José Francisco Velasco, but the irons surely had been known for more than a century by then. The account is not specific about the location of the irons, other than to say they were in a canyon called Puerto de los Muchachos in the Sierra de la Madera (Santa Rita Mountains), near the present town of Tubac. More intriguing is the implication that there were several masses. These people had apparently discovered a field of iron meteorites. Velasco's account goes on to say that a medium-sized iron was taken to Tucson, "where it has remained in the plaza of said presidio."

And for what purpose was this "medium-sized iron" dragged to Tucson? To serve as an anvil, a priceless possession to the frontier blacksmith. This anvil is like no other. It is an oblong mass with a giant off-center hole roughly 2 feet across, making the mass a huge ring. The thinnest side is only 2.66 inches thick, and the opposite side measures 17.5 inches. The longest exterior dimension of the ring is about 4 feet. It weighs 1,513 pounds. Although the Tucson Ring meteorite weighs less than the other "great" meteorites, its ring shape is unique among iron meteorites of the world.

The meteorite was buried narrow-end down and a foot or more deep in the ground. This positioned a natural flat area on the thick side nearly horizontal to the ground, allowing it to be used as an anvil. It remained in the presidio for many years, serving the Mexican military until the Gadsden Purchase of 1853.

Artist's rendition of the Tucson Ring meteorite being used as an anvil in the Mexican garrison at Tucson before 1856.

The Carleton meteorite, with members of the California State Geological Survey, photographed shortly after it arrived in San Francisco in November 1862. –Arizona Historical Society

The Carleton Meteorite

The Tucson Ring was not the only meteorite from the Santa Rita Mountains to be used as an anvil. There was a second mass, a flat, kidney-shaped meteorite weighing 633 pounds. Another blacksmith, Ramon Pacheco, used it as an anvil in his front yard until 1862.

The Mexican military left Tucson by 1856. The Tucson Ring meteorite, which had been used in the Mexican garrison, was abandoned. For six years it was forgotten, slowly rusting within the crumbling walls of the presidio.

In February 1862, Confederate troops occupied Tucson, but not for long. On April 15, Union troops under General James H. Carleton fought and won a brief battle at Picacho Peak, 40 miles west of the town. On May 20, General Carleton arrived in Tucson, proclaimed himself military governor, and declared martial law. He had heard of the Tucson Ring meteorite's being used as an anvil. By that time the Ring had already been removed, and Carleton's men confused Pacheco's meteorite with it. Without making any civil arrangements with Pacheco, who had personally hauled the heavy iron from the Santa Ritas to Tucson in 1849, Carleton confiscated it and sent it to California by wagon train. He wanted to give the meteorite to the city of San Francisco as a memorial to his march into Arizona. When it arrived in San Francisco, it was christened the Carleton meteorite. For a time it occupied the mayor's office but eventually found a home at the museum of the Society of California Pioneers. There it remained on display until interest waned and it was relegated to a basement storeroom in the museum for the next three-quarters of a century. In 1939 it was "rediscovered" by Dr. E. P. Henderson, who arranged for its purchase by the Smithsonian.

The Carleton and Ring Cross Paths

While the Carleton was still in Pacheco's possession, the Tucson Ring meteorite had met with a different fate. It had been abandoned in the old presidio in Tucson when the Mexican military turned Tucson over to the United States Army in 1856. In November 1857, a young army surgeon, Lieutenant Bernard John Dowling Irwin, arrived in southern Arizona Territory to join the Seventh Infantry at Fort Buchanan, 40 miles southeast of Tucson. Irwin was something of a naturalist and collected specimens for the Smithsonian Institution during his tenure in Arizona. Sometime in the spring of 1860, at the suggestion of his Smithsonian correspondents, Irwin located the Ring where it had been abandoned, "lying in one of the bystreets, half buried in the earth, having evidently been there a considerable time." Irwin had the meteorite removed and prepared it for shipment to the Smithsonian.

Irwin hired Augustin Ainsa, a Tucson freight broker, to take the meteorite to Guaymas, Sonora, on the Pacific Coast. From there it was to be shipped by steamer to San Francisco. The meteorite left Tucson in February 1861. But Ainsa had relatives living in El Alamito, Sonora, and he had never intended to go straight to Guaymas. Instead, his journey ended in El Alamito.

During the next two years, the Ainsa family "rewrote" the history of the Ring's discovery and recovery, mentioning the Ainsas but omitting Irwin. Their story was totally contrived. Augustin's brother Santiago finally set out for Guaymas with the meteorite. There he boarded a steamer for San Francisco and arrived in June 1863.

The Ring was displayed on the steps of the Custom House for several weeks. Only then did scientists in San Francisco and at the Smithsonian finally realize there were actually two meteorites from Tucson. This explained discrepancies of size, shape, and weight in the reports by Carleton, Irwin, and others. The Carleton and the Ring had finally crossed paths, but only briefly. They were destined to be separated for another seventy-six years.

While the meteorite remained on display in San Francisco, Santiago notified the Smithsonian that the meteorite would soon be on its way to Washington. Smithsonian scientists were delighted, and having no reason to question Ainsa's version of the story, they named the great iron the Ainsa meteorite. This was later confirmed by Joseph Henry, secretary of the Smithsonian, in the 1893 annual report:

> As the [meteorite] was first brought from the mountains north of Tucson by the great-grandfather of the gentleman to whose exertions in transporting it to Washington the Institution owes so much, it is proposed to call it the "Ainsa Meteorite."

When Irwin heard about this dedication, he wrote a scathing letter to Joseph Henry in which he "corrected" the Ainsa historical notes and insisted Ainsa's only role was to transport the meteorite, for which he was paid $50. Irwin explained that he had personally retrieved the meteorite from the presidio and sent it via Ainsa to the Smithsonian as a gift to the nation. "As the officers of the Institution have failed to recognize my services by bestowing my name upon it, it should be known as it was until I brought it within the reach of the scientific world–i.e. the Tucson or Arizona Meteorite." Irwin further concluded, "The fact that Mr. Ainsa had it carried for me a certain distance is no valid reason why it should bear his name."

Finally recognizing their embarrassing error, the Smithsonian changed the name to the Irwin-Ainsa meteorite, hoping this action would appease both parties. It didn't. Irwin quickly wrote a short booklet entitled "History of the Great Tucson Meteorite donated by B. J. D. Irwin, Surgeon U.S.A." In this booklet he published his account of its history, Ainsa's claims, and all the correspondence with the Smithsonian. He distributed it to scientists and politicians in Washington and elsewhere, and sent a copy to Joseph Henry, again objecting to the name. But the name remained unchanged for another decade. Finally, after repeated correspondence between Irwin and Henry, the Smithsonian relented and removed both names, leaving only the name Tucson meteorite, which it bears to this day.

Nearly all surviving references to the Tucson meteorite agree that somewhere in the Santa Ritas is a field of great irons. As at Port Orford, people have combed the mountains in hope of rediscovering the meteorite field that most certainly must exist. The saga of the Tucson meteorites continues.

A Personal Note

In 1975 I had the privilege of negotiating with the Smithsonian Institution for the temporary return to Arizona of the Tucson meteorite in commemoration of the opening of the Grace H. Flandrau Planetarium on the University of Arizona campus. As its founding director, I thought it would be an appropriate gesture, since Tucson had become a world center for research in astronomy, and meteorite research was a high priority at the university. Dr. Roy S. Clarke Jr., curator of the Division of Meteorites at the Smithsonian, agreed to the loan, and preparations were made for the great Tucson Ring meteorite to return home.

The planetarium is only a few miles from the spot where the Ring had originally been used as an anvil. During its six-month stay at the planetarium, I had four full-scale fiberglass-and-resin replicas of it made. One is on display at the Flandrau, one is at the Griffith Observatory in Los Angeles, one is at the planetarium of Mount San Antonio College in Walnut, California, and one is at the Center for Meteorite Studies at Arizona State University in Tempe, Arizona.

The World Abounds with "Great" Meteorites

Not all the great irons are in museums; the world's largest iron remains where it was found in Namibia, Africa.

China. China claims what may be the third-largest iron meteorite in the world. It fell in the Gobi Desert in prehistoric times and has been known since the nineteenth century as the "silver camel" by the local Kazakh people. In July 1965, the great iron was transported more than 300 miles from Chingho County to Urumshi, the capital of the Sinkiang Uighur Autonomous Region in northwest China. The Gobi meteorite's estimated weight is 33 tons, only a ton less than the Ahnighito meteorite.

Argentina. As early as 1576, tons of iron meteorites were known to exist in Chaco Province, Argentina. These are the Campo del Cielo meteorites. A 15-ton mass was recovered in 1813 and found its way to the British Museum. An 18-ton mass was found in a shallow crater as late as 1970. Other great masses exist, including one that was nearly transported to the United States in 1990; it might weigh as much as 37 tons and is still in Chaco Province.

Mexico. Mexico is home to one of the great iron meteorites. This mass, found in Sinaloa, Mexico, in 1863, measures 12 feet in its longest dimension and is estimated to weigh 22 tons. This meteorite, the Bacubirito iron, remained where it fell on a farm 7 miles south of the town of Bacubirito until 1959, when it was finally moved to the Centro Civico Constitution in Culiacan, Sinaloa, where it can be viewed today.

Australia. In recent years, Australia has proved to be a rich hunting ground for meteorites, especially in the Nullarbor Plains, Western Australia. This nearly featureless desert environment preserves meteorites for centuries. In 1967, geologists surveying the Eucla Basin near the town of Mundrabilla discovered two large irons, weighing roughly 8 and 11 tons. The larger still holds the record for Australia. Since the initial discovery, hundreds of smaller irons have been found in the Mundrabilla strewn field, and more are surely waiting to be discovered.

Harvey Harlow Nininger (1887–1986).
—Courtesy of Margaret Huss, American Meteorite
Laboratory

FIFTEEN

Harvey Harlow Nininger, Meteorite Hunter

Do with your life something that has never been done, but which you feel needs doing.

–H. H. Nininger

Dr. Harvey Harlow Nininger began his extraordinary career in a very ordinary way. In the early 1920s, he was a biology instructor at a small college in McPherson, Kansas. Biology was his profession but meteorites became his passion. In less than a decade he would wean himself from the security of college life for the dubious new "profession" of meteorite hunter.

Nininger found his life's work on the night of November 9, 1923. Witnessing a great fireball that lighted the landscape and left a trail of smoke in the sky, he immediately dropped to his knees and plotted the path of the meteor on the walkway. Scientific training guided his critical eye and analytical mind. He quickly concluded that the meteoroid had reached Earth, and he was determined to find it. Nininger knew he could not accomplish this alone–he needed to find other observers to help him. He immediately contacted newspapers around the state, requesting reports from other observers, and although within days his mailbox was overflowing, most of the reports were exaggerated or erroneous. Most people insisted the meteorite had landed within walking distance of their location. Untrained observers and a novice meteorite hunter did not make a good team. Relying on the reports of startled laymen was unsatisfactory and Nininger had much to learn.

The meteorite appeared to have landed near the town of Coldwater, Kansas, so Nininger talked to school and church groups in the area, offering a reward of one dollar a pound for any recovered meteorites. (This remained the standard price for meteorites for many years.) To his surprise, meteorites began to appear, but not from the fall he was pursuing. Rocks were brought to him that had been collected on farms during plowing and saved because they seemed different from other nearby rocks. Many old rocks that had been used to prop open a door or hold down the lid to an old well turned out to be meteorites. Nininger's lectures had rekindled memories and sparked a new interest among local residents. Though he never did succeed in finding the Coldwater meteorite, he did get two unknown specimens that were later named Coldwater. Two small meteorites from known falls were also recovered. He was encouraged by his early success.

More important than the recovered meteorites, Nininger saw what a bit of education, personal contact, and a modest reward could produce. He was developing methods he would use throughout his life, and his approach was unique. No one had ever tried to establish a search program by educating local populations. The plains states were perfect for it because, in farm country, fields were plowed yearly. The fields had long ago been cleared of rocks, some of which were meteorites. If new rocks appeared, there was a good chance they had fallen there as new meteorites.

Nininger's success at Coldwater led him to believe meteorites were much more plentiful than he had thought from his readings, and he reasoned that a diligent search over a large enough area should result in more finds. He desperately wanted to test this idea by establishing a field program for meteorite recovery, but the demands of family life and teaching prevented him from acting immediately. What he needed was financial support from a museum to free him from his teaching position. After he recovered his first meteorite, he had corresponded with Dr. George P. Merrill, curator of geology and meteorites at the Smithsonian Institution. Nininger presented his idea about a field program to Merrill and asked for the Institution's support. Merrill did not believe meteorites were as plentiful as Nininger suggested, but neither did he understand the determination that drove the man who had come to him for support. Merrill's response reflected this attitude: "Young man, if we gave you all the money your program required and you spent the rest of your life doing what you propose, you might find one meteorite." Merrill would live to see how very shortsighted that prediction was.

First Successes

Between 1923 and 1929, Nininger conducted his meteorite search-program while teaching at McPherson College. Every weekend and holiday he was on the road, investigating fireball reports and visiting farmers within a few hundred miles of home. He would ask to examine the rock piles that had accumulated on every farm. Occasionally, he discovered an old meteorite. His personal contact with farmers often netted meteorites they had plowed up and pitched against a fence, where they remained until Nininger retrieved them years later. During the summer months he extended his range to include all the plains states. He gave hundreds of lectures at schools and community groups, supplied meteorite articles to dozens of newspaper editors, and distributed thousands of leaflets.

Gradually, as they learned about meteorites, people began sending him samples of rocks they had found. Most were just that—rocks—but one in a thousand was the real thing. The people he talked to had long memories. Years after he had delivered a lecture in a small town, he would receive a request from someone asking him to look at an unusual stone. Sometimes this unusual stone proved to be a meteorite.

Going It Alone

Nininger tried and failed in his quest for funding, but his search for meteorites was proving successful. He wanted to devote his full attention to this work without the burden of teaching. His only hope was to make the sale of meteorites self-supporting. He perused rock and mineral catalogs, noting retail prices of the few available meteorites. Even in the 1920s prices were high, considerably beyond the dollar per pound he had been paying. There was a profit to be made. Perhaps science supply houses like Ward's Natural Science Establishment in Rochester, New York, would purchase meteorites from him, but first he had to build an inventory of enough specimens to guarantee a constant supply; only then could he finance his program and pay himself a salary.

His plan was to go where there were plenty of meteorites, and Mexico was the place. Many iron meteorites were found in the late eighteenth century in the state of Mexico, near the small town of Xiquipilco, near Toluca. Local people had known about these irons for centuries, and many generations of Mexicans had fashioned tools from the metal. Before the beginning of the twentieth century, both Ward and A. E. Foote had collected tons of Xiquipilco (now called Toluca) iron, but little was available by the 1920s. This seemed to be the ideal meteorite for his inventory—plentiful but not commonly available.

Nininger traveled to Xiquipilco, where he spoke with the town mayor through an interpreter. This gringo wanted iron rocks? The mayor disappeared, leaving Nininger to wonder if he had failed to accurately communicate his request. In minutes the mayor returned with a 20-pound meteorite specimen. After assuring himself this really was what Nininger wanted, he told him he could find many more specimens. By afternoon, people had heard that some crazy American was offering pesos for rusty iron rocks. They gathered in the town square, bringing as much of the iron as they could carry. Nininger purchased over 700 pounds, as much as his meager budget allowed. His hunch had been correct: The Toluca field was still strewn with meteorites. This iron would be his trading material, helping to launch his shaky new career.

The Toluca meteorites, along with many others he had collected, provided a substantial start toward independence, but it was not quite enough. His start-up fund required at least one major sale. This money rained from the sky on February 17, 1930, over Paragould, Arkansas. At 4:00 A.M. a huge fireball was seen over several states, followed by several explosions at the end of its trajectory that rattled windows and awakened people below. Within hours an 85-pound, broken fragment was found, nestled in a small crater in a farmer's field. Eyewitness accounts led Nininger to suspect that a larger specimen lay a few miles beyond the first. He plotted a likely direction and left it to the resident farmers to search their fields. They did, and they found it. The meteorite was more than Nininger could have imagined. It was an 820-pound stony meteorite, the largest ever from a witnessed fall. Nininger paid the finder $3,600 for the Paragould meteorite. At this price, an enormous amount in 1930 dollars, he could not afford to keep it. He resold it for $6,200. Profits from the Paragould meteorite gave him confidence. Now, with $2,600 profit in the bank and an ample supply of meteorites, he was ready to make his move. He resigned his teaching position, moved his family and their few possessions to Denver, Colorado, and naïvely but courageously became the world's first full-time, self-employed

Nininger cutting a large meteorite with a band saw.
−Center for Meteorite Studies, Arizona State University

meteoriticist. Meteorites were no longer just his passion, they were now his sole means of existence, and this at the beginning of the Great Depression.

There Is No Science of Meteorites

While traveling in search of meteorites, Nininger made a point of visiting geology museums along the way. His collecting excursions were always educational. He had thought that geology museums should be excellent places to learn about new meteorites. They weren't. He found that, almost without exception, museum directors had little interest in meteorites, and their knowledge in this area was wanting. Their small collections were hidden in storerooms or, if they were displayed, they were often mislabeled or unidentified. Nininger frequently offered to put their collections in order, correcting labels and descriptions, and often arranged for a trade.

College and university science curricula were just as neglectful. Nowhere did meteorites occupy even a single lecture in a geology course, and they were mentioned only briefly in astronomy survey courses. (The academic situation is not much better today.) A complete course of study in meteoritics was a fantasy. There was no research program in meteoritics in the United States in the early decades of the twentieth century. The mineralogy of meteorites had been investigated in the late nineteenth century and it seemed there was little to gain from any further study. There were more pressing and interesting problems to occupy a geologist's time.

Nininger felt more alone than ever in those early days. He found himself in an intellectual vacuum. He knew there was public interest, but he could not depend on the nation's colleges and universities to educate students in the fledgling science. For the present, he would have to provide that education himself.

A Society for Research on Meteorites

During the Depression years, there were a few other people in the United States who shared Nininger's enthusiasm. George Merrill at the Smithsonian and Oliver Farrington at the Chicago Field Museum were two. On the West Coast, Frederick Leonard at UCLA was one of the first to teach a formal course in meteoritics. Leonard was enthusiastic about organizing a meteoritical society. He and Nininger worked to bring together a small group of interested people, from collectors to museum curators. In August 1933 an organizational meeting was held at the Field Museum in Chicago, establishing the Society for Research on Meteorites. Leonard was its first president, and Nininger was secretary-treasurer. Only about a dozen people attended this first meeting but membership grew steadily. Within a decade, the society had become international, with a membership of over two hundred. After World War II, the society shortened its name to the Meteoritical Society and by 1953 was publishing a journal, *Meteoritics*.

Tracking Down Meteorites

Nininger was the first person to locate a meteorite from fireball reports. After his first unsuccessful attempt in 1923, he recognized the limitations of depending on eyewitness reports. "We learned to approximate all our judgments to take into account our dependence on the memories of untrained witnesses, unprepared for the phenomenon they viewed and unprepared to report it."

He looked for certain key observations that established the distance of the fireball from the witness. If the light from the fireball extended to the horizon or was seen behind a building or nearby hill, it was probably still hundreds of miles away, though the witness may have been sure it had fallen very close. If the light was extinguished before it reached the horizon, the meteorite was fairly close, perhaps 20 or 30 miles from the witness. If the witness heard a sound, the falling stone was even closer.

He learned to judge the direction of a fireball only after finding witnesses who had seen it from three sides, so he could triangulate the approximate direction.

The Pasamonte Fall

Nininger used these rules in successfully tracking the Pasamonte, New Mexico, fireball of March 24, 1933. This great meteor produced a fireball several cubic miles in volume and left an enormous smoke train estimated to be a mile in diameter. Shock waves from the descending fireball shook the ground. Nininger interviewed dozens of witnesses from Kansas to New Mexico, all of whom were certain it had landed "just over the hill." After nine months of investigation, he finally pinpointed the path of fall, feeling certain the meteorite had landed somewhere on the Pasamonte Ranch in Union County, New Mexico. When he visited the site, the rancher was surprisingly positive. Yes, he had seen the fireball. Yes, it had fallen just 2 miles from the ranch, and yes, he had some black stones picked up by a ranch hand. But the rancher had taken a hammer to them, thinking they contained something valuable, and they lay in many broken pieces. The meteorites were small, most not over an ounce or so in weight. About one hundred were found, scattered along a 25-mile path.

From the spectacular fireball and smoke train, Nininger had expected a sizable meteorite. The whole lot weighed under 8 pounds. Apparently, the initial mass of a meteorite could not be judged from the meteor display. The Pasamonte meteorites were eucrites, a rare form of calcium-rich achondrite, making the search well worth the effort. The lesson of Pasamonte was clear: Most of the mass of a falling meteorite was deposited in the smoke train. Perhaps only 10 percent made it to the ground. Fragile meteorites like the Pasamonte stones, which tend to crumble when handled, suffer even more.

"There's More Where Those Came From!"

Nininger learned from his experience with the Toluca irons that old strewn fields usually had not been thoroughly searched. He proved it again near Plainview, Texas. No one witnessed the fall of a stony meteorite there, but a young man from the rural area found a stone in 1917 that he suspected was a meteorite. He sent it to the Smithsonian Institution for verification. Dr. Merrill confirmed the young man's suspicions and asked him to see if he could locate others. Within two years, he had found more than a dozen stones. Apparently, Plainview was a multiple fall. This should have rung some bells at the Smithsonian, but Merrill was not inclined to investigate further.

In 1928, Nininger asked Merrill about the chances of finding more Plainview stones. Merrill was not encouraging, believing the area had been thoroughly searched. Nininger was not convinced. He put it on his "to investigate in the future" list. One day in December 1933, returning from an unsuccessful meteorite trip to Mexico, he decided it was time for a Plainview search. It was early evening when he put his well-practiced meteorite-recovery technique to work: hunting from farmhouse to farmhouse, making personal contacts with families, and stimulating their interest by offering a dollar a pound. Before the end of the evening, he possessed an 8-pound Plainview meteorite that for years had been used to hold down the roof of a chicken coop.

The next day's hunt netted him 26 additional specimens, totaling 152 pounds. They came from everywhere—retrieved from behind barns, brought up from storm cellars, used as weights to hold down all sorts of farm implements and prop doors and windows open during hot summer nights. Nininger's appearance in this rural community only a few days before Christmas during the depths of the Depression must have seemed like a miracle. Here was a stranger going door to door buying rocks. He was either a crazy man or an angel, sent from heaven at a time of great need; the farmers preferred to believe the latter.

Nininger visited Plainview often over the next fifteen years, collecting more than 900 meteorites. Among these were three that were different from the others. They represented different falls within the Plainview strewn field.

The Bonanza Years

Between 1933 and 1939, Nininger demonstrated his incredible skill as a meteorite hunter. At no other time in the history of meteoritics were so many meteorites recovered in such a short time. With the help of his wife, Addie, their son, Bob, and several friends, he found more than one hundred meteorites from different falls, almost all in the Midwest. The reasons for their success were clear. Meteorites were far more plentiful than anyone had ever suspected. Nininger was convinced that meteorites fell more than just occasionally and that they were distributed more or less uniformly on Earth. Their apparent rarity was due to a general lack of knowledge.

At the Smithsonian, there is a map of the United States showing locations of most meteorite falls and finds. There is an obvious concentration in Kansas, Nebraska, and Texas. Nearly all were found by the Nininger group during his bonanza years. Nininger attributed this skewed distribution to the "interest factor"—he had personally introduced people in these rural communities to meteorites, and with this new awareness they were on the lookout for meteorites in their fields.

During these years many meteorites now commonly displayed in museums worldwide were discovered. The list reads like a *Who's Who* of meteorites, and the following are only a few: Plainview, Texas (1933); Morland, Kansas, 632 pounds (1935); Hugoton, Kansas, 749 pounds (1935); Potter, Nebraska, 600 pounds (1937); Johnstown, Colorado, (1939); Grady, New Mexico, 30 pounds (1933); Beardsley, Kansas (1931–1933); Melrose, Texas (1933).

This prolific period was capped by the recovery of the famous Goose Lake iron meteorite in 1939. In October 1938, three deer hunters found an iron meteorite resting on top of an old lava flow. The flow covered a mesa overlooking the western shore of Goose Lake in northeast California near the California-Oregon border. With considerable difficulty, they removed a small sample, which eventually reached meteoriticist Frederick Leonard at UCLA. By this time, Nininger had received a letter from Dr. Earl Linsley, director of the Chabot Observa-

Clarence A. Schmidt, one of the discoverers of the Goose Lake meteorite, and H. H. Nininger (right) pose next to the meteorite still resting where it was found on top of a lava flow near Goose Lake, California.
—Courtesy of Clarence A. Schmidt

tory in Oakland, about a suspected iron meteorite newly found in Modoc County, California. Nininger visited Leonard in Los Angeles the following April and confirmed that Leonard's specimen was meteoritic.

Nininger needed no further evidence. There was an iron meteorite weighing over a ton near Goose Lake and he was determined to recover it. Through Linsley, he contacted Clarence Schmidt, one of the hunters, who agreed to lead Nininger to the meteorite.

After only a day of searching, they located the iron exactly as Schmidt had described it. Nininger contacted Leonard and Linsley, and recovery operations got under way. The complete story is told in Nininger's autobiography, in which he captures the human drama of this unique discovery. Especially amusing was Leonard's reaction when he first saw the iron. Leonard always presented a dignified appearance, even on field trips. He was never without his double-breasted suit and tie, and he was no different that day, in the backcountry of Modoc County:

> We reached the mesa top and then proceeded across the two miles or so of boulder-strewn, muddy mesa dodging ponds and fording small streams of spring run-off, until we approached the meteorite where it lay on reasonably dry, boulder-paved terrain.
>
> Frederick Leonard had shed most of his academic dignity by this time. When we came within sight of the big iron the pudgy little professor ran on ahead, placing his hands lovingly on the great meteorite, bent and kissed it. Then he lifted his hands skyward and turned to face us.
>
> "This is the greatest day in meteoric astronomy!"

After a brief sojourn at Treasure Island, where it was displayed, the meteorite was sent to the Smithsonian Institution. It had been found on federal land and belonged to the United States government.

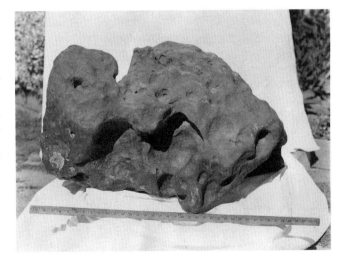

The Goose Lake meteorite at the Chabot Observatory in Oakland, California, where it was taken shortly after being recovered near Goose Lake in the spring of 1939.
—Courtesy of Clarence A. Schmidt

Where Are Earth's Meteorite Craters?

Nininger's phenomenal success recovering meteorites raised some interesting questions. Meteorite falls were apparently much more common than previously thought. This had to also mean that much larger bodies must be striking Earth on a more or less regular basis, and those bodies must be producing sizable craters. The apparent lack of meteorite craters puzzled Nininger.

The Haviland Crater. Ranchers and farmers on the Kansas plains were aware of the heavy, black rocks scattered throughout Brenham Township in Kiowa County. It was excellent farm country, partly because there were so few rocks to clear from the fields. But there were some black rocks, mostly buried, that farmers occasionally struck when plowing. In the late 1890s, Frank and Mary Kimberly had homesteaded their land. They too noticed the black rocks. Frank saw them as a nuisance, since he frequently hit them with his plow and had to remove them. Mary found them intriguing. It seemed to her they didn't belong there. They had a curious composition—iron with yellow, glassy crystals. They reminded her of a large meteorite she had seen when she was a schoolgirl. She began collecting them as quickly as Frank pulled them out of the ground. Soon she had well over a ton of "iron" rocks in a pile outside the barn. She was so obsessed with the rocks that her husband and neighbors began to ridicule her. After several years of collecting, she finally persuaded a geologist from Washburn College to look at them. Mary had the last laugh: The geologist identified them as stony-iron meteorites and purchased several hundred pounds on the spot. The meteorites were called Brenham pallasites, named after the township. The Kimberly farm that "produced" them was known as the "Kansas Meteorite Farm." With the proceeds of the sale of the first meteorites, the Kimberlys purchased an adjacent farm. This farm, too, was covered with pallasites. As it turned out, the Kimberly farm near the tiny town of Haviland was the site of the largest pallasite find in the United States and the strewn field lay squarely across their land.

Nininger became close friends with the Kimberlys when he conducted his meteorite campaign in rural Kansas in the mid-1920s. During a social visit in 1929, Frank Kimberly just happened to mention a large meteorite he had found in a shallow depression he called a buffalo wallow, near the farmhouse. The depression was an elongated hole 36 feet by 55 feet, with a depth of about 11 feet, and it had a rim projecting above the level of the plowed field. Nininger was intrigued. Frank took him out to the wallow, and Nininger dug a trench along the rim. Within minutes he found rusty-looking nodules containing olivine crystals. He interpreted them as completely oxidized meteorite fragments. He was certain the buffalo wallow was a meteorite crater.

Four years later he returned to the Kimberly farm to search for further evidence of a crater. He excavated the depression, uncovering hundreds of rusty nodules. Large nodules weighing up to 85 pounds were found within the crater. It probably was once a single mass weighing nearly a ton. The Haviland Crater did not have the impressive dimensions of Arizona's Meteor Crater, but it was the first in which meteorites were found.

By 1933, only five craters were considered probable meteorite craters. Even after Nininger published his results, the Haviland Crater was not immediately added to the list. There was still confusion about just what makes a crater meteoritic. Oddly, the existence of iron meteorites around a crater was not considered proof of an impact origin. Meteoritic material was found around all five craters but, unlike the Haviland Crater, no meteorites were found in them. Also, all the suspected meteorite craters were round, while the Haviland Crater was oblong. This eventually led to the realization that there were two types of meteorite craters, differentiated by the energy of impact. Large bodies with great mass and high impact velocity produced craters like those in Arizona and Odessa, Texas. Their energy was so great that an explosion occurred on impact, and most of the terrestrial rock and meteoritic material was either vaporized or excavated away from the crater. A crater formed by explosion rather than mechanical excavation is always round, regardless of the angle at which the meteorite strikes Earth. The Haviland Crater was also meteoritic, but it was formed by a much smaller body. The body lacked sufficient energy to produce an explosion crater and excavated only a small crater by percussion alone. These are called percussion, or penetration, pits. Such craters are usually not round and meteoritic material is frequently found in them. Often in multiple falls, both types of craters are produced. The Henbury Craters in Australia and Sikhote-Alin Craters in Siberia are examples of both types.

Odessa Crater. Encouraged by his work at Haviland, Nininger decided to excavate the Odessa Crater. In 1932, he fashioned an electromagnetic device similar to a gardener's rake that he used to comb the soil around the crater. With this contraption, he successfully recovered about 1,500 small irons. This showed that iron meteorites were distributed at shallow depths around the crater, much as they were around Meteor Crater.

Nininger was after the large masses he believed lay somewhere within the crater. Three years later he took on the crater again, this time with a new device called a magnetic balance, which he borrowed from its inventor, G. L. Barnett of Oklahoma. It was the ancestor of the modern metal detector. With this cumbersome device, he and his son found larger meteorites below the surface. He searched inside the crater but found nothing. He assumed the detector was simply not sensitive enough to pick up large masses, which must be buried too deep to detect.

Nininger's work at Odessa did not prove the crater was meteoritic, but it made history. For the first time, a metal detector was used to recover iron meteorites. He left the crater still convinced that large masses would be found. A year later, his faith in this idea was badly shaken. The explosive nature of meteorite craters was brought to his attention in 1936 by the mathematician F. R. Moulton. Nininger, like many other scientists at the time, believed meteorite craters should contain large masses. In 1929, D. M. Barringer had announced that his drills had struck a large mass beneath the floor of Meteor Crater. Nininger was reluctant to give up the idea, and his meeting with the great mathematician only sharpened his resolve to search for evidence that would answer the question one way or the other.

Meteor Crater. If Moulton was right, the evidence must be in Meteor Crater, where material had been explosively ejected. In 1939, with a magnetic rake attached to the rear of his car, Nininger made a survey of meteoritic material within a 2.5-mile radius of the crater's center. The magnetic rake picked up over 12,000 small iron fragments. When he plotted the distribution of these fragments, he saw that they tended to form a radial pattern centered on the crater, suggesting that the fragments had been thrown out of the crater by an explosion.

The War Years

World War II interrupted Nininger's work on the crater. For the next five years he worked for the war effort, first as a salvage investigator, searching for scrap iron in mining towns, and later as an oil-company investigator, negotiating land leases in New Mexico. These temporary jobs suited Nininger well, since they guaranteed him a weekly paycheck and required extensive travel throughout the Southwest, giving him an opportunity to search for meteorites.

His research work on meteorites continued but on a smaller scale. He longed to return to Meteor Crater to continue his surveys, which had barely begun before the war intervened. As the war came to a close, the Niningers had to make important decisions. The brief respite from self-employment was at an end. Nininger was approaching the age of sixty, and he knew he soon must find a more stable means of support and security. For over two decades he had financed his research program through the occasional sale of meteorites. His family constantly lived with the fear of financial collapse. He tried often to find financial support from government and private agencies, but was seldom successful. When any support was found it was only a few hundred dollars and usually tied to a stipulation that the funds be matched by another source. However, he had managed to accumulate one of the largest private collections of meteorites in the world. He wanted to place this great collection in an institution dedicated to research on meteorites, where he could spend his declining years doing what he loved best.

Ruins of Nininger's meteorite museum near Meteor Crater.

But that was not to be. Once again, the Niningers had to depend totally on their own resources. What they lacked in support funds, they more than made up for in lofty goals. They were doggedly determined to create their own institute.

The American Meteorite Museum

Forty miles east of Flagstaff along the now-abandoned Highway 66 stands a crumbling stone tower, a lonely sentinel on the plain overlooking Meteor Crater. I stood there on a cool afternoon in early spring forty years after the building had been abandoned. The wind whistled through openings where once there had been doors and windows. Here and there were fragments of oxidized meteorites, a rusting tool, a few pieces of broken pottery. The walls had been patched by the occupants many times as they fought a losing battle against the severe elements of the cold, high, northern Arizona desert. Graffiti adorned the walls. No longer could you ascend a staircase to the tower, to view the crater

6 miles to the south. Only ghosts could do that now. The museum had held such promise.

"Our institution would be a Meteorite Museum; nothing else, just meteorites; no snakes, no skeletons, no stuffed birds, no wild animals, and no curios—just meteorites." Nininger's emphatic statement was a reference to the many curio shops and "museums" scattered along Highway 66. His would be the world's first museum dedicated solely to meteorites. It would feature one of the world's great collections and would be on a "high educational plane." It would be called the American Meteorite Museum. These were the lofty principles upon which Nininger's museum was founded. He never varied from those principles.

Nininger's museum was to be located near Meteor Crater, Arizona. It was not a very practical choice. In 1946 the Niningers moved to Flagstaff and waited for an opportunity. Soon a building on Highway 66 just opposite the crater became available. It was situated on a low bluff of brick-red stone. The building was made of the same stone, a mudstone from the Moenkopi formation of early Triassic age. It lacked some amenities of modern civilization, like electricity, but its location seemed nearly perfect for a meteorite museum.

Nininger was so taken by its proximity to the crater that he did not pause to consider the business problems it would create. They were located on a lonely stretch of road 20 miles west of Winslow and 40 miles east of Flagstaff, both towns with small populations. The museum would have to depend entirely on the tourist trade, which meant the summertime traffic. There was plenty to attract tourists along Highway 66. Curio shops lined the highway. Other shops offered cold drinks, Indian trinkets, and collections of reptiles. Then there was the "Mystery Thing of the Desert" that everybody simply had to see. Signs advertised the "Thing" for 50 miles or more on either side of the new museum. There were no highway signs advertising the meteorite museum. Nininger hated those despicable signs. Would people even know what a meteorite museum was? They had to drive by on their way to visit the crater; the Niningers hoped people would make the connection.

They leased the property and returned to Denver to take the collection to its new home. They knew that a new highway, part of the interstate system, was already under construction, but no one could say exactly where it would go or when it would reach the area. Surely not for years, they hoped. They could not imagine that in less than three years it would open, skirting the old highway by a half mile and leaving the world's only meteorite museum marooned on the hostile, northern Arizona desert.

The new museum opened on October 9, 1946. That night, as if to celebrate the event, the sky was filled with meteors from the Draconid meteor shower.

The Commercial Enterprise

Nininger's primary reason for opening his museum near the crater was to enable him to continue his research there. But on the practical side, he had to make a living and he needed funds to finance his work. The crater was prominently featured at the American Meteorite Museum in displays describing the research being conducted there.

The museum housed more than five thousand meteorite specimens from 526 falls. They were available to anyone who wished to study them. Daily lectures were presented in an ongoing program of education. In essence, the museum accepted the role of educational institution, informing the public about meteorites and their importance in Earth's geology. They were filling a void that had been ignored by America's schools and universities. Frequently, the museum was visited by scientists from around the world. They were both delighted by what they saw and amazed by Nininger's tenacity. They were astonished to discover that no tax base or granting agency supported this museum and its research laboratory.

The museum charged an admission fee of twenty-five cents, which met with mixed reactions from the public, as many people had the odd notion that, since it was an educational museum, it should be free.

Nininger's meteorite museum in its heyday, around 1948.
−Courtesy of Margaret Huss, American Meteorite Laboratory

Once inside, tourists could purchase meteorite specimens and books. In the first year of operation the Niningers began to create jewelry with small, mounted meteorites from Meteor Crater. Oxidized fragments of meteorites were also used. The rarity of meteorites generally precluded them from use in jewelry, but thousands of small irons had been recovered during Nininger's survey and it did not seem unreasonable to mount a few of them tastefully. As jewelry, the meteorites were probably handled with greater care and given more attention than if they had been sold as the raw, rusty pieces of metal they were when found around the crater.

Nininger was constantly criticized in academic circles, especially by some members of the Meteoritical Society, for "abject commercialism"—selling meteorites was considered a form of prostitution. Anyone, especially a scientist, who sold or even owned such precious items of scientific interest was committing scientific blasphemy. "Legitimate" scientists simply did not become involved in such ventures, and academics in their tenure-secured ivory towers could not allow themselves to be soiled by association with commercialism. This arrogant attitude was a major reason why Nininger seldom received financial support from educational institutions and granting agencies; unfortunately, it still prevails today.

The paleontologist does not dare own a fossil, nor should a meteoriticist covet a personal meteorite specimen. Yet the Niningers sold thousands of beautifully prepared meteorites from hundreds of falls to museums and educational and scientific institutions worldwide, and scientists were more than happy to purchase them with tax-based or granting-agency funds. It was far easier and less expensive to buy them from that renegade than to go out and search for them. The science of meteoritics could progress only if new specimens were found and disseminated. Almost every museum and institution collecting and studying meteorites has representative samples of Nininger's materials.

Nininger was fond of pointing out that past mineral dealers, such as Dr. Henry Ward of Ward's Natural Science Establishment of Rochester, New York, and A. E. Foote of Foote Minerals of Philadelphia, were primarily concerned with the commercial aspects of meteorite collecting. He considered his commercial venture a necessary evil, allowing him to continue his research. Unfortunately, his research was never really taken seriously by scientists, leading to denials of support from granting agencies.

With Nininger it was a simple matter. He had a living to make and a family to support. He correctly saw little difference between a tax-based paycheck in return for teaching and research services and receiving payment for the services rendered by the American Meteorite Museum. The jewelry proved especially annoying to scientists. In addition, Nininger self-published several booklets dealing with meteorites and

*Nininger's booklet
A Comet Strikes the Earth.
Note the "meteorite" sample
in the round window.*

tektites, which he sold at the museum. One booklet had a piece of oxidized meteorite fragment attached to its cover. In early editions he called the specimen a meteorite fragment, but technically this was incorrect, for the specimen had been terrestrialized by weathering and contained little free iron. He corrected the caption in later editions, but it remained an annoyance to purists.

A Time of Crisis

The museum's first year of operation was a success. The summer of 1947 brought thousands of paying visitors. The museum made enough money to allow Nininger to hire a few helpers. But in the second summer, attendance fell off. The meteorite collection was the Niningers' only real asset. If the museum did not succeed financially, they would be forced to give up the dream and sell the collection.

As their third summer arrived (1949), attendance increased as expected at the beginning of the summer tourist season. Then it dropped abruptly. Interstate 40 had opened. Most tourists skirted old Highway 66 and took little notice of the building with the tower on the desert. From then on, attendance and income, barely sufficient in the past, was cut in half. This period of reduced attendance had its plus side: The reduced workload gave Nininger more time for his crater studies, and he made his most important discoveries during this period.

Return to the Crater

Nininger was anxious to continue his research. He had collected hundreds of soil samples during his magnetic rake survey in 1939 and stored them for future examination. Sifting through the soil, he found millimeter-sized spherical particles made of cemented sand and soil grains. In their centers were metal cores that tested positive for nickel. He interpreted the metal cores as condensation products from the cloud that had formed over the crater when the nickel-iron meteorite was vaporized: Nickel-iron droplets had rained out of the cloud as it cooled. The droplets had formed in the absence of oxygen, since the atmosphere above the crater had been briefly forced aside. They still retained their metal composition, with only a thin veneer of metallic oxide that cemented the sand and soil grains surrounding the metallic droplets.

Nininger found other metallic particles in the soil that were completely oxidized. He interpreted these as droplets from the edge of the cloud that had mixed with oxygen in the atmosphere before falling to Earth in an oxidized state.

At the moment of impact, when the meteorite was vaporized, a great deal of terrestrial rock must have been vaporized also. This material would have condensed out of the expanding cloud and may have trapped tiny metallic particles on their fall to the ground. Fused silica glass with entrapped metallic particles called impactite had been recovered at the Wabar and Henbury Craters a dozen years earlier. Many scientists had pointed to the absence of fused silica at Meteor Crater as evidence against an explosion. But no one had really looked for it. This was Nininger's moment of discovery.

On the northwest side of the crater rim, the highway department had dug a gravel pit. There, in 1952, Nininger found glassy, greenish-yellow slag distinctive from materials he had examined earlier. He ground a piece flat to examine its interior. Bright, silvery metallic particles appeared, embedded in the glass. The metal tested positive for nickel. This was meteoritic material. He had found the "absent" impactite. Moulton had been right—the great iron meteorite had vaporized on impact.

A New Beginning

Despite the hardships the Niningers endured in the museum on Highway 66, their goals were undiminished. They attributed the failure of the museum primarily to the highway change. What they needed was a new location and a modern building. In their seven years on the Arizona desert, electrical power had never reached the museum. Despite the advantages of being near the crater, they finally decided it was

American Meteorite Museum in Sedona, Arizona, about 1958.
—Center for Meteorite Studies, Arizona State University

better to move closer to civilization. They found some property in Sedona, Arizona, in beautiful Oak Creek Canyon a few miles south of Flagstaff, and constructed a small brick building facing Highway 89. They moved the meteorite collection to its new home in September 1953.

The first year in the new museum was encouraging but not as busy as it had been on Highway 66. There was no Meteor Crater to attract attention, just a sleepy town of craftspeople and artists. Still, the Niningers were busy enough to ask their daughter Margaret and her husband, Glenn Huss, to come to Arizona to help. Fortunately for the Niningers, their son-in-law had a keen interest in meteorites and soon became an indispensable addition to the staff. Glenn and Margaret continued as custodians of the museum until its closing in 1959.

In 1958, the Niningers finally acknowledged that the museum could not pay for itself. For several years, Nininger had been sending out "feelers" to institutions in the United States and abroad about the sale of his collection. Finally, in 1956, Dr. Max Hey, curator of meteorites at the British Museum, visited him. Hey examined the collection and expressed the museum's interest in purchasing pieces cut from many of the specimens. A firm offer would have to wait for the acquisition of funds. During the waiting period, both the Smithsonian and Arizona State University expressed interest in purchasing the collection. Neither of these two institutions had the ready cash to make a firm offer. The race was on.

285

Glenn Huss at the American Meteorite Laboratory in Denver. The Acuna iron meteorite, foreground, was recognized by Huss in 1981.
—Courtesy of Margaret Huss, American Meteorite Laboratory

Nininger had always wanted to sell the collection intact, but it was too large for a single institution to buy. He reluctantly agreed to split the collection. He was especially reluctant to see the collection leave the United States, but his financial situation was critical. In June 1958, before the American institutions could act, the British Museum offered a firm bid of $140,000 to purchase 21 percent of the collection, amounting to 1,200 specimens from 276 falls. This left 79 percent intact and gave the Niningers the first financial freedom they had ever experienced. With their bills paid, they were free to build a new home and travel for pleasure for the first time in their lives. They went to the Far East for six months and returned to Sedona in May 1959. The Husses had been operating the museum for them, and the situation was no different. The museum had barely survived financially. It was time for a difficult decision.

The Final Sale

In the fall, Arizona State University contacted Nininger to reaffirm its interest in purchasing the remaining collection. Nininger had inventoried the specimens: 1,300 meteorites weighing 7.5 tons, which he would sell for $530,000. In the spring of 1960, with funds from the

National Science Foundation, the Arizona State University Foundation, and an anonymous donor, the university made an offer of $275,000 for most of the remaining specimens, barely half its worth. Nininger accepted, pleased that the bulk of his collection would stay in the United States and, most appropriately, in Arizona. Today, it is the largest university-owned meteorite collection in the world. The collection became the nucleus for a new university-sponsored institution, the Center for Meteorite Studies, where research is currently being conducted on those meteorites. Nininger's lifelong wish was finally granted.

A Personal View

The sale of the collection brought an era to an end. For fourteen years, the American Meteorite Museum had struggled to exist through the dogged perseverance of one man and his family. Their sacrifices had been great, and during that brief period they had touched the lives of more than one-half million people who had browsed through the museum's unique collection of celestial bodies and heard lectures from the meteorite crusader.

I was one of those privileged visitors. In the spring of 1958, on his annual trek to Meteor Crater, Frederick Leonard took me to the museum in Sedona to meet Nininger. He didn't lecture to me. He just let me look at the specimens. I was too naïve to realize I was in the company of two pioneer meteoriticists. What a history they had shared! I left the museum with a small stony meteorite from Dimmit, Texas, a gift from Nininger to Leonard's aspiring pupil. Two years later, the Niningers closed their museum forever. As if to lament its closing, my favorite professor, Frederick Leonard, passed away shortly thereafter, leaving the Meteoritical Society as his legacy.

The American Meteorite Laboratory

Glenn Huss carried on his father-in-law's meteorite work. He worked to keep the spirit of the American Meteorite Museum alive after the Sedona museum closed. In the summer of 1960 he and his family returned to Denver. They took the remaining specimens, still a sizable collection, and set up the American Meteorite Laboratory, determined to provide meteorite specimens for the world. This Glenn Huss did admirably until his death on September 28, 1991.

Harvey H. Nininger lived to his ninety-ninth year and died March 1, 1986. Who would fill the void? Who would continue the search for meteorites? Months before Nininger died, he was visited by a young man full of fire—meteorites were his passion. Nininger looked at him and saw himself in his youth. He encouraged the young man to take to the field, to pursue the heavenly rocks wherever they might take him. This he did. His name was Robert A. Haag, a self-styled "meteorite man."

Robert A. Haag. –Jeff Smith photo

SIXTEEN

Robert A. Haag,
the "Meteorite Man"

Late one evening in the spring of 1992 as I was watching a popular TV program, *Missing–Reward,* a young man from Tucson, Arizona, appeared on the screen. He was casually handing a $200,000 check to a lucky man from Argentina. The Argentinean had something the young man wanted–a meteorite from Patagonia. This was no ordinary meteorite; this was the world's largest intact, privately owned stony-iron meteorite, the 1,250-pound Esquel pallasite. The young man was Robert A. Haag, the "Meteorite Man." I immediately checked the *Catalogue of Meteorites* and found this notation for the Esquel pallasite: "Main mass, Buenos Aires, in the possession of the finder." No longer, I thought. Haag had struck again.

Our paths had crossed briefly ten years earlier. I had been browsing in a Tucson shopping mall when a wild-looking fellow in a silver jumpsuit caught my eye. He was standing in front of an enormous picture of the Moon, delivering his pitch. He was selling "space passports" and people were captivated. "Sounds like a con artist," I thought.

With a halo of curly blond hair and a fine physique, he had the look of a rock star and the smile of a television evangelist. His enthusiasm was overpowering and his message was simple. He was convinced that people soon would be flying into space aboard commercial space shuttles and would surely need a space passport. He would take your picture in front of the Moon backdrop and attach it to a passport, all for only $5.00. He was charming and people seemed to accept his gimmick in good humor. As I wandered away, I heard him say to a young woman as he handed her a passport, "Don't leave Earth without it!"

The Visit

The following spring a curious advertisement asking for meteorites appeared in the newspaper. I dialed the number. The voice on the phone was young, enthusiastic, and impatient (where had I heard that voice before?). He wanted meteorites–it didn't matter what kind–and he wanted them NOW! He'd be over within the hour. Forty-five minutes later I answered a knock on my door. There, without his silver jumpsuit, was the flamboyant Robert A. Haag. A sling around his neck supported his right arm, recently broken in a hang-gliding accident.

I had laid out a dozen or so meteorites in the family room. One look and he was speechless. I handed him an iron and he took it as though it was the Hope diamond. Then he exploded with obvious incredulity. Other than a few specimens he had seen at the Flandrau Planetarium—which had been inaccessible behind glass—these were the first meteorites he had ever been able to hold. The air was electric. I handed him a chondrite, then a pallasite. Here was a man possessed. His reaction was far beyond anything I had experienced with countless other people to whom I had shown meteorites over the years.

Haag had always been close to minerals. His parents were commercial rock and mineral dealers, and his brother Zee had sources for gigantic quartz crystals from South Africa. Haag had no taste for academic pursuits. He took classes at Pima College, in Tucson, but student life was not for him. He was impatient. School was slow and laborious. He tried his hand at surveying mineral claims and made a scant living drilling mining claims for validation purposes. He bought and sold minerals with his parents and brother, but none of these activities suited his fireball personality.

As a child, he had shown a keen interest in space. He was twelve when astronauts first walked on the Moon. Indeed, he was born the very year the first satellites were launched into space. Space occupied his waking hours. His interest intensified when, like Nininger a half century before, he witnessed a brilliant fireball from the beach in Sonora, Mexico. He desperately wanted to find that meteorite and naïvely spent hours searching the shoreline. He simply didn't know enough to find the prize. It was 1969, the year of the great Allende fall. Like Nininger, he didn't find the meteorite, but the seed had been planted. To own a piece of another planet became a recurrent dream.

Haag's dream almost became reality when he was twenty-one years old. He was astonished to see an iron meteorite for sale at a gem and mineral show. It was expensive, far more than he could pay. Someone else could afford it, though, and purchased the dream. He vowed never to let another opportunity slip through his fingers.

A year or so later, Haag noticed an ad in a lapidary magazine; someone was asking for meteorites. The ad was signed "Jim Dupont of New Jersey." Jim Dupont was a millionaire and owner of one of the largest private meteorite collections in the world. Haag placed a fateful call. Dupont wasn't selling meteorites, he was buying them, but he took an interest in the young man and encouraged him to go after his dream aggressively. From that moment, Robert Haag began his search in earnest. Dupont urged him to run classified ads for meteorites in his local newspaper. "You never know who's using one as a doorstop," he said.

Haag had done just that and it brought him to my house. He had finally realized his dream.

A Time to Learn

Haag came to buy meteorites knowing little about them, but I had a feeling he wouldn't stop with the few small specimens I could furnish. It was time to learn and he was a willing student. He had a voracious appetite for the subject, and for weeks he came by the house every few days to talk meteorites. He soon learned that an understanding of meteorites required knowledge in many scientific fields: mineralogy, astronomy, petrology, and chemistry. Meteoritics is a hybrid science, derived from them all. He read every book he could find on the subject, and there weren't many—you could hold them all in one hand.

His pursuit of bigger and better meteorites continued and his ads began to pay off. A 10-pound Canyon Diablo iron appeared from Nogales, Mexico. This specimen had Nininger's label on it.

In February 1982 the Tucson Gem and Mineral Show opened, and I asked Haag to sell several specimens from my collection. His brother had booth space at a motel, and he planned to sell them there with his mineral specimens. We priced the meteorites at seventy-five cents per gram (0.035 ounce), an exorbitant amount, thinking we might possibly get fifty cents per gram. They all sold in less than two days at the higher price. Haag was convinced: People wanted meteorites, and there weren't many available. Dozens of people heard about mine but were disappointed to arrive too late to buy them. No one at the show knew much about the meteorite market, if indeed there was one. Most important, except for the more common Canyon Diablo and Odessa meteorites, no one knew where to get them. A market without a product is no market at all. The opportunity was there if only meteorites could be found. Haag remembered Dupont's words: "If you want them, go out and find them." This was just the challenge Robert Haag was looking for.

Haag Searches for Meteorites

The best place to find meteorites is a place where they have been found before, and Meteor Crater was Haag's first target. He bought the best metal detector he could find and headed for the crater. It was essentially virgin territory. Although Nininger had searched there twenty-five years earlier, there had been enormous technical advances in metal detectors since then. Modern detectors can find iron meteorites 3 or more feet deep. Only a few hours' searching on the west side of the crater uncovered dozens of meteorites, from a few grams to several pounds in weight. It was like fishing with an irresistible lure.

There was a time in the 1960s and 1970s when searchers could scour the area around the crater with the rancher's and crater curator's consent. They asked only that all holes be thoroughly filled in, to prevent cows from stumbling into them. Ranch hands patrolling the area often saw Haag out in the sage, and greeted him with a wave and a reminder to cover the holes. He was careful to stay far to the west of the crater, 3 miles out beyond Canyon Diablo, so he would not interfere with the ranch business.

Those were wonderful days for Robert Haag. Searching the plains by day and camping under the stars at night gave him an increased awareness of his destiny. These rocks from space were beginning to play a major role in his life.

A spirit of cooperation with the ranch prevailed through the first few years of Haag's explorations, but ended when other, less conscientious hunters left their holes unfilled. Inevitably, permission to search the land around the crater was revoked and remains so today.

Haag's success at the crater gave him a stock of iron meteorites he could sell at mineral shows and by mail. More important, it was material he could trade for other kinds of meteorites.

His initial success was not without controversy. The problems commercial fossil dealers have encountered for years soon plagued the meteorite hunters. The commercial collection of meteorites produced "heartburn" for some academics who insisted this "harvesting" of meteorites degraded the science. These vague criticisms do not withstand scrutiny. If anything, the opposite is true. The meteorites Haag and others rescued from the plains were carefully cleaned and painstakingly prepared for distribution and display. Today, Canyon Diablo meteorites are the most plentiful irons, gracing collections in public institutions and the private sector. These are often the first meteorites that students of astronomy are permitted to handle. In this respect, the collectors have contributed to public education. The alternative is to allow the meteorites to continue rusting away in the ground, for the benefit of no one.

Some time after the crater became off limits to hunters, Haag filed for a 10-square-mile mining claim around the crater. His petition was flatly denied. Scientists had objected to any mining claims within several miles of the crater; they abhorred the collecting of meteorites on state land.

To the Strewn Fields

Haag realized that meteorites from one fall alone did not guarantee a successful business. He had to offer collectors and educators a variety of specimens from many locations, the more exotic the better. The Haviland Crater on the old Kimberley farm near Brenham, Kansas, was his next target. This was the location of the Brenham pallasite strewn field. Few meteorites had been found on the farm since the 1930s. Things had changed since Nininger first recognized the crater in 1929. The old farm had burned down and its remains had been plowed under, scattering rusty nails and metal plumbing fixtures through the prime meteorite field.

Haag and two companions, one of whom was David Baker, located the property and began searching with metal detectors. In less than a minute they found a 20-pound pallasite and believed they were in for a real turkey shoot, but in the next two weeks they found nothing more. On the last day of their search, they followed a lead from a local farmer who claimed he consistently hit a rock when plowing. They searched all day but could not find the rock. They returned to Tucson all but empty-handed. Haag could only conclude that, through the years, the Brenham field had been searched out.

Three weeks later David Baker showed up with a 360-pound Brenham pallasite. (Baker was good at finding the big ones. A year earlier he had located a 125-pound meteorite under a massive limestone rock in Canyon Diablo.) He refused to reveal exactly where he had found the specimen (it was probably the rock the farmer kept hitting), but Haag didn't care. He bought the pallasite immediately—a wise move. Weeks later beautiful slabs of Brenham pallasites magically appeared on the market. Everybody wanted one. Haag had been right. There was an incredible market for meteorites.

The sale of more than 300 pounds of Brenham gave Haag the capital he needed to investigate other strewn fields. His next stop was the Allende field. Most of this rare meteorite had been collected by the Smithsonian Institution within days of its fall in February 1969, but that did not discourage Haag. Using the Smithsonian's map, he began contacting ranchers along the major axis of the strewn field ellipse, which proved to be a slow and inefficient process. Fluent in Spanish, Haag spread the word on the local radio station, offering a reward in American dollars for meteorites. Out into the desert the local residents went, and within days, meteorites began to appear—buckets of them. Haag later described his adventure:

> This place is like Brigadoon. It seems far more natural to go about on horseback than in automobiles. The people of Allende are just tremendous; they took me in, fed me and encouraged me. I loved it there. It seems that if ever a place deserved to be "blessed by heaven," Allende is that place.

A week or so later, Bob Haag walked into my house with boxes of Allende carbonaceous chondrites. He put more of these rare meteorites on my kitchen table than had existed in all the museums of the world before Allende fell. He had collected more than 100 pounds.

At Allende, Haag learned the most efficient way to recover meteorites—recruit local residents to do the searching. Next he drove to the Nuevo Mercurio strewn field in Zacatecas, Mexico. Again his method was tested, and his success was stunning. The entire town showed up for a meteorite hunt. Thousands of little Nuevo Mercurio meteorites were recovered, weighing a total of over 90 pounds.

About this time a very special meteorite was making headlines. Organic molecules, amino acids among them, had just been found in the CM2 carbonaceous chondrite from Murchison, Australia. Murchison, more than any other carbonaceous chondrite, seemed to hold clues to the origin of life. Suddenly, Murchison was on every collector's wish list. Researchers at many universities across the United States and abroad wanted samples of this pristine piece of the early solar system.

Off to Australia Haag went, not knowing if there would be any specimens left after more than ten years. More than 1,500 pounds had been recovered initially, but the meteorite was so fragile it could not survive long in nature. In Murchison, Haag went door to door, asking for meteorites. Most people shook their heads. A few did not. Betty Maslin, a Murchison resident, vividly remembered that Sunday morning in September 1969 when she and her father heard loud explosions and hissing noises. She especially remembered a strong smell like that of methyl alcohol that filled the town. She sold Haag a sealed jar of meteorites. It was like opening a good bottle of wine after a decade of anaerobic rest. A powerful smell of alcohol and ether was released, organic molecules not of this Earth.

Again Bob walked into my kitchen, this time carrying the impossible—several pounds of Murchison meteorites. He was good enough to let me have my pick of them before they went on the market. His Murchisons carried a message to the world: If you want a meteorite—any meteorite—Robert Haag will find it for you. Murchison commanded a price of $20 per gram (0.035 ounce) then and even at that, they didn't last long. The sudden availability of this rare meteorite made the world of rock and mineral dealers suddenly take note. Who was this Robert Haag? How did he manage to acquire the unobtainable?

"I'll Go Anywhere in the World for a New Meteorite"

Nininger scoured the United States and Mexico for meteorites. In his time, the world beyond North America was far less accessible. But the world would be Robert Haag's collecting field. By the mid-1980s, his young business was producing annual receipts of $100,000. The stars seemed to be shining on him.

However, traveling the world in search of meteorites was expensive business, and every trip was risky; he often returned with nothing more than a reduced bank account.

He decided to go to Namibia, to the Gibeon iron-meteorite strewn field. Although tons of Gibeon had been collected through the years, little had reached the United States and almost none was available. The strewn field was enormous. When Haag arrived in Gibeon, he went directly to the principal of the local school—what better place to recruit some enthusiastic searchers. He showed the principal a few rusty Canyon Diablo irons and offered a reward to any of the students who could find some Gibeon irons.

> The next thing I knew he—the principal—was on an intercom calling all of his students out of their classes, all four hundred of them! They formed a single line and each student walked past the rusty iron meteorites, looking at them carefully and handling them. Within the next three weeks they had found ten Gibeon meteorites. Fabulous specimens! The biggest was around 500 pounds.

From that contact the Gibeon harvest began. When the children took money home to their parents, the parents also began searching. Then uncles, aunts, and cousins. It became a family affair. Then a tribal affair. The searchers recovered more than thirty Gibeons. The largest weighed over 1 ton, larger than any previously found. (Since that time, other collectors have recovered literally thousands of Gibeons and distributed them worldwide.)

Haag learned valuable lessons from these experiences. First, he learned never to accept the experts' estimates. If the *Catalogue of Meteorites* says one hundred meteorites fell in an area, most certainly two or three times that many actually fell. Second, an incentive is required. Most people don't realize that meteorites have monetary value. Once a reward is offered for meteorite recovery, people's attention is captured. This is especially true in Third World countries, where wages are low. Finally, the more people who can be recruited, the more likely it is meteorites will be found. The most inefficient way to search for meteorites, especially in unfamiliar territory, is to do it yourself. Nininger used the local farmers. Haag uses whole towns and villages.

Robert Haag distributes a beautiful, full-color catalog on meteorites, which he calls *Field Guide of Meteorites*. The best specimens in his collection are beautifully illustrated, and the booklet is a curious mixture of information about meteorites and his personal adventures. He describes his experience in La Criolla, Argentina, when he was pursuing La Criolla stony meteorites. On January 6, 1985, hundreds of stones fell on the small town of La Criolla, smashing through the roofs of houses and scaring people. They were soon identified as L6 olivinehypersthene chondrites. This story from Haag's catalog demonstrates his enthusiastic approach to meteorite hunting and illustrates again why he has been so successful.

> Back in Tucson, Arizona, I was reading a report of the new fall at La Criolla in a Geophysical newsletter. I was interested, but didn't think much about it until a month later, when I read a second report saying that "tens" of stones had been found in the area. A bulb lit up over my head: if there were "tens" of stones, there were just as probably hundreds of stones, and my chances of finding some were good. I made immediate preparations to go.

In Buenos Aires, I showed a map of the area to someone in a local travel agency and they got me on a flight to a town called Concordia, which was within 10 kilometers of La Criolla. . . .

In La Criolla, we stopped at a small grocery store and started asking questions. The storekeeper remembered the fall vividly. Over big cups of tea, I mentioned purchasing some meteorites. She went into her back room and returned with a 50-gram stone. I bought it from her on the spot and asked if she knew where I might find more. She said to try "El Policia."

"El Policia" turned out to be the town's only constable, a friendly man who knew the community well—the perfect contact. When I explained what I wanted he asked to go along. First, we stopped at the home of the Mayor, a Mr. Silva. I asked him if he knew where I could find more stones. He excused himself and came back with the most beautiful, fresh meteorite I had ever seen. It was a 15.5-pound stone with scalloping over two sides and a clean break showing the interior. I was ecstatic. I offered him the equivalent of a year's wages, which he

Haag and the mayor of La Criolla, Argentina, holding a 15-pound La Criolla stony meteorite. –Courtesy of Robert Haag

accepted. Furthermore, he knew of many others! So we all piled into the taxi: the driver, the long-haired gringo, the policeman and the mayor, and went in search of treasure from heaven.

Just up the road, we stopped at a small farm house. Over more tea, I asked the woman there all the same questions and she told me that there was a piece in the barn, right beside the oil and tools. It turned out to be a 15-pound complete specimen. Then, as if that weren't enough, her husband brought out another 5-pound specimen from his father's house.

The news of the crazy man buying rocks spread like wildfire, and soon people were coming from all over carrying black stones. After eight days, I had recovered fifty-four stones with a combined weight of some 66 pounds.

Haag returned twice that year and bought an additional 26 pounds of small stones. These meteorites are now in collections scattered throughout the world.

The One that Got Away—And Almost Took Haag with It

Argentina is a gold mine for meteorite hunters. It is best known for the Campo del Cielo crater field and the associated great irons. The *Catalogue of Meteorites* lists a 15-ton iron, with many others in the 1-ton class. Robert Haag knew there was a larger one estimated to be 37 tons—larger than the great Cape York iron.

In Tokyo in 1989, when Haag was displaying and selling meteorites at an international gem and mineral show, he met an Argentine mineral dealer, Miguel Fernandez. Fernandez knew the location of an enormous iron meteorite in the Campo del Cielo strewn field. It was on private property outside the small village of Gancedo, in Chaco Province, about 600 miles northwest of Buenos Aires. If Haag was willing to pay a finder's fee, Fernandez would make arrangements with the rancher to sell the iron. To Haag, "on private property" meant the meteorite was the property of the landowner, as in the United States.

Fernandez would also arrange to truck the meteorite to the coast for transport by ship to the United States. Haag agreed to pay the rancher $200,000 for the meteorite. He paid Fernandez $40,000 in advance and an additional $8,000 to get a crane for lifting the meteorite onto a tractor-trailer rig.

In January 1990 all was ready. Fernandez assured Haag that the rancher had agreed on the price and he could come and get his meteorite. Haag and his photographer, Jeff Smith, left for Argentina. Fernandez met them in Gancedo and they headed to the ranch with an eleven-man crew and lots of heavy equipment. The next day they began the task of loading the monster onto the truck. The operation went smoothly, except when the crane nearly sank into the mud. Finally, the meteorite

Haag with the Campo del Cielo iron meteorite, estimated at 37 tons. –Jeff Smith photo

was loaded and ready to roll. Down the highway they went, the tractor-trailer with the meteorite, the truck with the crane, and one passenger car. They hadn't been on the highway long when they were stopped by police and arrested near Charata, in northern Chaco Province. The charge–stealing a national treasure.

They were taken to Charata, a city of about seven thousand people, and jailed. Jeff Smith pleaded innocent. As he was only along to document the expedition, he was released after four days for lack of evidence. Haag remained in custody for nearly a month, but the experience was hardly one of deprivation or penance. He was treated more like a celebrity than a suspected criminal and was never behind bars. He was free to walk around the jail grounds, eat steak, and enjoy wine with his captors. When his sister called the jail from her home in Tucson, she was surprised when Haag was the first to answer the phone.

After nine days in custody, Haag paid $20,000 bail. He remained in Charata, waiting for a hearing and confident charges would be dropped. In an interview with the Associated Press after his release on bond Haag said, "I had nothing to do with any theft. I came here with over $50,000 in cash, television cameras, my own personal photographer, a crane and a semi truck and trailer. How could I have gotten away with it? I didn't come to steal the meteorite, I came to buy it. But instead I got set up."

The judge told Haag that meteorites in Chaco Province are the property of the province, not the landowner, and it was not for sale. He had been duped. He spent the next three weeks negotiating with the Argentine government, trying to get the meteorite back, but the negotiations were fruitless. He was free to leave Argentina after paying bail, and he reluctantly returned to Tucson.

If he does not go back to Argentina for trial, he forfeits the $20,000. For most people this would have been a disaster, but for Robert Haag it was a bonanza.

"The exposure in the media was terrific for my business. It more than doubled," Haag said, after returning from his ill-fated trip. Then he added wistfully, "I still want to get that meteorite." There are those who want to help him, too. He received letters from Argentine lawyers who say they will have it delivered to him for $1 million dollars!

An Australian Lunar Meteorite

In early February 1991, I wandered into Robert Haag's room at the Desert Inn in Tucson. This is where he conducts business during the Gem and Mineral Show each year. He wasn't there but his assistant, Mark Carlton, motioned to me. He reached into the display case and picked up something in his cupped hand. Taking my hand he opened his, and onto my palm dropped a small, black stone no bigger than a walnut. With a satisfied smile he proudly revealed its secret. This was a rock from the lunar surface, and there I was, without museum guard or alarm system, holding a priceless piece of the Moon. "Bob wants people to hold it," Mark said. "Where else would this be possible?" I thought. Then Mark told me the story of the lunar rock.

Every meteorite hunter knows that the best place to hunt meteorites is where the soil is sandy, where there are few natural rocks, and where the weather is usually clear and dry—desert country. Australia has the best desert for meteorite hunting. The Nullarbor Plain near the south coast of Western Australia is well known for large Mundrabilla irons and many other smaller finds.

In October 1960 (the exact day is unknown) a fall of a rare type of stony meteorite startled sheep ranchers on the Millbillillie Station homestead, about 7 miles from the town of Wiluna, Western Australia. The meteorites were eucrite achondrites, worth plenty on the meteorite market. They fell over a large area, but fortunately the meteorites had a shiny black crust that was easily distinguished from the desert rocks. A gold prospector named Harry Redford heard about the meteorites and the prices museums were paying for them, and headed for the strewn field to find his fortune. He did, too. In a few days of searching, he found several dozen small stones. But it was tough work and he was alone. So, adopting Haag's technique, he offered rewards to the local Aborigines. Soon he had an army of searchers finding stones by the bucketful.

Haag purchased hundreds of these interesting meteorites from Redford and planned to sell them worldwide. Before he could sell them, however, he had to examine each one for verification. He was looking through a pile when he picked up a small stone that "felt different." He put it aside, intending to look at it more carefully later and continued sorting:

> When I looked at the stone again, I could see that this specimen had tiny gas vesicles on the fusion crust, something so rare that I've only seen it in a few other meteorites. The crust had a slight greenish tint, unlike the Millbillillies. I ground off a tiny corner to look inside, and again, it was different from the Millbillillie eucrite. It contained tiny white clasts I'd seen in photos of Moon rocks.

Barely able to contain his excitement, Haag rushed the little meteorite over to Dr. William Boynton and Dolores Hill, meteorite experts at the University of Arizona's Lunar and Planetary Laboratory. One look was enough to excite them, too. Boynton called for an analysis. Haag and the scientists waited for the results with the anxiety of expectant parents. Then the official word came: "We have analyzed the new meteorite by neutron activation analysis and believe it to be of lunar origin based on comparison with other meteorites and samples returned from Apollo landing sites."

The news was out. Robert Haag had done the impossible again. He now possessed a Moon rock, something NASA had paid dearly to acquire two decades earlier. It was a piece of lunar crust from an area not sampled by either American manned or Russian unmanned missions, making it unique. Haag committed six grams (0.21 ounce) for scientific analysis and kept the rest in his collection.

This was not the first Moon rock found on Earth. Researchers had discovered eleven on the Antarctic ice over the last several years. For five years, this 19-gram specimen remained the only Moon rock in the world in private hands. On March 23, 1997, a private collector recovered a second, much larger Moon rock in the Libyan Sahara. The rock, a polymict anorthositic breccia from the lunar highlands, weighs 531 grams and was named Dar al Gani 262.

To own a piece of the Moon was another dream come true for Bob. His response to the news was predictable. With his usual enthusiasm and optimism, he saw far beyond the Moon rock. "If you can find Moon rocks on Earth, I'll bet you can find rocks from Mars, Venus . . ."

On the Trail of the Mars Rocks

The SNC meteorites are volcanic rocks with characteristics suggesting an origin on Mars. The best chance of finding meteorites is to search

Haag's 0.7-ounce moon rock, which measures 1.35 inches across. Note the white anorthosite (calcium plagioclase) fragments common in lunar breccias.

where they have been found before, in known strewn fields. If the strewn field is depleted, the next best place to find meteorites, especially rare ones, is in a museum. Haag tried both sources in his quest for an SNC meteorite. He decided the Nakhlite meteorite from Nakhla, Egypt, would be his best bet.

Arriving in Cairo in the fall of 1984, Haag hired a taxi to take him on the six-hour drive to the Nakhla strewn field, an agricultural area in the Nile floodplain. Farmers there still worked their fields with horse-drawn plows and used oxen to lift buckets of water from irrigation canals. The people were living much as their ancestors had four thousand years ago. Haag showed them stony meteorites and they looked at him in disbelief when he told them they had fallen from the sky. He offered money for any black stones they could find, but money didn't interest them—they had no need for money and would have preferred to receive camels or goats.

The methods he had used so successfully before did not work in Egypt. There was little interest in helping "this crazy American," who told such ridiculous stories. So he returned to Cairo, this time to the Cairo Museum, where most of the Nakhla meteorite was located. For eight days he negotiated with museum officials. He had brought a suitcase full of prize specimens that any other museum would have exchanged its director for, but they offered only 1 gram of Nakhla. They said they had been offered $1,000 per gram for their piece of Mars. Haag had reached his limit and declined the meager offer.

Haag with several villagers on the site of the Zagami meteorite fall. Zagami rock is in the background. —Courtesy of Robert A. Haag

Not all Haag's meteorite adventures end successfully, that is, if success is measured in weight of meteorites acquired. Persistence is the rule in his business. Another chance arose later, and he bought a 120-gram (4.3-ounce) Nakhla meteorite for $350 per gram, thirty times the price of gold. Even this outrageous price turned out to be a bargain. He had some Nakhla to sell, and the price was $500 a gram. Besides possibly being a Mars rock, Nakhla is a treasure for another reason. When the stones fell on June 28, 1911, one hit and killed a dog—one of the rare witnessed accounts of a death caused by a falling meteorite.

Haag's quest for an SNC meteorite did not end in Nakhla. On October 3, 1962, another Mars rock, this time a shergottite, fell in a field near Zagami Rock in Katsina Province, Nigeria. A single meteorite weighing about 40 pounds was recovered and taken to a museum in the provincial capital of Kaduna, where it was put on display. It remained intact until Robert Haag "picked up its scent." In 1988, Haag took a trip to Nigeria to search for something truly exotic—the Zagami shergottite.

Kaduna, in north-central Nigeria, is a small place. The museum has few visitors. Haag walked in, asked to see the director, and told him he was a private collector. He said he would like to trade specimens for any Nigerian meteorites, purposely playing down his lust for the Zagami shergottite. He opened his suitcase and revealed a substantial collection of meteorites any museum would be proud to display. The director was astounded. He could offer only a few specimens in trade. One was a rare silicated medium octahedrite iron from a witnessed fall called Udei Station. He was willing to cut off a piece but they had never cut an iron meteorite and did not realize how difficult it would be.

While the meteorite was being cut, Haag hired a Land Rover and driver and went to Zagami, hoping to find someone who might have picked up meteorites during the fall. He met an old man who had been in the very field where the meteorite hit—it missed him by only a few feet. Then he met the village chief, and through an interpreter asked if anyone in his village had Zagami meteorites. This time the technique failed. No one had any. SNC meteorites were proving to be elusive.

Back at the Kaduna museum, they finally succeeded in slicing a 700-gram piece (1.6 pounds) of the silicated iron, but not before wearing out several blades on their cutoff saw. They were embarrassed and frustrated by their small offering, compared with Haag's more substantial collection. Would he be willing to accept a piece of a stony meteorite? They pointed to the 40-pound shergottite. His heart pounded but he maintained self-control. Yes, of course he would accept a piece. They cut off more than 13 pounds and, after making sure that was enough, added it to their small pile of Nigerian meteorites. The trade was complete.

Haag finally had his Mars rock—if he could just hang on to it. He left Kaduna the same day and headed southwest to the city of Lagos on the Atlantic coast. There he was stopped cold by an official who was checking baggage. The meteorites set off the metal detectors. "What is this rock?" the official asked. Haag told her it was a meteorite given to him by the museum director in Kaduna. "That can't leave the country," she said coldly and set it aside. Haag grabbed it back, determined to "walk across Africa barefoot or swim the Atlantic Ocean" to get it to Tucson. The main problem was that he had no papers from the museum releasing the meteorite. The line behind him was filled with impatient travelers, and the plane was scheduled to leave momentarily. The official was under pressure. "Well, I've got these papers," Haag said, and thrust all his Nigerian money on the table. "I have something back here," he continued and turned around to retrieve a bag. In that instant the "papers" vanished. With a smile and a thank-you, the official motioned Haag to the waiting plane.

This meteorite had a price on the world market of $1 million. Returning to Tucson, he sent letters to his clients, offering small slices of Mars. In a few weeks he paid off his house, bought a new Corvette, a new boat, and more meteorites.

Looking to the Future

Some people might call Robert Haag just a good businessman, and he is. He turned his passion for meteorites into a million-dollar business and became the most important meteorite supplier in the world. But he is not an H. H. Nininger. He has never claimed to be a scientist, although meteorite scientists have reluctantly come to respect his tenacity and drive. He has furnished rare meteorites to laboratories around the world, and his efforts have changed the world of meteorites. Millions of people have been touched by his enthusiasm and passion for these rocks from space. He has been featured in *National Geographic* magazine and appeared on television in the United States and abroad.

Haag knows the world is rich in meteorites. The Sikhote-Alin strewn field in Siberia awaits him with his metal detector and television crew; the Antarctic ice cap is strewn with fresh meteorites; Wabar Crater in Saudi Arabia has never been searched with modern metal detectors. But one goal stands out from all others. The Tucson Ring meteorite tolls for Robert Haag. Its strewn field is practically in his backyard, only 30 miles south of Tucson. Somewhere on the slopes of the Santa Rita Mountains, below the great telescopes of Mount Hopkins Observatory, more iron masses must be buried, as history claims. People are out there looking. Haag has sent them. A $100,000 reward has been offered, and I have every expectation that Robert Haag will be successful once again.

"Space shuttle to Haag. We have your meteorite in tow. Where do you want it?"

The strewn field of the Imilac pallasites in the low hills of the Atacama Desert, Chile. The hard desert pavement and lack of vegetation make this area ideal for hunting meteorites.
—Courtesy Peter Larson, Black Hills Institute of Geological Research

SEVENTEEN

Hunting for Meteorites

Meteorites are where you find them. They can be anywhere. What are the chances of finding a meteorite after a diligent search? It all depends on the rate at which they fall and their lifetimes on Earth. Years ago, Nininger was convinced there should be at least one meteorite for every square mile on Earth. More recent studies suggest meteorites are reaching Earth at a much greater rate than five hundred per year, the number Nininger used. Canada and the United States have conducted studies using wide-angle automated cameras to record the flight of meteorite-producing meteors. One goal was to determine where meteorites come from and how many land on Earth every year. Data from the Canadian study (the Meteor Observation and Recovery Project) showed that about 23,930 meteorites, ranging in size from 4 ounces to 20 pounds, land somewhere on Earth every year. Of these, nearly three-fourths are lost in the oceans, leaving about 7,500 over Earth's entire land area. This works out to about one meteorite for every 10,000 square miles per year. Assuming a lifetime of a million years and a uniform distribution over Earth's land area, there would be one hundred meteorites per square mile landing in a million years of time. We should be able to find at least 10 percent, or ten meteorites in every square mile—ten times Nininger's original estimate.

This is only a crude statistical estimate based on assumptions that are not true for all places on Earth. Meteorites probably do not fall uniformly everywhere in a million years, and the terrestrial longevity of a meteorite depends on its type and the local climate. In humid areas, weathering and erosion destroy most meteorites within a few thousand years. They could conceivably last a million years or more in an arid desert. Most meteorites found long after they have fallen show substantial disintegration of their original structure.

You are most likely to find a meteorite if you go where they have been found in the past. There are still meteorites in many well-known areas, but hunting may be restricted, as it is at Arizona's Meteor Crater. You might find a meteorite at the Odessa Crater; Haviland Crater, Kansas; Pueblito de Allende, Mexico; Plainview, Texas; or Toluca, Mexico. As Nininger proved, strewn fields are never completely searched out.

H. H. Nininger uses a magnet attached to a stick to pick up iron meteorites on the surface around Meteor Crater. –Courtesy of Margaret Huss, American Meteorite Laboratory

Magnet and Magnifier

Before rushing out to start your search, consider the tools you will need. Besides the obvious geologist's pick and shovel, a 10X magnifier is a must. And because most meteorites contain enough elemental iron to be attracted to a magnet, carry a small, powerful magnet with you. If you are hunting for surface specimens in a known iron-meteorite strewn field, try attaching your magnet to a stick about a yard long. The magnet-on-a-stick tool is a meteorite hunter's best friend. You cannot count on your eyes alone to locate meteorites in old strewn fields, as they are often coated with minerals and look like the normal rocks in the area. Probing with this tool will quickly reveal the iron.

Small stony meteorites can also be retrieved with a magnet, as Nininger learned when he combed anthills near the railroad yard in Holbrook, Arizona, and extracted dozens of tiny stone meteorites. For testing larger specimens it is better to hold both meteorite and magnet in your hand, so you can feel the slight pull of the magnet.

Metal Detectors

In large strewn fields of irons, meteorites may be buried by as much as a foot or more, making a metal detector essential. All metal detectors work on the principle of inductive balance. A circular head 8 or 10 inches in diameter on the end of a long pole induces a magnetic field in a coil of copper wire embedded in the head. When the head is held within a few inches of the ground, the induced field penetrates the ground. Any conductive material, such as metals buried in the ground, changes the field's phase and amplitude, distorting the shape of the field. The electronics near the handle end of the detector sense this distortion and send a signal to a meter that measures the strength of the distortion, which is a rough indication of the size and depth of the metal source. Most detectors emit an audible signal, loudness being a measure of the signal strength.

Steve Howard, a customer service manager at White's Electronics, demonstrates the use of a modern electronic metal detector.

Top-of-the-line "Spectrum" metal detector made by White's Electronics, showing the visual display and control center. —Courtesy of White's Electronics, Inc., Sweet Home, Oregon

Many meteorite hunters use electronic metal detectors made in Sweet Home, Oregon, by White's Electronics. Models 5900 and 6000-DI have been the workhorses in the field for years. They can locate iron meteorites in fairly compact soil to a depth of about 2 feet. The detectors can be adjusted to any sensitivity. On the high sensitivity setting, they can find small surface specimens that weigh as little as 3 or 4 grams, and small subsurface specimens that weigh as little as 10 grams. Unfortunately, they will also pick up every bullet, nail, and every rusty can that has lain there for decades. Seasoned meteorite hunters usually select a less sensitive setting to avoid the trash. This problem has been essentially solved with White's new Spectrum series detectors. These top-of-the-line detectors have a visual display that replaces the older meters, providing precise information on the detected signal. The system can be custom programmed to discriminate between meteorites and manmade trash, determining the nature of the metal by the way in which the field is distorted. They are even capable of detecting the iron in stony meteorites to a depth of a foot or more if the ground itself is not too conducting. Unlike the earlier models, the Spectrum is not affected by, nor does it interfere with, other detectors being used nearby.

310

A bigger head sweeps out a larger area and penetrates much deeper. These are more cumbersome to handle, however, and usually skip over the smaller surface and subsurface meteorites. Large heads are used to locate large specimens at depths of 4 or 5 feet, which could be missed by the smaller heads.

Choosing Search Areas

You can find meteorites in many areas far removed from well-known strewn fields. Don't select your search area at random. Obviously, it is far more difficult to look for meteorites in an area where there are many dark rocks, or where there is a thick growth of vegetation. You are far more likely to find a meteorite in an area that is somewhat free of this natural camouflage.

Desert Dry Lakes. Desert dry lakes, or playas, seem to be natural meteorite hunting grounds. Playas are flat areas in the center of small valleys where water from adjacent mountains drains. They are commonly found in the Basin and Range country in the western states. Here Earth's crust has been broken into parallel-running faults, and great crustal blocks have been uplifted and tilted westward. The east sides of the blocks that trend north-south are steep and rapidly drain water into the basins between the mountains. The lakes are usually dry, with water accumulating briefly only during the rainy season.

Landsat *satellite image of the Basin and Range country around Las Vegas viewed from a 570-mile altitude. White mineral deposits cover dry lake beds between the mountains. Lake Mead is at upper right.*

311

Dry lake beds are a good place to find meteorites for several reasons. They are essentially flat, featureless, vegetation-free catchment basins. They can be crossed quickly on a bicycle or motorbike, and meteorites stand out against their generally light-beige surfaces. You can spot dark-colored rocks hundreds of feet away. One drawback is that many dry lake beds in the West contain a sprinkling of small, dark, basalt rocks, especially around the edges. Also, old rocks frequently have a light mineral coating on them, deposited by evaporation. Meteorites will also become coated with minerals in time, making them more difficult to find.

In 1936, several small chondrites were found on the surface of Muroc Dry Lake in the Mohave Desert in Kern County, California. Four years later, a single chondrite weighing about one-quarter pound was found a few miles west on Rosamond Dry Lake north of Lancaster. This single specimen was identical to those found on Muroc Dry Lake and probably is a stone from the same fall.

The idea of hunting for meteorites on dry lake beds was tested by Ronald Hartman in 1963. He selected Lucerne Dry Lake east of Victorville, California, also in the Mohave Desert. In one day he found a nearly complete chondrite with only a small chip missing. The

Lucerne Dry Lake meteorite. –Courtesy of Ronald N. Hartman

Udall Park stony meteorite (black rock in center) found in desert pavement near Tucson, Arizona. Note the lack of soil in the area.

meteorite was found in a group of small, black pebbles in a clearing near the edge of the dry lake. Winds had apparently blown away the fine soil from a small area surrounded by sagebrush, leaving behind the pebble field, which happened to have a meteorite in it. The missing chip was later found near the access road to the lake. The same year, Ronald Oriti organized a search party of one hundred students from Antelope Valley High School. The students maintained a distance of 6 feet between them and walked the lake all day, searching for additional meteorites. What amounted to a total of eight hundred man-hours of searching bagged one small additional specimen. All three meteorites had fresh fusion crusts, suggesting they had been on the lake bed no more than a year.

Although stony meteorites generally contain some iron metal and will respond to a metal detector, it doesn't make much sense to drag a detector around on a dry lake bed; let your eye be the main instrument. Metal detectors should be reserved for known meteorite strewn fields.

Blowouts. In the dry deserts of the western United States, rocks are often sorted from small sand and soil particles by the wind. In desert environments, topsoil is thin or absent altogether, being transported by winds and flash floods. Left behind is a hardpan surface with cemented small rocks called desert pavement. Among the countless millions of rocks making up this pavement, there undoubtedly are meteorites, but locating them is nearly impossible. However, in less rocky terrain, winds often remove soil and leave behind hollowed-out areas called blowouts, in which buried rocks have been excavated. If the area has few natural rocks, there is a chance of finding meteorites in these blowouts.

Gordon Nelson of Tucson, Arizona, first brought this to my attention. He found some meteorites in desert blowouts in Valencia County, New Mexico, about 25 miles west of Albuquerque. In 1979, Gordon was searching for interesting rocks on the surface of a bowl-shaped area about the size of a house, where the thin desert soil had been removed by the winds. Left behind were many small stones that had once been covered by soil but were now exposed. He noticed one that looked different from the rest. It was smooth and light-colored. Nickel-iron flakes were evident and chondrules were present. Gordon had found an H5 chondrite. The surface was weathered, but the internal structure was still intact. Once he knew what to look for, he was able to find more. Five months and twenty trips later, he had recovered more than two hundred specimens, ranging in size from 10 grams to 200 grams. These are the well-known Correo meteorites, named after the nearest post office.

The story doesn't end there. The Correo strewn field was small, about a mile long and one-quarter mile wide. Within that field were some areas where sand had not been completely removed by the wind. There Gordon found another meteorite, only partially covered, but from an entirely different fall. This specimen was much larger than any of the Correo specimens; it weighed more than 4 pounds. The new specimen was an L5 chondrite. This very different meteorite was given the name Suwanee Springs, even though it was found in the Correo field. It is the only Suwanee Springs meteorite known.

Interestingly, this area is well known to archaeologists for its Indian artifacts. For several years, scientists had roamed unknowingly through the Correo strewn field, looking for artifacts and missing the meteorites. Gordon suggests the searcher set his mind to a mental picture of what a meteorite might look like. "Don't look for a boulder-sized meteorite when it's likely to be walnut-sized," he told me. "The eye tends to see what the mind expects." He pointed out that six months later about 30 miles east of the Correo field, near the small town of Los Lunas, he had been searching another blowout, looking for Correo-sized meteorites. There, he found the Los Lunas H4 chondrite.

No doubt there are countless more Correo meteorites waiting to be discovered as the wind removes the thin desert soil. Since the location of this multiple fall is well known, it might be a good place to use a metal detector, especially in the sandy areas on the edge of the blowouts near the dunes.

The Farm Belt. It has been more than fifty years since Harvey Nininger successfully searched the farms of the Midwest for meteorites. It's time to retrace his footsteps and try again. Surely more meteorites have fallen in the Midwest in the last half century. Farmlands in Michigan, Illinois, Ohio, Iowa, and Indiana were barely touched by Nininger's search. Any place where land is plowed on an annual basis is potential meteorite country. "New" rocks are noticed by farmers. Following Nininger's example, advertisements placed in local newspapers in farm communities will surely bag a few meteorites.

The Antarctic Dream

Meteorites are most easily recovered on flat, featureless terrain where there is little natural rock and no vegetation; terrain where meteorites are not only preserved, but where some natural mechanism is slowly bringing them to the surface. It is helpful if the surface is light in color, so the meteorite's dark fusion crust stands out in stark contrast. Such an ideal place exists—in Antarctica. What better place to find a meteorite than on a blue-white ice surface.

Meteorites were the furthest things from the minds of a group of Japanese glaciologists when they came across the first of nine meteorites lying on the Antarctic ice cap near the Yamato Mountains in December 1969. At first, the scientists assumed they were all from the same fall, but later examination of the specimens yielded surprising results. They were from at least four different classes of stony meteorites, which meant they were from different falls. Yet they were found together, just inland of the Yamato Mountains. Although different meteorites have occasionally been found within the strewn field of a major fall, to find four different meteorite classes clustered together was too extraordinary to be a coincidence. There had to be a natural mechanism that somehow concentrated meteorites from widely scattered landing points on the ice shelf.

GLACIOLOGY. Glaciers are nature's moving platforms. As snow accumulates in cold climates and fails to melt during the warm seasons, it goes through a transformation and becomes a solid slab of ice, a glacier. When enough ice has accumulated, the glacier spreads out under its own weight and begins to move slowly, carrying with it rock and soil excavated along its path. When the glacier begins to melt and recede, it leaves behind tons of rocky debris, forming glacial moraines along its margins.

Glaciers in Antarctica operate somewhat differently. The entire continent is capped by a glacier, which forms in the center of the continent and slowly spreads to the edges at a rate of from 3 feet to 30 feet per year. Most of the glacial ice reaches the sea, where it calves, or breaks apart, by wave action, forming icebergs. Unfortunately, glaciers containing trapped meteorites eventually find their way to the sea when the glaciers break apart and melt. The sea floor along the continental margins may be rich in meteorites that will probably remain lost forever.

Fortunately, there are obstructions along the way. Antarctica has mountain ranges around its periphery that temporarily impede the ice's progress. The ice is so thick that little natural rock finds its way through it from below. Meteorites landing on the surface of the glaciers are covered by snow, which eventually turns to blue ice. The meteorites are never buried too deeply. When the glacier pushes upward against the mountains, downslope winds scour the ice, slowly ablating and evaporating its surface as the glacier, acting like a conveyor belt, continues to press against the mountain barrier. This continual motion of the glacier carries meteorites from widely separated areas to the mountain barriers. There they are exhumed, accumulating at the base of the mountains on top of the ice.

Meteorite Rush. The first Antarctic meteorites were found inland at the base of the Yamato Mountains, where the advancing glacier is

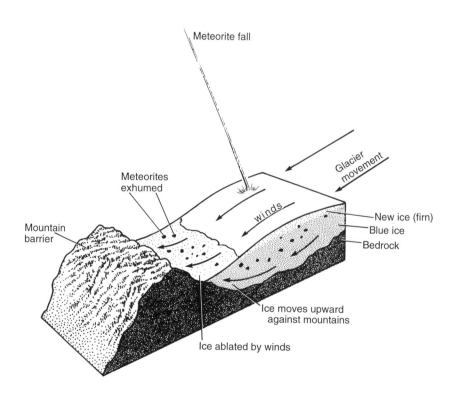

Natural glacial movement accumulates and exhumes buried meteorites against mountain barriers in Antarctica.

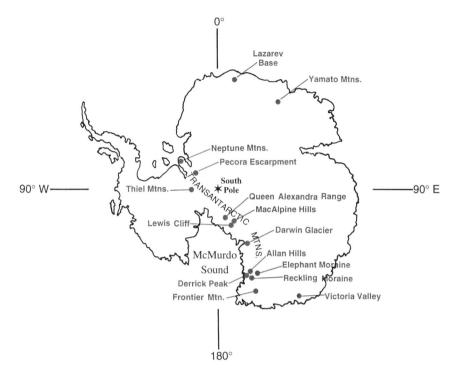

Meteorite finds in Antarctica. Note that most of the meteorite locations are along the continental perimeter, where mountains are located.

rapidly eroded by winds. The Japanese were quick to recognize the natural sorting and excavation mechanism and sent a meteorite field party back to the Yamato Mountains in 1973, where they found more than a dozen new meteorites. In 1974 they netted 663 specimens and the next year 307. Never before had such a rich accumulation been found. The meteorite rush was on!

Allan Hills Find. In 1976, the Americans entered the picture. William Cassidy of the University of Pittsburgh and Edward Olsen of the Field Museum of Natural History in Chicago traveled to McMurdo Station on the edge of the Ross Ice Shelf. McMurdo Station is diagonally opposite the Yamato Mountains. For several weeks they searched the base of the Transantarctic Range near Mount Baldr, but found only a few meteorites. After six weeks they gave up the search and headed back to McMurdo Station. On the way their helicopter pilot suggested that they try searching on bare ice near the Allan Hills at latitude 76°S and longitude 156°E. Within minutes of landing, they found a meteorite. Many other finds quickly followed. They had made a meteorite strike,

William Cassidy with the first meteorite discovered at Allan Hills, Antarctica, in 1976. A small meteorite lies on the snow near his hand. —Courtesy of Ursula Marvin, Smithsonian Astrophysical Observatory

and it was a rich one—between 1976 and 1980, more than eight hundred meteorites were collected from the Allan Hills area.

Meteorite Bonanza. In 1979, a team from the National Institute of Polar Research in Tokyo returned to the Yamato Mountains and recovered an astonishing 3,300 meteorite specimens. This phenomenal success continued, and by 1988 more than 10,000 meteorites had been collected by the Japanese and American teams. This probably represents about eight hundred different falls, an increase in the total number of known falls by more than 25 percent. As expected, most are ordinary chondrites, but all meteorite classes and types discussed in this book have been found in Antarctica. Included are rare ureilites, achondrites, and carbonaceous chondrites, as well as a few that seem to have characteristics intermediate between known classes (see color plate VII). Two shergottites were found in Victoria Land in 1979, due north of Allan Hills. Only seven were known before these were found. One has dark-green glass inclusions, containing krypton, argon, zenon, and nitrogen gases. The gases may be samples of the atmosphere from its home world. They were found in the same relative abundance as those occurring in the Martian atmosphere, measured by the *Viking* landers in 1976.

318

Beautifully shaped iron meteorites were found on the slopes of Derrick Peak, north of Darwin Camp, also in Victoria Land. Sixteen specimens were recovered, with a total weight of 750 pounds. They all have deep "thumbprints" on their surfaces with protruding schreibersite inclusions and are classified as coarsest octahedrites.

They Are All Old Meteorites

Isotopic studies of the Antarctic meteorites reveal the length of time they have spent on Earth. While in space, meteorites are constantly bombarded by cosmic rays that interact with elements in them, producing radioactive isotopes. The isotopes slowly decay, but since the cosmic radiation in space is constant, new isotopes continually take their place and the isotopic content remains constant. As soon as meteorites reach Earth, where cosmic rays can no longer bombard them, their radioactive isotopes decay slowly over time. By measuring the isotopic content in these meteorites and knowing their rate of decay, scientists can calculate how long they have been on Earth. The terrestrial ages vary from 10,000 to more than 1 million years. Antarctic meteorites have the oldest terrestrial ages of any known meteorites. Most of the meteorites in collections around the world were picked up

A beautiful hexahedrite iron (ALHA 81013) weighing nearly 40 pounds lies on the Antarctic ice before being retrieved. The number held above the meteorite serves to identify it in the field. –Courtesy of Ursula Marvin, Smithsonian Astrophysical Observatory

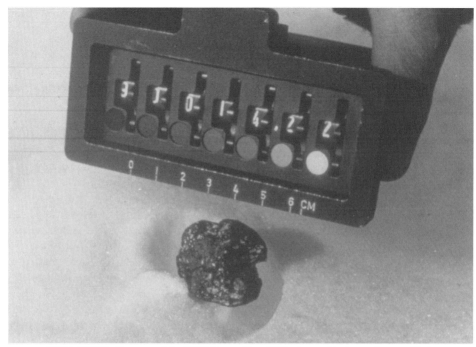

ALHA 81005, discovered in 1982 from Allan Hills, proved to be a glass-welded soil breccia from the lunar highlands. This is the first meteorite from the Moon ever recognized on Earth. Specimen is about 1.2 inches high. —John Schutt photo, courtesy of Ursula Marvin, Smithsonian Astrophysical Observatory

The interior of ALHA 81005 lunar meteorite shows a brecciated texture typical of lunar highland rock. The white rock fragments are plagioclase feldspar. —Smithsonian Institution

soon after they fell and have terrestrial ages averaging one hundred to two hundred years.

The meteorites in Antarctica provide an excellent opportunity to test whether meteorites in the world's collections are typical of meteorites that have fallen over the past million years. Of the eight hundred or so meteorite falls recognized at Antarctica, the ratio of chondrites to achondrites, irons, and stony-irons is close to the ratios in museum collections worldwide. This means that the sampling of meteorites is probably a true representation of the real meteorite populations in space.

Antarctic meteorites provide another unique test. If all meteorites come from the same sources, then the Antarctic meteorites should have compositions identical to those in collections around the world. Some Antarctic meteorites are new types never seen before or new specimens of rare but known types. Recent research has shown that Antarctic meteorites have significant chemical differences, suggesting that meteorites falling to Earth do change with time. This may mean that meteorites come from different sources over time.

Lunar Rocks

A brecciated meteorite that proved to be a sample from the Moon was recovered in 1982 in the Allan Hills area. This rock resembles rocks from the *Apollo 16* landing site in the lunar highlands. It contains large rock fragments cemented in a dark glassy matrix. Apparently, impacts by meteorites on the Moon dislodged lunar rocks, sending them into Earth-crossing orbits. After the first discovery, ten more lunar rocks were recovered including several collected by Japanese scientists in the Yamato Mountains. They are mostly highland breccias with similar mineral compositions. Two are mare basalts. They represent at least two or three different lunar impact events, resulting in fragments that made it to Earth. They all probably came from areas on the Moon not explored by American astronauts or Soviet lunar robots.

A Unique Collecting Site

Antarctica is international, open to scientific groups from all nations; it is not open to amateur collectors. (A few meteorites from Allan Hills have appeared in the commercial and private collector's inventory.) The U.S. Antarctic Meteorite Program is funded by the National Science Foundation. Through grants to American scientists, meteorites are collected in the field with the same careful collection techniques developed by NASA's Apollo Program for collecting lunar samples. The meteorites have been buried in ice for thousands of years in one of the most sterile environments on Earth—they are the least contaminated meteorites ever collected. (Even though all are quite old, they show little weathering. Ice burial has protected them well.) Collection is made with instruments, never by hand, and each specimen is photographed in situ, given a temporary field number, then stored in sealed plastic bags. They are shipped frozen to the Meteorite Processing Laboratory at the Johnson Space Center in Houston. NASA, the National Science Foundation, and the Smithsonian Institution are responsible for cataloging the specimens and disseminating them to qualified scientists throughout the world.

How to Catalog Ten Thousand Meteorites

A simple scheme for cataloging meteorites was in operation before the Antarctic meteorites were discovered: Meteorites were named for the town or geographic feature nearest the place of discovery. So the Pultusk meteorites were named for the Polish town nearest the fall, and the Quinn Canyon, Nevada, meteorite, a superb iron weighing 3,190 pounds, was named for the canyon in which it was found. Oddly enough, irons around Meteor Crater are not named after the crater, but are referred to as Canyon Diablo irons, after a small meandering canyon 2 miles to the west. If more than one meteorite is found in a common area and they are distinctly different, they are given the same name distinguished by the letters *a* and *b*. A stone meteorite was found 7 miles from Forrest Station in the Nullarbor Plain, Western Australia, in 1967 and another about 20 miles away in 1980. The first, an H5 chondrite, was named Forrest (a). The second, an L6 chondrite, was called Forrest (b). An exact longitude and latitude of the recovery point accompanies each name.

Antarctic meteorites presented the unthinkable situation of too many meteorites and not enough distinctive geographic features. A new system had to be invented to deal with the large numbers. Actually, NASA had solved the problem years earlier when lunar rocks were collected. ALHA 80133 is the designation for a CV3 carbonaceous chondrite found at the Allan Hills site (ALH) by field party (A). The five-digit number following the abbreviation of the site lists the year (1980) and the number 133. These last three digits mean that it was the 133rd meteorite found at Allan Hills in the 1980–1981 collecting season.

Other Collecting Sites?

With the phenomenal success of the Antarctic Meteorite Program, scientists are beginning to look at other glacial locations. The Arctic is probably not a good place, since it is covered with pack ice with no continent underneath. The Greenland ice cap seems to be a good prospect, and plans for searches there are under way.

Scientists examine newly discovered iron meteorites found during the 1977–1978 collecting season at the base of Derrick Peak on Hatherton Glacier, Antarctica.

This 5.24-pound iron meteorite was one among seventeen irons found in December 1978 in fractured sandstone on the slopes of Derrick Peak, Victoria Land, Antarctica. This specimen is a coarsest octahedrite with 6.6 percent nickel and measures 6.5 inches long.

Part IV

ORIGINS

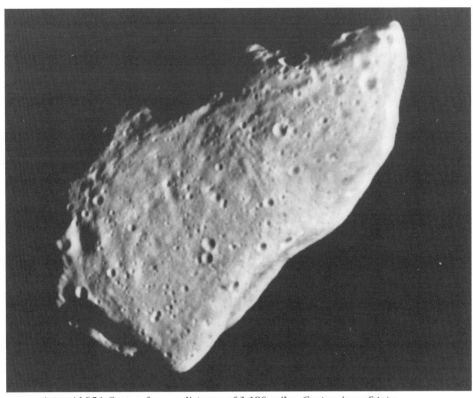

Asteroid 951 Gaspra from a distance of 3,180 miles. Gaspra is an S-type asteroid shaped like an 11-mile-long potato. More that 600 craters mark its surface, attesting to its violent past. Gaspra splintered off its larger parent body about 200 million years ago in a collision with another asteroid. –NASA

EIGHTEEN

Asteroids –
"Parents" of the Meteorites

In a broad zone between Mars and Jupiter lies the asteroid belt. Here great chunks of rock ranging in size from dozens of feet to hundreds of miles across orbit the Sun, just like the planets. They are a diverse group, suggesting the complexity of the processes that formed them nearly 5 billion years ago: These are the parents of meteorites. Simply put, meteorites are chips off these little worlds. The first challenge for scientists studying meteorites was to recognize where they came from. The second challenge was to learn to read and interpret the information locked within them. After the sciences of chemistry, physics, and geology were developed, analytical tools and procedures came into play to tease information out of these alien rocks. Along with our Earth-based technologies, instruments and techniques of the space age have revealed a new solar system unlike anything imaginable only twenty-five years ago. In this new solar system, meteorites were promoted from the level of mere curiosities to being the solar system's Rosetta stones. Today, meteoriticists stand at a crossroads. We are finally smart enough to read the language of the stones and can, with some confidence, begin to tell the story of our solar system's origin. Some gaps need filling in and some of the details will certainly change in time; these refinements will surely come as our ability to read the messages of stone and iron progresses.

The Solar Nebula—A Stellar Incubator

Our Milky Way Galaxy is loaded with great, turbulent clouds of gas and dust collectively called the interstellar medium. These clouds, called nebulae, are typically dozens of light years across. The gas is primarily hydrogen, but heavier gases such as nitrogen and oxygen are mixed with the lighter gases. The dust is made up of tiny, solid mineral grains, probably a mixture of carbon, silicates, and iron spewed out of massive stars during their evolution. The grains form the cores of dust particles, which are encased in ices of methane, ammonia, and water, and the dust is mixed with the nebulae.

Most of the gas is cold and dark, but if there are young, hot stars present within a nebula, the gas will fluoresce as it reacts with the ultraviolet light from the stars. The dust grains become visible when they reflect the blue light from hot stars (see color plate VII).

Among these gas clouds are massive, dark clouds full of molecules. In recent years, astronomers have found increasingly complex molecules in them. The growing inventory now includes organic molecules composed of more than fifty atoms found in the densest, darkest clouds. From these observations, we have reached an important conclusion: Organic molecules are common in interstellar space. Stars born of interstellar dust and gas are born within a rich organic medium, and planetary systems born with their mother stars are equally immersed in organic molecules. These molecules may act as "starter" molecules for the origin of life on planetary bodies.

In the densest parts of the interstellar medium, the stage is set for the birth of stars. A nebula is a stable mass of gas and dust until something compresses it. Once compressed, turbulent regions appear within the nebula, causing it to fragment into pieces with densities higher than the average density of the cloud. These fragments appear as dark, roughly spherical clumps about 10 light years across silhouetted against the bright nebula. Nebula fragments act like cocoons or

Lagoon Nebula in Sagittarius. A number of small black fragments can be seen against the brighter nebula. These are stellar cocoons, where stars are born. –National Optical Astronomy Observatory

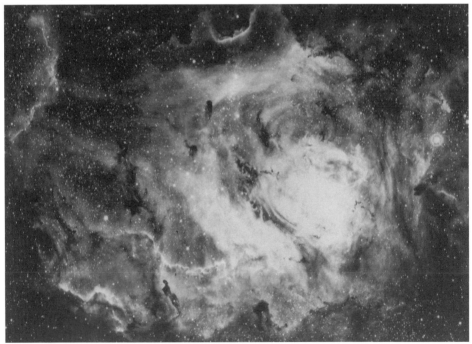

A section of the Rosette Nebula, showing bright young stars, members of an open star cluster that formed together within the nebula. —National Optical Astronomy Observatory

stellar incubators, in which matter is infalling and continues to be compressed through its own internal gravity.

Astronomers have tried to discover how compression of nebulae might take place. Cosmic collisions provide one way. Dark molecular clouds frequently collide with bright gaseous nebulae, setting up shock waves that compress the nebulae and spawn star formation. Collisions of molecular clouds with nebulae typically produce star-forming regions that result in multiple births of stars in one area. Probably as many as 80 percent of these stars are members of binary stars. They did not form as single stars but as twin, triplet, or quadruplet stars all orbiting around a common center of mass. These galactic clusters have limited lifetimes and eventually break apart, scattering the member stars, but many binary stars remain intact and continue to travel around our galaxy. Planetary bodies that may have formed in such multiple star systems cannot maintain stable orbits and ultimately are ejected from the system to wander as dark, cold bodies among the stars.

A Supernova Birth

As far as we know, our Sun was not born a member of a binary system, though it may have been a member of a galactic cluster early in its lifetime. It probably experienced a more solitary birth in a rather different way, though the final result is the same. There is excellent evidence that the Sun was born out of the death throes of a mighty star.

Five billion years ago a star many times more massive than the Sun happened to be close to a nebula that contained the material from which the Sun would be born. Throughout this star's lifetime, it had been using hydrogen fuel in its core at a prodigious rate, converting it to helium. (Cores of stars are giant thermonuclear furnaces that produce the energy necessary to keep them shining and, more important, to keep them from collapsing under the weight of their overlying masses.) There was sufficient hydrogen in the star to last only 10 or 20 million years. Once the hydrogen was completely consumed, its thermonuclear furnace shut down, causing its core to collapse as it responded to the tremendous gravitational forces within it. Its core, now pure helium, squeezed down by the overlying mass, heated rapidly until it reached 100 million degrees. At this incredible temperature the thermonuclear furnace reignited, this time using the helium in the core as a fuel to create still heavier elements such as carbon, oxygen, and silicon.

Once such a massive star begins to produce elements heavier than helium, it evolves rapidly and goes through several unstable stages creating layers of progressively heavier elements surrounding the reacting core. In the process of building heavy elements, the star passes quickly through a series of episodes in which the envelope of gas surrounding the layers expands to an enormous size. The star becomes a red giant. It passes through several of these red giant stages, growing to millions of miles in diameter. Throughout most of its relatively brief lifetime, the star releases heavy elements into space that condense to form dust grains. These grains continuously enrich the adjacent nebula.

The massive star followed this process until the inevitable end came. The final core reaction created the element iron, the end product in fusion reactions of massive stars. Iron cannot fuse to form heavier elements, so the core was no longer able to continue the nuclear reactions necessary to maintain the star's stability. The core experienced a final, rapid collapse to a super-dense state and then rebounded, sending a violent shock wave through the star's surrounding envelope. The shock wave compressed the envelope as it made its way through the star, creating many new heavy elements as it progressed. Then, as if to announce its glorious death to the universe, the shock wave penetrated the surface, ripping the star apart in a gigantic explosion and spewing out its newly created heavy elements like a flower releasing its seeds to the wind. These heavy elements quickly spread through the surrounding nebula.

Nearly simultaneously, the great shock wave invaded the nebula, producing turbulent regions that fragmented the nebula into prestellar

Crab Nebula in Taurus. This is a supernova remnant created when a star exploded into a supernova in A.D. *1054.* –National Optical Astronomy Observatory

clouds or cocoons. One of these cocoons, perhaps a hundred times the size of the present solar system, now called the solar nebula, was destined to give birth to the Sun. Gas and dust from the surrounding nebula rapidly gravitated to the dense center of the cocoon, forming a core of hot, dense gas. This core had a voracious appetite. As more material collapsed to the center of the nebula, the more it compressed and the hotter the core became. Eventually, the nebula's core began to emit a feeble light–the first rays of a star's birth.

As matter continued to accumulate in the center of the solar nebula, compressing and heating it, another reaction to this collapse occurred. Like all bodies in space, the original cocoon was rotating slowly. All rotating bodies possess angular momentum, which is always conserved. The total angular momentum depends upon the mass of the body, its radius, and its rotational velocity. A change in any of these three quantities results in a proportional change in one of the other two. As the solar nebula collapsed, it radius was changed. To conserve the body's angular momentum, it had to rotate faster, which is exactly what the solar nebula did during its earliest stages. (The nebula was also gaining mass at the same time, which tended to somewhat reduce but not stop the increase in rotation rate.) Even though strong internal gravitational forces acted on the cloud to compress it into a sphere, the rapidly spinning sphere resisted the inward motion of material and began to flatten out, reacting to strong centrifugal forces at right angles to its spin axis. A great disk of material spread outward from the core. The solar nebula became a spinning disk in which the planets and asteroids were destined to form (see color plate VII).

The Solar Disk

The initial conditions in this solar disk gave birth to the events that followed. The center of the disk was the densest and hottest part. It continued to collapse under its own gravity while material from the surrounding nebula flowed through its rotational axis, adding to its mass. Outside this center and above the disk plane, heavy-element dust rained down like tiny hailstones. Compression of this dusty material along the disk plane raised the disk's temperature but actually only slowed the cooling of the disk as the heat rapidly radiated into space.

A temperature gradient existed across the disk. Near its center, where the Sun was forming, the temperature was highest, close to 3,600 degrees Fahrenheit, but the temperature along the disk fell rapidly away from the center until it reached minus 446 degrees at the edge, typical of the temperature at the outer fringes of the solar system today. This steep gradient produced very different chemical conditions along the disk. Close to the center, where the temperature was above 2,900 degrees, the interstellar dust grains were vaporized. Turbulence in the nebula thoroughly mixed the interstellar atoms with the rest of the elements of the disk, producing a homogeneous gas. Farther out on the disk, temperatures were too low to vaporize and mix the interstellar grains and the grains survived, destined to be incorporated into calcium-aluminum–rich inclusions (CAIs) and chondrules.

Isotopes Tell Their Story

Survival of the original interstellar grains became strikingly clear to meteoriticists who analyzed CAIs found in the Allende carbonaceous chondrite (see chapter 10). The white, irregular grains contain isotopes that could not have formed in the solar nebula. Aluminum-26 (^{26}Al) is one of many isotopes that astronomers believe are formed during supernovae explosions. ^{26}Al has a rapid decay rate with a half-life of only 720,000 years, a short period of time in the calendar of solar system history. All the ^{26}Al has long ago disappeared from these primitive meteorites. ^{26}Al was initially incorporated into the crystal structure of the CAIs and chondrules but decayed to its daughter isotope magnesium-26 (^{26}Mg), which took aluminum's place in the crystal lattice. It is this magnesium-26 in the Allende meteorite that speaks loudly of a supernova event having triggered the formation of the solar system. Collapse of the solar nebula, signaling the beginnings of the solar system, must have taken place soon after the supernova occurred, within less than a million years, otherwise the ^{26}Al would have decayed away before being incorporated into the CAIs.

Other isotopes tell a similar story. Oxygen has three stable isotopes: ^{16}O, ^{17}O, and ^{18}O. These isotopes differ only by the number of neutrons in their nuclei. In terrestrial minerals there are specific, known ratios of ^{18}O to ^{16}O and ^{17}O to ^{16}O. The CAIs in Allende have considerably more ^{16}O than other minerals in the same meteorite. Like ^{26}Al, ^{16}O is produced by supernovae explosions. The other two oxygen isotopes are destroyed in the violent process.

Discovery of these isotopes reveals not only the presence of a supernova that may have started the process, but also that the disk was not uniformly homogeneous; the inner disk close to the developing Sun was homogeneous, while the outer parts of the disk remained heterogeneous. Thus, meteorites or parent bodies that formed outside the homogeneous boundaries would have a different isotopic composition. This important discovery will tell us where to look for the parent bodies of the meteorites.

Formation of the First Minerals

As the inner solar disk cooled, condensation of minerals began to take place. For a short time, the inner disk was at a temperature above 3,270 degrees Farenheit. At that temperature, all elements are in a gaseous state, even elements such as aluminum and iron. Minerals differ widely in their melting and vaporization temperatures. If two minerals are mixed in the liquid state, as many minerals are in a magma, they can be easily separated by simply cooling the liquid until one of the two minerals crystallizes and drops out. This is called fractional crystallization, or fractionation. Fractionation is common in nature. A good analogy is the formation of ice when the atmosphere cools below the condensation and freezing temperatures on a clear winter night. Much the same process occurs in a rapidly cooling elemental mixture of gases. Elements that condense at high temperatures, such as aluminum, magnesium, or iron, are called refractory elements. They are the first elements to form solid grains when a gas cools. Elements that condense at low temperatures are called volatile elements, since they remain gases at temperatures at which the refractory elements condense. The volatile elements form molecules such as water, methane, and carbon dioxide. When these molecules condense, they form ice in the outer regions of the solar system, where the temperatures are quite low.

The first elements to condense from the inner solar disk were metals such as aluminum, titanium, calcium, and magnesium. These metals reacted with surrounding hot gas to form metallic oxides. The

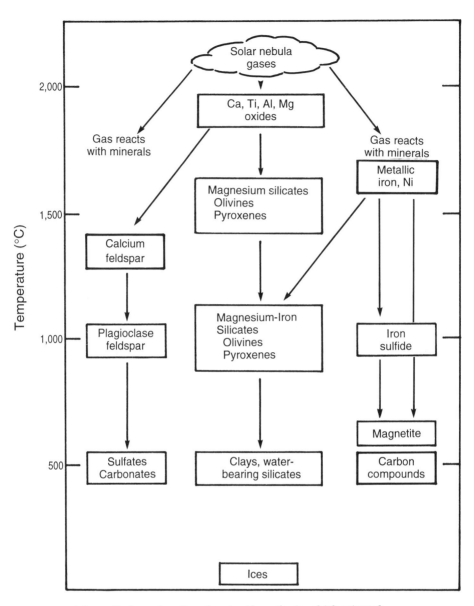

A theoretical construction showing the order in which minerals condense out of a cooling gas in a solar nebula. Oxides appear at about 1,800 degrees Centigrade (3,270 degrees Fahrenheit). Arrows indicate those minerals that continue to react with the nebula to form other minerals at lower temperatures.

appearance of silicon (Si) at about 2,900 degrees Fahrenheit allowed the first pyroxenes, olivines, and feldspars to form: diopside, forsterite, anorthite, and enstatite. All of these are important constituents of chondritic meteorites and all are rich in magnesium, calcium, or aluminum.

Metallic iron alloyed with nickel condensed with the first silicates, probably forming the metal flakes found in chondrites. Iron-bearing olivine, pyroxene, and troilite did not form until the gas reached about 1,800 degrees Fahrenheit. The claylike minerals with water in their crystal structures formed when the temperature reached 932 degrees. This occurred first at the outer edge of the present asteroid belt. Carbon compounds also formed at this temperature and together with the clays provided the material for the carbonaceous chondrites. Finally, near 212 degrees, water began to condense and react with other volatile gases such as ammonia and methane.

This condensation sequence suggests the composition we should expect to find for bodies forming at different points along the solar disk. It's important to remember that it is the minimum temperature at all points on the disk that determines the composition of the condensed bodies.

We would expect the ratio of metallic iron to silicates in the planets to vary from the Sun outward. Close to the forming Sun, we would expect to see planets made up primarily of the highest refractory minerals, namely metallic iron and lesser amounts of silicates that form at the highest temperatures with no volatile gases such as hydrogen or helium or, in fact, water. Mercury, at 0.4 AU from the Sun (1 Astronomical Unit is equal to the mean distance of Earth from the Sun—92.9 million miles), fits this composition rather well. Relative to its diameter, it has the largest iron core of any of the terrestrial planets (the four inner planets, Mercury, Venus, Earth, and Mars). Outward from Mercury scientists could expect to find less metallic iron and more iron sulfide, along with more magnesium and iron silicates composing the planets. Indeed, the iron core of Earth is smaller relative to its diameter than Mercury's. Water is still a gas at 1 AU and should not be an important constituent in the makeup of these planets. Mercury is predictably volatile-poor but Earth, formed farther out, is more volatile-rich than we would expect. Earth's oceans seem to be a knot in the theory, but much of its water may have come from cometary bodies after its formation. Its rich oxygen atmosphere was generated much later through metabolic processes of plant life. Venus lies somewhere between with most of its volatile materials in its atmosphere. High surface temperatures forced all of its water into the atmosphere, where it reacted with sulfur dioxide to form sulfuric acid clouds.

At Mars's mean distance from the Sun, 1.5 AU, more volatile elements were available. Most of the iron was combined with oxygen, so Mars probably formed only a small core of iron or iron sulfide. It has a thick, iron-rich silicate mantle and distinct crust. Of the four terrestrial planets, we would expect Mars to have the lowest density, and it does.

Farther out we would expect to find both silicates and ices of the volatile gases mixing to form solid bodies. This is certainly the case for the satellites of the outer planets.

From Minerals to Rocks to Planetesimals

On the solar disk, tiny, dust-sized mineral grains were beginning to rain out of the solar nebula as they condensed under cooling conditions. They accumulated along the central plane of the disk, forming a thin dust layer. This occurred within a short period of time, only a few tens of thousands of years. Turbulence in the disk mixed the grains, causing them to collide and stick together, as snow crystals stick together to form flakes. In fact, the first grain clumps may have been the consistency of fluffy snowflakes. The fluffy flakes of minerals accumulated rapidly, building boulder-sized bodies much as a rolling snowball gathers snow and increases in size. In the case of iron and magnetite, the solar magnetic field may have provided magnetism to hold the minerals together. Within 10 million years rocky bodies were formed, ranging from the size of large boulders to dozens of miles across. The largest of these bodies, now called planetesimals, eventually became massive enough to create a substantial gravitational field around themselves. They continued their growth process by pulling in more solid particles with their increasing gravity.

Occasionally, planetesimals with slightly intersecting orbits collided. In some cases this may have disrupted the planetesimals, but since they were all orbiting the infant Sun in more or less the same direction and with the same speed, the collisions would be at low velocity, so the pieces probably gravitated together, resulting in a fragmented but larger body. Other colliding planetesimals simply stuck together without fragmenting (see color plate VIII). This accretion process continued as the growing planets, now called the protoplanets, orbited the Sun, tunneled through the rocky debris, and added substantially to their masses. Accretion of the terrestrial planets took about 100 million years to complete.

Planets farther out than Mars also accreted rocky material from the condensing nebula, but they were in a region where the temperature was low enough to condense water, so their compositions are part rock and part ice. The satellites of the outer planets have rocky-ice compositions. The giant planet Jupiter accreted a sizable rocky core that gravitationally attracted the most volatile gases, hydrogen and helium. At Jupiter's distance from the Sun, these volatile gases were plentiful. Reactions of hydrogen with carbon formed methane, and hydrogen with nitrogen formed ammonia. These, along with hydrogen and helium gas, formed the huge atmospheres of Jupiter and Saturn. Jupiter and Saturn probably reached their present form within the first 10 to 20 million years. This early formation had a profound effect on the solar system.

From Protosun to Sun

While mineral grains were condensing, the central part of the solar disk continued to pull in material from the surrounding nebula. Rapid collapse of the roughly spherical body continued to increase its temperature. For a star of the Sun's mass, less than a century is required for the star to begin to emit heat and then a feeble light. When that occurred, the central condensation became a protosun. Increasing heat from the developing protosun probably acted to reduce somewhat the rate of cooling of the disk.

As the collapse continued, the temperature rose steadily. The protosun continued to collapse for 50 million years. For a period between the collapse phase and the beginning of nuclear reactions in its center, the protosun increased in brightness and violently ejected material into space, forming a high-speed wind. Stars with masses similar to the Sun's experience this period of instability during their formative years. This period is called the T-Tauri stage, named after a star in the constellation Taurus, the first observed to exhibit this activity. Solar winds quickly swept the inner solar system clean of volatile elements and unused mineral grains, leaving behind only the growing planetesimals and protoplanets. This violent activity continued until the protosun's core temperature reached 27 million degrees Fahrenheit. At this high temperature, nuclear reactions occur in which hydrogen fuses to form helium. Enormous gas pressures built up in the protosun's core that stopped the gravitational collapse. At this point, the protosun moved to the ranks of a stable star much as it is today—in equilibrium, with its inward-directed gravitational force exactly balanced by its outward-directed gas pressure. The Sun as we know it had finally arrived.

The Asteroid Belt

The asteroids are failed planetesimals that were never incorporated into a planet-sized body. They are the parent bodies of meteorites, which are pieces chipped off by collision. Meteorites show great diversity in composition and structure, and apparently they do not all share the same history. Many remain as primitive bodies, never melted or changed since the solar system began. Others have been heated. Some even melted.

Early Asteroid Population. Why did the asteroids tend to cluster within a zone between Mars and Jupiter? They probably didn't at first. Originally, asteroids formed throughout the inner solar system. This may have placed them as close as 20 million miles from the Sun. On the other extreme, there were probably sufficient refractory minerals forming as far out as the present orbit of Jupiter, nearly 500 million miles from the Sun, where more volatile elements mixed with the refractory elements to produce water-bearing planetesimals. Therefore, in the early solar system, the asteroids were distributed more or less uniformly within the inner solar system. When the planets began to form, the situation changed dramatically. Much asteroid material was swept up by the forming planets, leaving zones around their orbits devoid of asteroids. Those asteroids that did not meet this fatal end were perturbed by the gravity of these growing worlds. Their orbits gradually changed and became more elongated, sending them across the orbits of the inner planets and through the present asteroid belt. This is like crossing several lanes of traffic on a freeway. The chances of colliding with an oncoming automobile increase appreciably. It was only a matter of a few hundred million years before the asteroids that formed in the inner solar system were essentially wiped out.

The Jupiter Effect. Enter the great planet Jupiter, with a mass 317 times Earth's mass. It presents an enormous gravitational cross section in space that dramatically affects the motions of small bodies. Jupiter formed early in the solar system's history, millions of years before any planets could accrete where the asteroid belt now lies. The formation of a massive planet at Jupiter's distance from the Sun had a profound effect on this zone. Planetesimals formed there much as they did throughout most of the solar system, but gravitational perturbations from Jupiter prevented the planetesimals from coalescing into a single planet-sized body. Originally, there may have been sufficient planetesimals to form two Earth-mass planets, but Jupiter's gravity accelerated many from their initial orbits, sending them out of the solar system forever.

The remaining asteroids, if consolidated, add up to a body considerably less massive than Earth's moon. Thus, our solar system was robbed of an Earth-sized planet beyond Mars. The small size of Mars, only half that of Earth, may have been an additional effect of Jupiter's interference with the accretion process.

Discovery of the First Asteroid

The first asteroid was an accidental discovery on New Year's Day, 1801. Giuseppe Piazzi, then director of the Palermo Observatory on the island of Sicily, was measuring star positions when he came across a star in the constellation Taurus that was not on any star charts. Piazzi tracked the position of the object over several nights and found it was moving slowly, with a speed that suggested it was between Mars and Jupiter.

This discovery did not take Piazzi completely by surprise—an unseen planet had been predicted at that distance as long ago as the late sixteenth century, when astronomer Johann Kepler suggested in 1596 that for metaphysical reasons, one *ought* to be found.

Kepler, a contemporary of Galileo, later discovered a mathematical relationship between the orbital period of a planet and its mean or average distance from the Sun. Once the planets' periods of revolution around the Sun were determined through observation, their mean distances could be calculated. The mathematical equation required the

Saturn 9.54 AU

Jupiter 5.20 AU

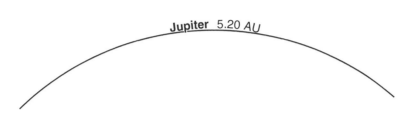

Missing planet 2.80 AU

Mars 1.52 AU

Earth 1.0 AU
Venus 0.73 AU

Mercury 0.39 AU

Sun

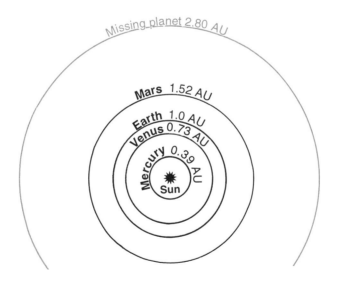

*Planetary orbits out to Saturn with their distances in AUs
derived from their orbital periods. Note the remarkably even
spacing of the planets with the single exception of an obvious
blank space between Mars and Jupiter.*

THE TITIUS-BODE RULE

Planet	Titius-Bode Rule	Actual Distance (AU)
Mercury	0.4	0.39
Venus	0.7	0.73
Earth	1.0	1.00
Mars	1.6	1.52
Missing Planet	2.8	2.77 (Ceres)
Jupiter	5.2	5.20
Saturn	10.0	9.54
Uranus	19.6	19.18
Neptune[a]	38.8	30.06
Pluto[a]	77.2	39.52

[a] The rule fails for both Neptune and Pluto.

periods of the planets to be expressed in terms of the Earth year and their mean distances from the Sun in terms of Earth's mean distance. Kepler could not give a physical reason for the even spacing of the planets, nor could he offer an explanation for the gap between Mars and Jupiter. He could only conclude that "though invisible to the eyes now, a large planet existed in this region."

More than a century later, in 1766, German astronomer Titius von Wittenburg found a curious mathematical tool that showed the spacing of the planets as a mathematical progression. He began a sequence of numbers beginning with zero, then 3, then doubled each number thereafter: 0, 3, 6, 12, 24, 48, 96. . . . In this progression, he added 4 to each number and then divided the resultant numbers by 10. The sequence became 0.4, 0.7, 1.0, 1.6, 2.8, 5.2, 10.0. . . . These numbers represent almost exactly the mean distances of the planets in AUs from the Sun to the last planet then known, Saturn, at 10 AU. Titius's "rule" clearly showed a vacant space between Mars and Jupiter, at 2.8 AU, exactly as Kepler had noted.

Titius's rule is often called Bode's law, after the German astronomer who used Titius's rule to argue for an unknown planet between Mars and Jupiter. It remained a curiosity and was not taken seriously until the English observer William Herschel, then an amateur astronomer, made the first discovery of a major planet beyond Saturn in 1781. Its distance was 19.18 AU, the planet Uranus. Titius's rule placed the new planet at 19.6 AU, remarkably close to the real value. This seemed to validate the rule, and astronomers began to take more seriously the possibility of an undiscovered planet between Mars and Jupiter.

Piazzi was certainly aware of Bode's law and the prediction of an unseen planet. He continued to observe the object until February 11, when poor weather intervened. By the time Piazzi made his position observations known so that an orbit could be calculated, the object was lost from sight. Fortunately, it was recovered exactly one year later and hasn't been lost since. Piazzi gave the object the name Ceres, after the Roman goddess of agriculture. The asteroid Ceres was found to have a mean distance of 2.77 AU, exactly where a planet was predicted to be.

A Belt of Asteroids. Scarcely had the excitement over Ceres' discovery subsided when another starlike object was located near Ceres' position. Heinrick Wilhelm Olbers, a German physician and astronomer, found a second asteroid, Pallas, on March 28, 1802. It was after this second discovery that William Herschel gave the generic name asteroids to these obviously small bodies. He referred to them as starlike, or "asteroidal" in appearance.

With a respite of only one year, a third asteroid, Juno, was found on September 1, 1804, and after two and a half more years, Olbers discovered Vesta on March 29, 1807. It was becoming increasingly apparent that one large planet did not form at 2.8 AU but, instead, many small planets. There the issue rested for the next thirty-eight years, due in part to war waged between Germany and France, which interrupted observations until 1845. In that year a German amateur found asteroid number five, Astraea. From this time forward the asteroid bottle was uncorked, and more and more asteroids were discovered in rapid suc-

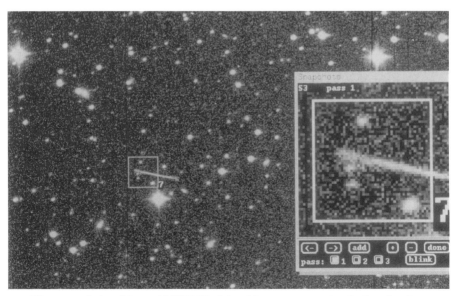

Discovery photo of Asteroid 1992 JD. Asteroid appears as a streak, showing its motion among the stars. The electronic image is enlarged in the box at right.
—Courtesy of Tom Gehrels, Spacewatch project, University of Arizona

cession. By 1890, three hundred asteroids had been found both by professional and amateur astronomers, using only their eyes at the telescope.

Thirty years earlier, the photographic process had been successfully applied to the telescope to produce pictures of the brightest astronomical objects such as the Moon, planets, bright stars, and nebulae. By the late 1890s, increasing film sensitivity allowed astronomers to replace the eye at the telescope with a camera, and the real asteroid harvest began. Photographic plates exposed over a period of a day show a passing asteroid, seen as a starlike object, moving among the stars. Today, thousands of asteroids are known, and orbits for more than four thousand of these have been calculated and cataloged. Most are main-belt asteroids staying within the confines of the belt. A few pass inside the orbit of Mars with at least two known to pass inside Mercury's orbit. One, at least, has recently been found beyond Saturn, but the most interesting are those that pass across Earth's orbit.

Gaps in the Asteroid Belt. As Jupiter's gravity confined the asteroids into a more or less restricted area, it also dictated their locations within that belt. Plots of their mean distances from the Sun revealed gaps in the belt where no asteroids seemed to exist. Especially prominent were gaps at 2.50, 2.83, and 3.28 AU. In 1866, the American astronomer Daniel Kirkwood provided an explanation for these gaps, known today as Kirkwood gaps. He noticed that the distances corresponded to orbital periods that are simple fractions of the orbital period of Jupiter (11.86 years). For example, the gap at 2.50 AU represents an orbital period of 3.95 years, which is exactly one-third of Jupiter's period. The

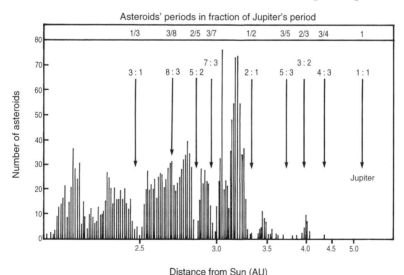

Kirkwood gaps. Plot of asteroid distance versus orbital period shows obvious gaps in the asteroid belt produced by Jupiter's perturbations on commensurate asteroids.

significance of this is that an asteroid at 2.50 AU from the Sun will line up with Jupiter every three asteroid years, thereby falling under the influence of Jupiter's strong gravity. To be sure, other asteroids at various random distances also pass by Jupiter, but not nearly as often as those with orbital periods commensurate with Jupiter's. Asteroids at 2.83 and 3.28 AU have periods of 4.76 and 5.94 years, respectively, which when compared with Jupiter's period give ratios of 5:2 and 2:1.

The constant repetition of asteroid-Jupiter alignments over millions of years takes its toll. Periodic gravitational perturbations create a phenomenon astronomers call resonance. Most people experience resonance in some form almost daily. Anything that vibrates has a resonant frequency. A clock pendulum is a good example. As long as the pendulum maintains a constant swing period, the clock operates normally. Energy is imparted to the pendulum by the clock's mainspring to give the pendulum just the right amount of push at the right moment and direction in its swing to keep the pendulum oscillating at the same frequency. But if the pendulum is interrupted at midswing, forcing it to vary its period, the clock acts erratically or ceases to function altogether. Conversely, if the push increases at regular intervals, the period will lengthen, again causing the clock to malfunction.

Like the orbits of planets, the motions of asteroids are controlled by the Sun's gravity, the mainspring of the solar system. Jupiter's gravity acts as an additional force upon the tiny bodies. This periodic pull slowly increases the eccentricity of the asteroid's resonant orbit, causing the asteroid to change its position and swing either outward or inward from Jupiter, leaving behind a gap where it used to be. Asteroids or meteorites entering the Kirkwood gaps are perturbed into orbits that allow them to leave the confines of the belt altogether. The gaps are open doors through which meteorites and asteroids can invade the inner solar system and, indeed, come to Earth.

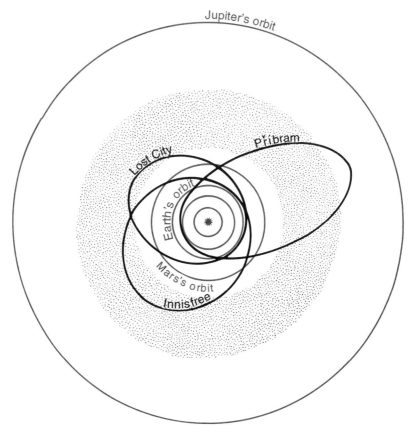

Orbits of the Příbram, Lost City, and Innisfree meteorites, all of which had their aphelions within the asteroid belt.

NINETEEN

Searching for Meteorites' Lost "Parents"

Little more than thirty years ago, a few courageous scientists put forth the argument that meteorites come from asteroids. This fact seems obvious to us today, but it was a long "rocky" road to establish that idea. Asteroids were known to be small bodies, too small to have had thermal histories like those of the planets. It was hard to imagine that even the largest asteroid was large enough to retain sufficient heat to melt and differentiate. Most scientists agreed that asteroids were primitive bodies, unchanged since their formation, and because most meteorites showed a history of melting, they must have encountered very different conditions.

Scientists would have given all their graduate students to get their hands on samples of these primitive bodies, the asteroids. Enormous sums of money and hundreds of thousands of hours were expended to reach the Moon, where scientists thought rock samples could be collected from the "beginning." Meanwhile, meteorites already lay in collection drawers around the world, waiting to yield their precious cargo of solar-system history. Where are the monuments to those scientists who were the first to realize that meteorites represented the very samples they wanted?

Tantalizing evidence that meteorites came from the asteroid belt actually did exist during the Apollo years. Ideally, astronomers wanted to know the orbit of a meteorite before it entered Earth's atmosphere, and it would be nice if the meteorite itself was on hand for examination. This unlikely combination came about in 1959 when an accidental photograph of a brilliant fireball was captured by several cameras set up in Czechoslovakia to record satellite tracks. From these photographs, taken from widely separated locations, the orbit of the object was accurately determined. And as a bonus, four meteorites were found near the town of Příbram, not far from Prague. All were H5 chondrites. For the first time in history, the orbit of a recovered meteorite was known, and its aphelion was in the outer reaches of the asteroid belt.

Encouraged by this success, the Smithsonian Astrophysical Observatory set up cameras in an attempt to photograph incoming fireballs.

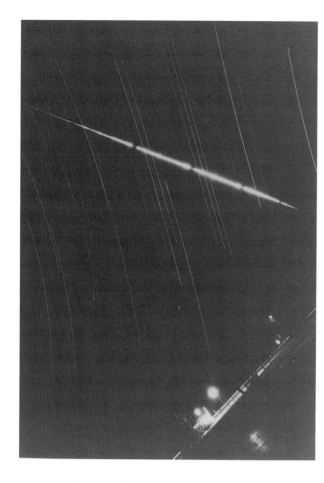

Lost City fireball from Prairie Network cameras. The camera was tilted with respect to the ground. Arcs are star trails, and the meteor train is segmented due to a rotating blade operating over the camera lens.
−Courtesy of Smithsonian Astrophysical Observatory

They hoped this would lead to the recovery of meteorites and a calculation of their orbits. A battery of cameras called the Prairie Network, consisting of sixteen stations centered on southeast Nebraska, was set up in 1964. It extended 300 miles in all directions. The cameras operated for ten years. In that time hundreds of fireballs were photographed, but only one yielded a meteorite. Four meteorite fragments totaling 37 pounds were recovered near Lost City, Oklahoma, in January 1970. It turned out to be another ordinary H5 chondrite, and, more important, calculation of its orbit determined that, once again, the aphelion placed it within the asteroid belt.

One additional meteorite was recovered in 1977 by a similar Canadian photographic network. This meteorite fell near Innisfree, Alberta, and was a rarer LL5 amphoterite. Its orbit placed it in the middle of the asteroid belt. The orbits of these three meteorites provided strong evidence that meteorites were pieces of asteroids.

Now the pieces of the puzzle are falling into place—scientists have the tools and knowledge to begin searching for the parents of these meteorite orphans. Using texture and chemical composition, meteor-

Recovery of the Innisfree meteorite through a photo survey by the Canadian photographic network.
—Courtesy of Hal Povenmire

iticists are trying to match meteorites with their parent asteroids at various locations in the asteroid belt.

It is a tricky business to make correct assumptions about conditions in the early solar system—what did the original parent bodies look like? How were they formed? What happened to them after they were initially formed? Meteorites yield at least some of the answers and provide the materials to test the hypotheses.

Chondrite Parent Bodies

Chondrites are the most common stony meteorites and show the greatest diversity. Meteoriticists are still arguing about how the enigmatic chondrules formed. With the exception of the CAIs, chondrules are the oldest structures in the solar system. Chondrules in the CV3 carbonaceous chondrites are the best preserved and show the least amount of heating or no heating at all. The ordinary chondrites show various amounts of heating. This suggests that those asteroids which formed closest to the Sun (ordinary chondrites) were heated much more than those formed farther away (carbonaceous chondrites). What was responsible for the early heating of the asteroids? The Sun was much brighter during its formative stages, just before it stabilized, but

347

significant heating did not come from an overly bright Sun. Another heat source tied either to the Sun or to a source in the solar disk close to the Sun must have been available.

Heating may have occurred during the T-Tauri stage of the protosun. At that time, the Sun ejected huge quantities of charged particles, forming a violent solar wind that spread throughout the solar system and was particularly intense close to the Sun. The charged particles flowing around the planetesimal surfaces produced electric currents that heated the surface to various states of metamorphism. At the same time, the Sun was rotating many times faster than it does today, creating a strong magnetic field. Magnetic field lines anchored within the Sun spread outward across the solar disk. The Sun dragged the field lines through the solar disk as it rotated, winding up the magnetic field lines into a spiral. As the Sun's strong but varying magnetic field swept across the forming planetesimals, they induced electric currents within the bodies. Electric currents produce heat much as electric currents induced by magnetic fields in an induction stove heat cookware made of conducting iron. Depending upon the conductivity of the asteroid, sufficient heat could have been produced to thermally alter the chondritic structure to the very core of the asteroid.

Heating by electromagnetic induction was short-lived on the scale of solar system history—about a million years. Once nuclear reactions began in the Sun's center, the Sun stabilized. Its luminosity and solar winds subsided to their present level. During this stabilizing period, the Sun's rotation slowed as it lost most of its angular momentum to the escaping solar-wind particles. This altered the Sun's magnetic field, reducing drastically the induced electric currents in the disk.

The effectiveness of this heat source depended upon the body's distance from the protosun. The carbonaceous chondrites probably formed at the outer edge of the asteroid belt. Several observations lead to that conclusion. Carbonaceous chondrites show no thermal metamorphism. Their position farther from the Sun may have protected them from the solar wind and induction heating. They are far enough out to be within the zone of ice condensation. Careful study of C1 carbonaceous meteorites shows structural changes due to the presence of water. There must have been water on or within the parent bodies circulating over a sufficient length of time to alter the original minerals to claylike minerals. C2 meteorites show much less aqueous alteration and C3 meteorites show no alteration at all.

All of this leads back to the same conclusion: Ordinary chondrites must have formed closer to the Sun than carbonaceous chondrites, where they were subjected to more solar heating. The C1 meteorites formed farthest out and the C2 and C3 meteorites progressively closer to the Sun, where less water was available, but not close enough to be affected by the solar wind.

Oxidation of elements in the chondrites is another criterion that suggests specific placement of the parent bodies in the asteroid belt. Ordinary chondrites are classified as H, L, and LL, according to the amount of total iron (which includes iron combined with oxygen as well as free iron metal). Iron as a refractory element appeared early in the condensation sequence but at a time when the inner solar system was being rapidly depleted of volatile elements. One of those volatile elements was oxygen. Chondrites in which most of the iron is in the unoxidized metal state or in combination with sulfur imply that these meteorites formed in a nearly oxygen-free environment. This would place them on the inner edge of the asteroid belt less than 2 AU from the Sun. The enstatite (E) chondrites seem to fill this niche, since they have almost no oxidized iron but plenty of metal. Some meteoriticists believe that the E-chondrites may have formed inside Mercury's orbit.

How to Make a Chondrite Parent Body

Chondrite parent bodies were the first to form in the inner solar system (comets were probably the first to form in the outer solar

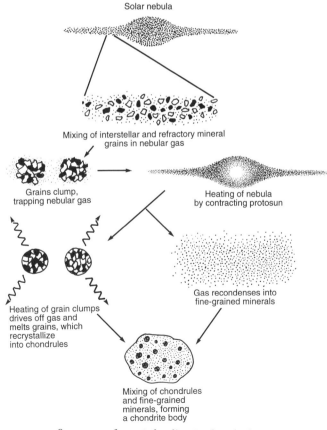

Sequence of events leading to chondrule formation and a chondritic parent body.

system). To make a chondrite body, chondrules first need to form. Gas pressure in the solar nebula was not high enough to allow liquid mineral droplets to condense from the gas. Instead, solid grains of refractory minerals probably condensed and mixed with the original solid interstellar grains that existed in the nebula when it was contracting. These grains clumped together (perhaps by magnetic forces), trapping volatile elements within them. During periods of intense heating by the protosun, the grains were heated to their melting points, producing liquid silicate droplets and driving off the more volatile materials. These droplets quickly solidified into chondrules, and some of the grains that may have remained unmelted were trapped. The chondrules then mixed with other fine-grained, solid minerals of similar composition condensing from the nebula. This mixture formed the first chondrites. By accretion, these small chondrite bodies gradually came together to form a planetesimal of chondritic composition.

A heat source within the body began to heat the interior. A sufficiently large body, probably between 20 and 50 miles in diameter, provides an insulating blanket of rock to retain the heat. The greatest heating occurs in the core, and progressively less heating takes place farther out. This differential heating produces layers of chondritic material showing various stages of metamorphism.

Later, if the chondrite body is disrupted by impact, its pieces, representing a mixture of metamorphic states, could be scattered along the original orbit and possibly perturbed into new orbits that might send them to Earth. It is more likely, though, that the body would gradually reassemble itself into a rock pile of mixed types.

Differentiated Meteorites

Achondrites, stony-irons, and irons come from differentiated parent bodies of original chondritic composition. Analysis of their composition tells us that their parent bodies were heated throughout to the melting point, allowing differentiation to take place. Achondrites show compositions and textures that suggest that igneous processes took place in their interiors and on their surfaces. Their parent bodies therefore must have been large enough to retain considerable heat in their interiors to support such igneous activity. If the parent bodies of these meteorites were asteroids in the main belt, then a problem exists: How can some asteroids be heated to the melting point while others in the same belt show little or no heating? It's a matter of distance, heat source, and size.

Solar heating through induced electric currents was sufficient to heat asteroids close to the Sun. Whether this was enough to melt the entire asteroid is still debated, but an additional heat source probably worked in tandem with solar heating. Together, they could have melted an asteroid.

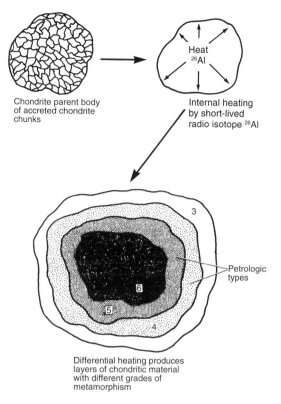

Chondrite parent body
of accreted chondrite
chunks

Heat
^{26}Al

Internal heating
by short-lived
radio isotope ^{26}Al

3

Petrologic
types

6

5

4

Differential heating produces
layers of chondritic material
with different grades of
metamorphism

Formation of different petrologic types of chondrite through differential heating in a chondritic parent body. The layers are labeled with their petrologic types from unmetamorphosed type 3 near the surface to highly metamorphosed type 6 at the core. The relatively small body cannot retain sufficient heat to completely melt and differentiate.

In the previous chapter, a supernova was described that spewed the isotope ^{26}Al into a forming solar nebula as it condensed. This short-lived isotope produces heat when it decays to ^{26}Mg. If a relatively large chondrite parent body contained sufficient ^{26}Al, internal heating would have occurred. Since rocky material conducts heat only poorly, a large enough body, say, over 100 miles in diameter, could retain sufficient heat to melt the body entirely. This requires that enough ^{26}Al was included in the original parent body, which implies that the body had to accrete early in the solar system's history—before the ^{26}AL decayed away. This would have been within the first million years, the same period of time in which we believe solar heating was at its most effective level. This, then, implies that differentiated parent bodies are very old, nearly as old as the chondrites, and since they formed from chondritic bodies, they should be found in the asteroid belt among the chondritic asteroids.

Matching Children with Parents

So far, our conclusions about meteorite origins have been best guesses. What we really need is a spaceship journey to the asteroid belt to sample the asteroids firsthand. It's something like testing blood types and DNA to determine the biological parents of lost children. Lacking that kind of close examination, is there any hope that museum specimen meteorites can ever be matched up with known asteroids? As impossible as it seems, that is exactly what is beginning to happen.

Planetary astronomers are a frustrated bunch. Those who study the Moon and planets are destined forever to study intangibles. They use every available technique to study the feeble light coming from these bodies to try to determine their nature. Recent space probes have provided tantalizing views of the Moon and planets and even four asteroids, but when at last the means is achieved to visit these other worlds and sample them, then the astronomer must move aside and make room for the chemist, geologist, and mineralogist; tangibles are not his domain.

We have not yet reached that point with the asteroids. At least for the present, we must turn to the astronomer to search for the parents of the meteorites.

Minerals that are transparent or translucent pass light selectively through their crystal lattices. Some wavelengths are easily transmitted, while others are absorbed in the crystal. Visually, this selective transmission and absorption does two things: It tends to color the mineral and makes it darker or lighter.

If asteroids are like meteorites, they are made up of silicate minerals. Tiny mineral crystals on the surface of an asteroid are exposed to light from the Sun. The light passing through the surface minerals is selectively absorbed. The remainder is reflected back into space from an asteroid, which is the reflected light seen through a telescope. What we observe is the reflection spectrum of an asteroid, extending from the violet end of the visible spectrum through the red end and into the infrared. A light-sensitive device called a spectrophotometer, attached to a telescope, scans the light returning from an asteroid and compares it with incoming light from the Sun. This is done for a number of wavelengths through the visible and near-infrared spectrum. The ratio of incident light from the Sun to reflected light from an asteroid is the albedo, or percentage of light returned from the asteroid.

Beginning in the 1970s, astronomers recorded reflection spectra from dozens of asteroids. Two facts immediately became apparent. Most asteroids have relatively low albedos, meaning they are very dark, reflecting only 3 percent to 7 percent of the light striking them. These are carbonaceous asteroids with dark carbon minerals on their surfaces. Asteroids also have various colors and exhibit absorption features in their spectra, implying different mineral compositions.

About this time some investigators realized that if meteorites are truly pieces of asteroids, they should exhibit similar spectral reflectance characteristics. To test this idea, they selected several basic kinds of meteorites, including ordinary chondrites, carbonaceous chondrites,

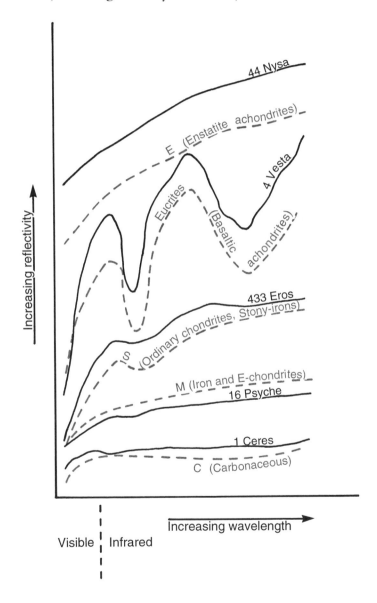

Reflection spectra of five asteroids compared with those of major meteorite types.

achondrites, irons, and stony-irons, and ground them to a fine powder. Then they obtained laboratory reflection spectra of the powders and compared them with those from the asteroids. When the reflection spectra are compared with laboratory spectra of meteorite samples, one must bear in mind that the spectrum is not from a single mineral but from several at one time. This tends to "muddy" the resulting spectrum, since it is really composed of several overlapping spectra. But for the moment, identifying the exact mineral composition at the surface of the asteroid is not as important as finding a close spectral match between asteroids and meteorite types.

The similarities between the asteroids and the meteorite samples are striking. Especially interesting is the spectrum of 4 Vesta, the first to be studied spectroscopically in 1970. (Asteroids are often designated with numbers preceding their names. This number indicates the order of discovery. Thus, Vesta was the fourth asteroid discovered.) Its spectrum matches the spectrum of a eucrite, an igneous, basaltlike meteorite. Vesta seems to be the sole source asteroid for these meteorites. Richard P. Binzel, at MIT, was able to match twenty small asteroids with a eucrite spectrum. Since these asteroids also have orbits similar to that of Vesta or derived from a Vesta-like orbit, Binzel believes them to be broken fragments of Vesta. The first asteroid discovered, 1 Ceres, has an albedo of about 6 percent. Its spectrum is very flat, showing equal reflectivity at nearly all wavelengths. Its spectrum falls into the broad area denoting the carbonaceous chondrites. These are collectively called C-asteroids (*C* for carbon). The asteroid 433 Eros falls within the zone of the ordinary chondrites. It is somewhat reddish in color and its spectrum shows a combination of pyroxene, olivine, and metallic nickel-iron as expected. Chondritic asteroids are called S-asteroids (*S* for stony). Asteroids with the highest albedo (40 percent) seem to have the composition of aubrites (enstatite achondrites). These are designated E-type asteroids (*E* for enstatite). The asteroid 16 Psyche has a flat spectrum with a moderate albedo of about 10 percent. Its spectrum shows no absorption features, indicating it is a metallic body, perhaps the nickel-iron core remnant of a differentiated asteroid. Nickel-iron is an opaque mineral, returning only reflected light. These are called M-type asteroids (*M* for metal). Interestingly, the enstatite chondrites have a nearly identical flat spectrum. They are almost pure magnesium pyroxene, with a high percentage of iron metal embedded in them. With their high metal content, it is no surprise that they exhibit an M-type spectrum.

The Last Word—So Far

The conclusions are tentative but encouraging. Having identified a few meteorites that seem to match some asteroids, we can begin to place them in their proper positions within the asteroid belt. As sus-

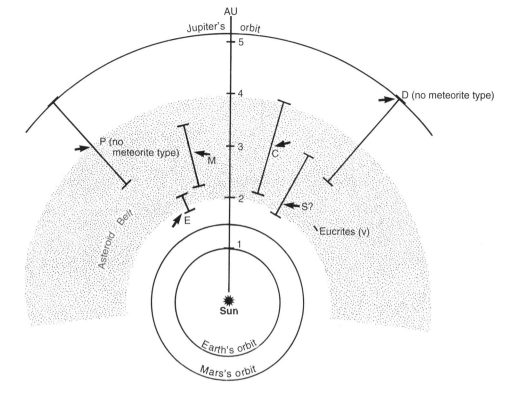

Distribution of asteroid types in the asteroid belt. The bars by each asteroid type show the extent of their range in the belt, and arrows indicate where the majority of each type is found.

pected, there is a direct correlation between compositional differences in the asteroids and their distribution in the asteroid belt.

More than 75 percent of the asteroids sampled were classified as C-asteroids. Of these, about two-thirds show evidence of water in their mineral structures, which shows up as an absorption feature in the ultraviolet part of the spectrum. These asteroids range from the middle to the outer edge of the belt.

The S-type asteroids are surprising. Despite the large number of ordinary chondrites in collections around the world, only about 16 percent of the S-type asteroids studied have chondritic compositions. The disparity between the carbonaceous chondrites, rare on Earth but plentiful in the asteroid belt, and the ordinary chondrites, common on Earth but relatively rare in the asteroid belt, indicates that chondrites probably came from one or at most a few parent bodies. The large numbers reaching Earth are apparently not indicative of large numbers of S-type asteroids in space. It seems we cannot rely on our meteorite collections to tell us the true ratios of asteroid types. The rarity of carbonaceous chondrites on Earth—by far the most plentiful asteroid type–probably illustrates the difficulty of removing them from the outer fringes of the belt, where they normally reside, to Earth-crossing orbits. S-type asteroids are confined to the inner belt, and only the E-type seem to be nearer the Sun but still within the belt.

Admittedly, there is some confusion about the S-type asteroids; they do not closely match the spectra of ordinary chondrites. They seem to have a larger complement of nickel-iron metal than that found in chondrites. To add to the confusion, the pallasites and mesosiderites also contain the three minerals pyroxene, olivine, and nickel-iron metal. S-type asteroids show a predominance of olivine over the normal pyroxene-to-olivine ratios in chondrites, so some of the S-type asteroids may actually be stony-iron parent bodies. The jury is still out on this one. Mineral analyses of asteroids via spectrophotometry do not provide the detailed results expected from the chemistry laboratory. It is crude by comparison and errors must be relatively large. So far, no large asteroid has been identified as the source body of the ordinary chondrites, though a few small bodies are suspect.

Wonderful Vesta

The most thoroughly observed asteroid is 4 Vesta, one of the brightest in the sky and at times just visible to the unaided eye. Careful study of its reflection spectrum shows changes over a period of hours. These changes are the combined results of its rotation and varying mineral composition. Apparently, it reveals a heterogeneous surface with different mineral regions passing into view as it goes through its 10.68-hour

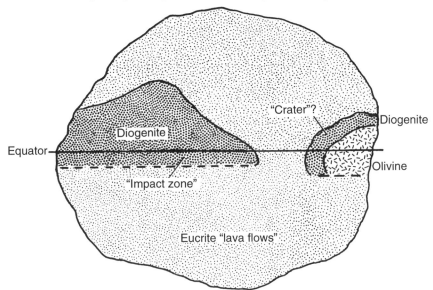

Interpretation of reflection spectra of 4 Vesta's surface mineralogy suggests eucrite "lava flows" and diogenite "crater fields." –Based on data from Michael Gaffey, University of Hawaii

rotation. These observations can be explained if we assume that Vesta is a differentiated body composed of layers of minerals. The observations made by Michael Gaffey, of the University of Hawaii, show a general covering of eucrite material (probably a basalt flow originally covering the entire surface), an irregular area of diogenite material (an igneous rock formed intrusively within the body), and a roughly circular area with a combination of diogenite material and an adjacent olivine-rich area. These areas are located on Vesta's equator and extend northward for some 30 degrees latitude. Vesta was struck several times in its history, and large impact craters mark its equator. The diogenite area is an impact zone where eucrite material was splintered off, revealing a diogenite mantle layer beneath. The second, more circular area is possibly an impact crater excavated through the eucrite and diogenite layers and on into a deeper, olivine-rich layer. Both layers are exposed in the walls and floor of the crater. Vesta is undoubtedly the source body for eucrite and possibly also for the diogenite meteorites. Not only are we able to see the scars left from the encounter, but amazingly, we have pieces of the broken ejecta from Vesta itself.

This hypothesis has recently been confirmed by Hubble Space

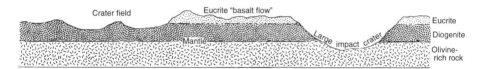

Theoretical cross section through the equator of 4 Vesta, showing a eucrite (basalt) covering, a diogenite exposed impact area, and an impact crater with eucrite-diogenite-olivine layers. –Based on data from Michael Gaffey, University of Hawaii

Telescope images of Vesta made in May 1997, which show an enormous crater more than 285 miles wide covering almost 75 percent of one side of Vesta. This is extraordinary since the asteroid is only 330 miles across. Cornell University researcher Peter Thomas concluded that the impactor was about 19 miles in diameter traveling over 3 miles per second when it hit Vesta. Though not quite large enough to break Vesta apart, it gouged a crater 8 miles deep and sent chunks of Vesta into space to orbit the Sun. The rocky debris from this collision continues to reach Earth as eucrites and diogenites.

How Do You Break Up an Asteroid?

The evidence confirming that meteorites come from asteroids raises new questions. How do asteroids shed their meteorites? Once shed, how do the meteorites manage to leave the asteroid belt? And even more puzzling, how do they find their way to Earth?

All asteroids are traveling in a counterclockwise direction around the Sun. The asteroid belt is like a giant freeway system with hundreds of traffic lanes representing orbits. An asteroid in the middle of the belt at 2.8 AU from the Sun is traveling at about 12 miles per second. Since asteroids are all traveling at about the same speed at that distance and in the same direction, like cars on a freeway, we would not expect energetic collisions to occur. But asteroid orbits are not all circular and nicely parallel to each other like freeway lanes. Most asteroids have elliptical orbits that carry them across the orbits of other asteroids. As if that weren't hazardous enough, some asteroid orbits change over time, as nearby Jupiter perturbs them with its strong gravity. Jupiter is responsible for stirring up the asteroids so their orbits carry them across each other at relative speeds of over 3 miles per second. During the lifetime of the solar system, virtually every asteroid has suffered numerous collisions. This seems to be borne out by observation. All but the largest asteroids vary in brightness as they rotate, suggesting that they are not round but irregular. These may be pieces broken off a larger body. The largest asteroids, such as Ceres and Pallas, apparently have not been disrupted, though they surely have been impacted many times and their surfaces must be heavily cratered.

What can be expected from an asteroid collision at 3 miles per second? The amount of damage asteroids suffer depends not only on their impact speed but also on their composition, total mass and size, and whether they receive a glancing blow or a direct hit. For example, if an asteroid 200 miles in diameter is struck by another 20 miles in diameter at 3 miles per second, the smaller asteroid would be at least partially vaporized as its kinetic energy is converted to heat. The remaining pieces would be ejected into space, but perhaps not very far.

Aftermath of a collision between two parent bodies with differing composition. Pieces of these bodies eventually gravitate together and intermix.
–Dorothy S. Norton painting

The larger asteroid might be broken into large pieces, but the energy of impact would be insufficient to scatter them, and before long the gravitational attraction of the pieces would draw them back together, forming a jumbled rock pile. Pieces from the smaller asteroid would also join the rock pile. Numerous impacts of this nature would thoroughly mix the rock fragments. This is how brecciated chondrites and achondrites form. Because mesosiderites are a jumbled mixture of broken mantle and lighter silicate rock, these meteorites most certainly were formed when mantle rock from a broken, differentiated parent body impacted another body, mixing surface rock with mantle rock.

Breaking up differentiated parent bodies is a bigger problem. Repeated glancing blows would successively chip off the lighter surface rock and then rocky mantle rocks beneath, exposing the nickel-iron core. Immediately surrounding the core would be core–mantle boundary material with a pallasite olivine-iron mix. This material, being harder, would be more difficult to break off than the rocky upper mantle, but the growing collection of pallasite meteorites provides convincing proof that the excavation was successful, not once but many times.

Left behind is the denuded core of the parent body. Among the known asteroids are several that are made of nearly pure metal. They are on average much smaller than chondritic asteroids, which suggests that they must represent such cores. The tensile strength of iron makes them difficult to fracture. They are better equipped to weather the grinding action of billions of years of impacts than are the chondrite parent bodies. Meteorite collections contain distinct iron-core material representing literally dozens of parent bodies. More parent bodies are represented by the iron and stony-iron meteorites than by the chondritic and achondritic meteorites. This may be explained by the much more durable nature of iron material.

Families of Asteroids

If two similar-sized asteroids collided—each about 50 or more miles in diameter and made of similar rocky material—they would not only be broken apart but the impact would be sufficiently energetic to throw the pieces into individual orbits closely paralleling the orbits of the original asteroids. Most of these pieces would range in size from between a fraction of a mile to 20 miles or more in their largest dimensions. All would be irregular in shape. They would probably look like the asteroid 243 Ida photographed by the *Galileo* spacecraft in 1993. The lighted side of Ida measures 35 miles.

Thus, a union of two similar asteroids results in a family of asteroids all traveling in roughly the same orbit. This idea of asteroid families became a recognized fact in 1918 when Kiyotsugu Hirayama in Japan showed that three groups of known asteroids have similar orbital char-

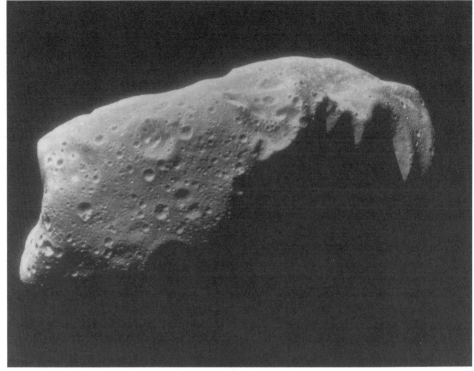

Asteroid 243 Ida from a distance of 2,100 miles. The densely cratered body is 35 miles long and is an S-type main-belt asteroid. A member of the Koronis family of asteroids, Ida is a fragment of a much larger parent body.
—Courtesy NASA

acteristics. He referred to them as families because he correctly believed that they were originally a single body that had suffered catastrophic disruption in the past. A recently disrupted family tends to orbit in a tight cluster, but as the family ages, the cluster members slowly separate from each other. Some families are so widely scattered that they must represent ancient collisions. This tells us that collisions among asteroids have occurred throughout the history of the solar system.

Planet-Crossing Asteroids

Asteroid impacts gradually grind asteroids down into smaller and smaller bodies. Very likely there were many more parent bodies in the early solar system than there are today. The largest known asteroids are among the few that just happen to have escaped catastrophic collisions in the past. During violent impacts, small bodies a few hundred feet to a mile or so across could be knocked into one of the Kirkwood gaps. These are unstable regions where no asteroid can remain long without its orbit being changed by Jupiter's gravity. As the asteroid fragment is

Asteroid 243 Ida from a distance of 6,755 miles. This remarkable image made by the Galileo *spacecraft on August 28, 1993, shows a mile-diameter satellite to the right of Ida. Dactyl, the first satellite of an asteroid to be found, is about 30 miles from Ida's surface, close enough to be under the influence of Ida's gravity.* Inset: *A magnified view of Dactyl, revealing a cratered surface and nonspherical shape.* —Courtesy NASA

pulled by Jupiter, its orbit is gradually elongated and it ventures outside the main belt and often inward toward the Sun.

Astronomers were slow to realize that some asteroids are not confined to the asteroid belt. The first indication came when American astronomer James Watson discovered asteroid 132 Aethra in 1873. He located the tiny body on June 14 and was able to follow it long enough to calculate a tentative orbit. Much to his surprise, this asteroid actually passed inside the orbit of Mars; after following the asteroid for three weeks, he lost track of it. Astronomers insist upon independent confirmation of any new discovery, without which the discovery remains classified as tentative.

Aethra remained the only hint that there may be Earth-approaching asteroids until 433 Eros was discovered in 1898. Eros was even more surprising. Its closest approach to the Sun took it to within 13 million miles of Earth's orbit. Eros confirmed Watson's suspicions about wandering asteroids. (Aethra was rediscovered in 1923.) The next discovery of a Mars orbit-crosser thirty years later, 1221 Amor, overshadowed

Eros. Amor had a perihelion—closest distance to the Sun—of 1.08 AU, only 8.5 million miles from Earth's orbit. It became the prototype and all others with Mars-crossing orbits were called Amors. All had orbits extending out to the main belt, showing that their origin was within the asteroid belt, but apparently they had been perturbed into elongated orbits that took them close to Earth.

Was it possible that asteroids from the main belt were actually crossing Earth's orbit? Astronomers didn't have to wait long for an answer. Asteroid hunter Karl Rienmuth at the Heidelberg Observatory photographed the trail of a faint asteroid on April 24, 1932. At the time of observation the object was about 7 million miles from Earth, but when its orbit was calculated, astronomers had their answer. This asteroid not only crossed Earth's orbit but also the orbit of Venus. Rienmuth gave it the name Apollo, after the Roman god of the Sun, because of its close approach to the Sun, but as with Aethra, astronomers let this asteroid slip away. Apollo was not rediscovered until 1973. Then a second Earth-crossing asteroid, Adonis, was found in 1936 to have passed within a mere million and a half miles of Earth just days before its discovery. It, too, was lost after only a few weeks of observation but was rediscovered in 1977.

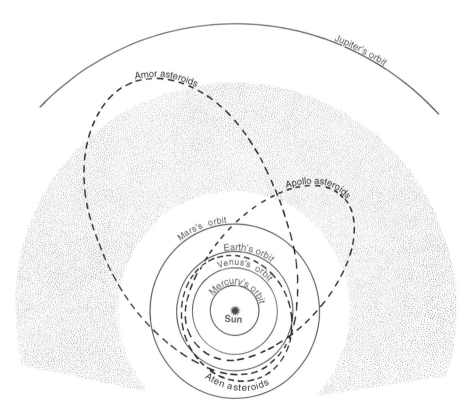

Typical orbits of Apollo, Amor, and Aten asteroids, all of which cross Earth's orbit or approach very near it.

Tom Gehrels, director of the University of Arizona's Spacewatch project, stands beside the 36-inch Spacewatch telescope located at Kitt Peak, Arizona.
—Courtesy of Carl E. Koppeschaar

A year later, in October 1937, an asteroid called Hermes, roughly one-half mile in diameter, passed within a half million miles of Earth—only twice the distance to the Moon. Unfortunately, like many of the faint asteroids, it too became lost. Hermes has not been seen again. (With some apprehension, I can imagine this "loose cannonball" somewhere out there, playing tag with the inner planets. Of the four inner planets, Earth is the most probable future "impactee.")

Apollo asteroids officially became the second asteroid class known to have left the asteroid belt. They are distinguishable from the Amor asteroids because their perihelion distance is inside Earth's orbit. Apollo asteroids are Earth-crossers.

A third asteroid class, perhaps the strangest of all, was discovered in 1976 and was given the name Aten. Aten has broken all ties with the asteroid belt. It orbits the Sun almost entirely inside the orbit of Mars and much of its orbit lies inside Earth's orbit. Aten asteroids have mean distances from the Sun of less than 1 AU, which means that they all overlap Earth's orbit.

Throughout this history, only a comparative handful of devoted astronomers searched the skies for asteroids. After the first four or five were found, it became obvious that no large planet existed between Mars and Jupiter and interest waned. Besides, there were more important things for astronomers to do. Large telescopes were being constructed primarily for the study of objects far beyond the solar system. There was a universe out there to explore. To most astronomers, asteroids simply got in the way during the long exposure times needed to photograph beautiful galaxies.

By 1970 interest in asteroid research reached its lowest ebb, or so it seemed. One astronomer, Dr. Tom Gehrels at the University of Arizona in Tucson, was determined to keep the study alive. Virtually single-handedly, with the aid of a few high-school science teachers, Gehrels began a research effort to determine the physical nature of asteroids. To him, they were more than just points of light in the sky. They were pieces of the early solar system and held the secrets to its origin. He gathered data from hundreds of observations of asteroids, but alone he could do little.

In 1971 he took a gamble and threw a party. He organized an international conference intended for any astronomers interested in the physical nature of asteroids—about which next to nothing was then known. He discovered that interest was latent but still alive. Nearly 150 astronomers showed up from around the world. Gehrels was encouraged and as a result of this conference, the first international scientific work on asteroids was published, *Physical Studies of Minor Planets,* edited by Gehrels. At the time, it was the most comprehensive work on asteroids. His conference uncorked the bottle, releasing the excitement that had been there all along. Interest in asteroids rose steeply. In 1979, Gehrels edited a second, much larger work, *Asteroids,* as an offspring of the second international conference, which superseded the former work and remains the most comprehensive book on asteroids. Thanks to the tenacity of Tom Gehrels, asteroid research was rejuvenated and interest in asteroids has remained high ever since.

Asteroid Surveys

The Apollo, Amor, and Aten asteroids, samples of which were undoubtedly in laboratories around the world, were especially exciting. What were their populations? Where had they originated? What were the sizes of these Earth-crossers? Most important, was anybody watching?

Two years after the 1971 conference, Eugene Shoemaker, a USGS geologist turned astronomer, and Eleanor Helin of the California Institute of Technology joined in a search effort to find Earth-crossing asteroids (the Palomar Earth-Crossing Asteroid Survey). Using a wide-field, 18-inch Schmidt telescope at Palomar Observatory in southern California, their systematic search added greatly to the known Earth-crossers and continues to do so today. Helin discovered the first Aten asteroid during this survey on January 7, 1976. The survey held the record for discovering the nearest asteroid until 1993. On March 31, 1989, they photographed a faint asteroid, 1989FC (identified, until it's verified, by year and letters instead of by a name and number), that had passed within 414,000 miles of Earth only a week earlier. Its diameter was estimated to be greater than 600 feet and possibly as large as 1,300 feet.

Meanwhile, Tom Gehrels joined the exclusive club of asteroid hunters with a survey of his own. For a long time, Gehrels had wanted a powerful telescope dedicated solely to finding new asteroids. He felt an almost urgent need to seek out these great, tumbling bodies for the very real hazards they present to Earth. His wish was granted in 1983 when a 36-inch telescope, housed atop Kitt Peak in southern Arizona, was commissioned as the Spacewatch telescope. This venerable old telescope had been built in 1919 for the University of Arizona's Steward Observatory. Originally housed on campus, it was moved to Kitt Peak in 1963. There it remained, used only occasionally, as it had been superseded by larger, more modern and glamorous instruments. Now revitalized with modern electronic systems, it automatically scans the sky. A computer compares the scanned areas with a star map in its memory. With this telescope, Gehrels and his group have found dozens of new asteroids, with the promise of hundreds more in the future.

That future came sooner than expected. In 1990, a sensitive new CCD electronic camera was coupled to the 36-inch telescope, allowing the researchers to detect much fainter, smaller bodies. Over the next two years, Gehrels and his coworker, David Rabinowitz, found eight new asteroids, which they estimated as measuring between 15 and 300 feet across, passing near Earth—and their search covered only a small area of the sky over a short period of time! The implication was that near-Earth space was populated by many times more asteroids than anyone had inferred previously. Gehrels and Rabinowitz estimated that at least fifty of these small bodies pass between Earth and the Moon each day. Apparently, this swarm of small asteroids, each with a nearly circular orbit, composes a miniature asteroid belt that lies within the neighborhood of Earth's orbit.

On May 20, 1993, the Spacewatch telescope picked up a faint asteroid leaving a barely visible streak on the discovery photographs. The object, $1993KA_2$, is estimated to be between 15 and 30 feet across. It is notable for being the smallest natural body ever detected beyond Earth. But its real fame lies in its nearness to Earth. Asteroid $1993KA_2$ just missed colliding with Earth, coming within 84,000 miles and beating all previous records.

Between the two search teams, near-Earth asteroid discoveries have reached thirty-five per year, and the rate is expected to climb in the near future. Many of these are large enough to cause irreparable damage to life on planet Earth. Lurking in the shadow of the Spacewatch project and the asteroid searches of Helin and Shoemaker is a persistent, driving question of profound importance to Earth's inhabitants. When can we expect the unthinkable to happen?

Dire Predictions

All asteroids that come near Earth's orbit are called, collectively, near-Earth asteroids. The discovery of Earth-crossing asteroids should have immediately set off alarm bells. We're talking about great mountains of rock and iron, from a few dozen feet to several miles across, intercepting Earth's orbit every year or two. Not counting Gehrels's mini-asteroids, more than two hundred near-Earth asteroids are known, and roughly half of these cross Earth's path. They are among the smallest asteroids, evident only because of their close approach to Earth. Their relatively small size is no cause for relief, however, for they still average 1 to 2 miles at their largest dimension (they are all irregular fragments). A few are more than 10 miles across.

One can only guess at the number of Earth-crossing asteroids in the 0.5- to 1-mile-diameter range that have so far gone undetected. From 1,000 to 1,200 is the usual estimate among asteroid experts. If this is extended to include the Amor asteroids, some of which could cross Earth's orbit sometime in the future, then the number increases to about 6,000: As many as 6,000 orbiting mountains playing tag across and near Earth's orbit. This is truly a cosmic shooting gallery—with Earth the largest target.

What everyone would really like to know is the probability of one of these asteroids hitting Earth—next week or next year or ten years down the road. Unfortunately, estimates are not yet very accurate. Reasonable estimates can be made of impact rates occurring over the past several billion years by examining craters on the Moon. Older craters can easily be distinguished from younger ones, even with a small telescope. Younger craters are easily spotted: Their white ejecta contrasts with the dark lunar maria. These craters formed after the large impact basins were flooded with basalts. The flooding of the lunar basins occurred between 3.9 and 3.3 billion years ago. Crater counts on the maria show there was a late heavy bombardment 4 billion years ago, but a steep reduction of impacts occurred between 3.9 and 3.3 billion years ago. Between 3.3 billion years ago and the present, the rate has been much slower, but steady. It is this steady rate that is of interest, because this is the rate that influences Earth today.

As researchers compile their crater counts, they arrange the craters into categories according to size, noting the number in each category.

Telescopic view of the Moon's Imbrium basin, showing numerous small impact craters formed after the basin was flooded with basalt.

Then each number can be extrapolated to include the entire surface area of the Moon. (This includes craters only in the dark maria, which occupies only a little more than one-sixth of the Moon's surface.) By dividing the number of craters of a given size into the average age of the maria, 3.5 billion years, it's possible to get an estimate of the Moon's impact rate for each crater size over the past 3.5 billion years.

The estimates are then applied to Earth—a considerably larger target. In the 30- to 60-mile-diameter size, about thirty impacts occurred on the Moon in the past 3.5 billion years. This would be about one impact for every 117 million years. For Earth, which has thirteen times the surface area of the Moon, the total number increases to about four hundred, or roughly one impact every 8.75 million years. An asteroid capable of producing a crater in this size range would be 3 to 5 miles in diameter. There are five times as many craters in the 15- to 30-mile-diameter range, which would result in nearly two thousand impacts on Earth in 3.5 billion years, or one crater 15 to 30 miles in diameter every 1.75 million years.

As the craters decrease in size, their numbers increase rapidly. This is not surprising, since there are many more asteroids in the 0.5- to 1-mile-diameter range. If we include all asteroids from 0.5 mile and larger, the number of impacts increases substantially, to about sixty impacts every 10 million years or one impact every 160,000 years.

As we continue to reduce the size of the impactor, the rate of impacts continues to rise. Asteroids in the 100- to 500-foot-diameter category, like the one that created Meteor Crater in Arizona, hit Earth about every 10,000 years. Finally, we can expect one impact a century like the Sikhote-Alin fall in eastern Siberia in 1947 or the blast over Tunguska, Siberia, in 1908. Fortunately, stony asteroids in the 30- to 300-foot range are completely destroyed during atmospheric flight, as over Tunguska. This is probably not true of metallic asteroids, however.

Fewer meteorites strike Mars and Venus, and half or more of these are ejected from the inner solar system altogether as their orbits are changed by repeated close passages to the inner planets. In total, about 10 percent of all the Earth-crossing asteroids are destined to hit Earth.

Orbital studies tell us that Apollo asteroids cannot have lifetimes of much more than 10 or 20 million years. However, the cratering record on the Moon tells a different story. Impacts have been going on at a steady pace for more than 3 billion years. There has never been a shortage of Apollo asteroids—there must be a constant resupply from the asteroid belt, furnished by continuous fragmenting of the existing main-belt asteroids.

How Often Do Asteroids Fragment?

There is a way to tell how long a meteorite has been separated from its original parent body and therefore how long ago the parent body fragmented. While the meteorite was still locked safely within the parent body, it was protected from high-energy cosmic rays that constantly enter the solar system from the galaxy. Cosmic rays are not true "rays," but highly energetic subatomic particles, primarily protons, alpha particles (helium nuclei), and some heavy nuclei. The most energetic are released during supernova explosions. When these particles encounter another atom, they alter it by knocking protons and neutrons out of its nucleus, changing the atom to an isotope of another element. Cosmic rays can penetrate only about 3 feet into solid rock; rock beneath that level is protected from cosmic rays. Once a parent body has been disrupted, however, those fragments a foot or two in diameter, now meteoroids, are suddenly subjected to the full effects of cosmic rays. This sets a cosmic-ray "clock" in the meteorite. Isotopes such as ^{3}He (helium), ^{21}Ne (neon), and ^{38}Ar (argon) begin to appear as the atoms in the minerals are attacked. Some of the isotopes of these new elements are radioactive, making them relatively easy to detect. The rate at which these isotopes form depends upon how long the meteorites have been exposed, assuming that the cosmic ray flux remains constant. Analyzing a meteorite for isotopic abundance reveals the cosmic-ray exposure age, or the total time the meteorite was in space.

Cosmic ray ages vary widely but they seem to be somewhat dependent on the type of meteorite. The H-chondrites show the youngest ages with a peak of about 5 million years. This suggests they must have all come from the same parent body at approximately the same time. Thus, a chondrite parent body suffered a major impact and fragmented 5 million years ago. Other chondrites range in age from 10 to 40 million years. Iron meteorites show by far the longest cosmic-ray exposure ages. Some exceed a billion years, while the youngest are around 100 million years, with an average in the 400- to 800-million-year range. This longevity can be explained by the durability and greater strength of the irons, which results in their being fragmented much less often and only during the highest energy impacts.

Meteorites that make it to Earth are young compared with their parent bodies. The cosmic-ray exposure ages of the stony meteorites compare reasonably well with the estimated lifetimes of Apollo asteroids. The impacts that fragmented the parent bodies may have been responsible for sending the fragments on orbits that led to Earth. One L-chondrite that fell in Farmington, Kansas, late in the last century turned out to have the youngest cosmic-ray exposure age of all—a mere 25,000 years. Studies of its orbit placed it among the Apollo asteroids.

The supply of meteorites seems to be endless, and Apollo, Amor, and Aten asteroids appear to be their parents. Earth is visited by thousands of these small fragments every year. It is only a matter of time before a really large fragment or even a parent body itself comes calling. This visit will change Earth as we know it forever.

Bumper stickers in space.

This hypothetical scene shows the Sun's red-dwarf companion star, Nemesis, passing through the inner Oort comet cloud as the star reaches perihelion. The star's gravity perturbs the icy comet chunks sending millions toward the Sun 30,000 AU away. The Sun is the brightest star on the lower left. —Painting by Dorothy S. Norton

TWENTY

When Worlds Collide

Fifty years ago, authors Philip Wylie and Edwin Balmer wrote a science fiction novel entitled *When Worlds Collide*. It is the story of a Professor Bronson's discovery of two planets that had entered our solar system from somewhere in the galaxy. Bronson Alpha, the larger of the two, was a gaseous body something like Jupiter and around it orbited a second world, Bronson Beta, similar in size to Earth. They both were headed for the inner solar system, and passed close by Earth as they rounded the Sun, raising enormous tides, reawakening dormant volcanoes, creating strong earthquakes, and knocking the Moon from the sky. Six months later they met Earth again on their way out. This second meeting doomed planet Earth to destruction, as the larger world and Earth collided. But released from the gravitational grip of Alpha, the smaller world took up Earth's orbit around the Sun, giving humans a second chance, with a new world full of hope.

As an impressionable teen, I devoured the book and spent many a waking hour wondering what it would be like to witness such an Armageddon. (Besides science fiction, I was an avid reader of astronomy books, and I knew the collisions of worlds just didn't happen; but it made a great story anyway.)

In the late 1950s, as an astronomy student at UCLA, I met my men-tor, meteoriticist Frederick C. Leonard. Next to his office was a tiny meteorite museum, full of wondrous rocks from space. Often I would wander into the museum just to be surrounded by alien things. It's odd that no one really thought much then about how those rocks got to Earth. Few investigators connected them with asteroids. I don't recall Leonard ever referring to them as pieces of larger bodies, though he must have recognized they were.

Later, as Leonard's teaching assistant, I took annual trips with him and his students to Meteor Crater in Arizona. Though this great, gaping hole a mile across had been blasted out by an iron body just a hundred feet or so across, no one referred to the object as an asteroid. We were blind to the message carried by such rocks from space. We were sleepwalkers, oblivious to the warning nature was sending us.

August 10, 1972, was a warm, clear day in the Grand Tetons. Vacationers at Jackson Lake were readying their boats for a day of fishing. A gentle breeze made lazy waves on the water. No one was aware that a nearly catastrophic event was silently taking place above the mountains. The blue sky was suddenly pierced by a blazing fireball traveling south to north, trailing a long smoke train. James M. Baker was alert enough to photograph it (see color plate VIII) and at least one motion picture was made. The object did not crash into Earth. It entered the atmosphere over the Central Rockies, traveling at 33,000 miles per hour. It remained in the atmosphere for 101 seconds, covering nearly 1,000 miles in that brief sojourn. Then it skipped out of the atmosphere, much as a stone skips over water. Estimates of its size vary from 33 feet to 260 feet across. If it was a stony asteroid, its mass was between 1,000 and 1 million tons. Earth and its inhabitants had narrowly escaped colliding with an Apollo asteroid. Still, Earthlings slept. . . .

The Great Awakening

I regret that I became interested in fossils rather late in life. Most of my professional years were spent looking up instead of down. After the Moon landings gave us our first look at Earth as a planet, against a jet-black void, I began to take more notice of home. I pored over geology books. I took university classes, talked to geologists, and visited all the great geological wonders around the western United States. Many

Early Cambrian trilobite, Olenellus, *from the Latham Shale of the Marble Mountains, Cady, California.*

of these sites contained fossils, which I learned were used to date the rock strata where they were found. To learn about fossils, I took paleontology classes, talked to paleontologists, and visited all the famous fossil beds in the western United States. I visited all the great paleontological museums in the country. During my personal renaissance of learning, I got the impression that there were distinct domains in science: Geologists studied the domain of rocks; paleontologists studied the domain of fossils; and astronomers owned the universe—minus Earth. An unwritten law prevailed: Thou shalt not trespass on the scientific domain of others.

The great awakening came when scientists reluctantly began to realize that dividing scientific knowledge into separate blocks creates an artificial, inefficient structure. Intricate interrelationships exist between the sciences, blending them into one vast kingdom of knowledge. The "big picture" requires knowledge about rocks and fossils and atoms and stars. At no time did this become more dramatically evident than in the decade of the 1980s, when dinosaurs once again came into vogue—big time! Solving the mystery of the disappearing dinosaurs required a marriage of the sciences.

Earth History

Thumbing through the pages of a paleontology book shows a world that doesn't exist today. It is a world of fantasy not even Disney could have imagined. Geologists divide all Earth history into four great blocks of time called eras. The first era, the Precambrian, covers a vast period beginning with the origin of Earth about 4.6 billion years ago to about 600 million years ago. The Precambrian engulfs by far the greatest block of time in Earth history. It is often divided into the early Precambrian, in which there was no life on Earth, and the late Precambrian, in which the first life-forms appeared and developed into soft-bodied, multicellular forms. The Precambrian came to a close when something extraordinary happened—soft-bodied animals living in the seas suddenly learned how to construct hard, calcareous coverings around themselves. This wonderful achievement not only protected them from predators, but after death the hard parts sometimes fossilized, leaving a record of their existence for future paleontologists to read.

This hard-part fossilization distinguished the beginning of the next great era, the Paleozoic, meaning early life. Geologists divide the Paleozoic into seven periods beginning with the Cambrian, 600 million years ago. Each period is distinguished by the appearance of new, more complex and diverse life-forms. Trilobites were among the most successful life-forms in the Paleozoic, dominating the seas in the Cambrian and living on for nearly 400 million years before suddenly dying out at the end of the Permian period. This dying marked the close of the Paleozoic era. Each geologic period was marked by the disappearance

Late Cretaceous ammonite, Hoploscaphites nebrascensis, *from the Fox Hills formation, Corson County, South Dakota.* –Black Hills Institute of Geological Research

of some species and the appearance of others, but the end of the Permian was especially disastrous to living things. More than 90 per-cent of all living species simply vanished.

The Mesozoic era is best known for the rise of the great reptiles, among them the dinosaurs. They developed through three periods beginning with the Triassic 248 million years ago, through the Jurassic, and ending with the Cretaceous 65 million years ago. Like the Paleo-zoic, the Mesozoic saw the rise and fall of many species of land and aquatic plants and animals. The end of the Cretaceous period also ended the Mesozoic era and represented another period of dying. This time the dinosaurs, which had successfully lived through 150 million years of evolution, suddenly met their end and went the way of the trilobites. The ammonites, wonderful spiral-shelled, mollusklike crea-tures that had survived the great Permian extinction, also joined the dinosaurs and over two-thirds of the existing species of animals.

The passing of the dinosaurs signaled the beginning of the last era of geologic history, the Cenozoic. It is divided into two periods, the Tertiary and the Quaternary. These in turn are divided into epochs. The Cenozoic is marked by the rise of the mammals. Even though fossils of early mammals are known from the Triassic, these animals remained in hiding during the reign of dinosaurs and were little competition to the great beasts. Only after the Cretaceous-Tertiary extinction was it possible for mammals to take the lead, filling the niches left by their precursors. What would the world have been like today had the dinosaurs lived on into Cenozoic times? Would there be mammals today? What about primates . . . and humans? What did in the great dinosaurs?

To Gubbio—And the "Mysterious Layer"

The history of science is strewn with examples of serendipity. Often scientists set out to find solutions to one mystery only to stumble onto observations that solve another. The "mysterious layer" is one example.

In the late 1970s, geologist Walter Alvarez, from Columbia University's Lamont-Doherty Geological Observatory, was studying rock layers at the Cretaceous-Tertiary boundary (K-T boundary) near the town of Gubbio, Italy. He was searching for clues to the great extinction. The rocks there were sedimentary beds made of limestone, deposited in an ancient sea during the Cretaceous period, when calcareous shells from countless billions of microscopic animals rained down onto the sea floor after their deaths. For millions of years the debris compressed and hardened, forming limestone rock above and below the K-T boundary.

Alvarez noticed that a curious layer of red-brown clay about half an inch thick interrupted the otherwise continuous limestone bed. This

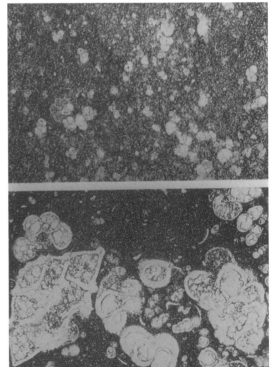

Top: *A single species of foramanifera appears in the Tertiary limestone immediately above the K-T boundary.* Bottom: *Fossil shells of numerous species of foramanifera in the Upper Cretaceous limestone near the K-T boundary.* —Walter Alvarez and Allessandro Montaneri, Lawrence Berkeley Laboratory, University of California

clay layer conveniently marked the end of the Cretaceous and beginning of the Tertiary (see color plate VIII). Looking closely at the limestone below the boundary clay, he could see the tiny shells of many microscopic, one-celled organisms called foramanifera. Forams were common in this period, having thrived for over 30 million years in the Cretaceous. It was obvious to Alvarez that the forams had been thriving right up to the end of the Cretaceous. Many species were represented as fossils in the lower bed.

Then the slow process of limestone deposition was interrupted by the clay layer. Immediately above the clay, the limestone continued but with a profound difference. The fossil forams had all but disappeared. Only one species in the sample had made it across the boundary clay.

This is not the way events normally happen in matters geological. Significant changes in Earth's geology occur at an excruciatingly slow pace. Human lifetimes are but the wink of an eye in geological terms. We cannot watch the deposition of sediments and formation of sedimentary beds—the process is far too slow for many human lifetimes. The limestone beds Alvarez was studying represented millions of years of deposition. The same can be said for the progress of evolution. Paleontologists can easily observe the rise and decline of species by studying the fossil record over millions of years of time, but changes over a few months or years either did not happen or could not be observed in the fossil record . . . until now.

It was obvious to Alvarez that indeed a drastic change had taken place that snuffed out the lives of many species of forams, and this change had occurred immediately before the deposition of the boundary clay. He thought the clay represented typical erosional debris off the continent, carried to the sea by rivers. Normally, this is a lengthy process—it might take centuries to lay down a half inch of sediment. Yet some catastrophic event temporarily stopped the limestone deposition, killed off the forams quickly, and deposited the clay. The boundary clay would tell him how long the event lasted. All he needed was to find out how fast the clay had been deposited.

Walter Alvarez was lucky. His father, Luis Alvarez, was a Nobel Prize–winning nuclear physicist at the Berkeley campus of the University of California. Walter went to Berkeley in the 1977-1978 academic year as a visiting professor with the possibility of leaving Columbia University and accepting a teaching and research position at Berkeley. After arriving at Berkeley, he never returned to Columbia. Walter showed Luis a sample of the Gubbio boundary clay layer he had encased in Lucite. The clay layer was bordered on both sides by the limestone. Using a microscope, Walter showed Luis the forams in the lower layer and the lack of forams in the upper.

Throughout human history there are great moments that change our lives and the way we think about things. Usually, they go unnoticed by the very participants involved. It is not until much later that the impact of those moments takes effect. From the moment Luis Alvarez looked at that sample from Gubbio, our perceptions of Earth would never be quite the same again. He immediately saw the importance of that sample and the need to determine the sedimentary rate of the clay layer. Intuitively, he knew that in his hands lay the answer to the death of the dinosaurs.

Iridium

The sample from the clay layer only reluctantly gave up its secrets. Much work had to be done and many false leads had to be abandoned. What was needed was a standard by which to measure the rate of

Luis Alvarez (left) and Walter Alvarez at the tilted limestone beds near Gubbio, Italy, where the K-T boundary clay was discovered. –Lawrence Berkeley Laboratory, University of California

sedimentation. That standard came from space. Roughly 10,000 tons of meteoritic dust rain down on Earth every year, and it is constant and uniform over the Earth. All one has to do is measure the amount of meteoritic dust particles accumulated in the clay to determine how long it took the layer to form. Sounds simple, doesn't it? But counting the dust particles is impossible—they are too many, too small.

Luis knew that the elemental composition of meteoritic material is similar to that of Earth's crustal rock, but not identical, since most meteorites are primitive bodies that have not been differentiated into core, mantle, and crust as Earth has. When Earth differentiated and the iron made its way to the center to form the core, it carried down with it nearly all of the platinum-group metals. These include such rare metals as gold, platinum, osmium, and iridium, collectively known as the noble metals, little of which is found in Earth's crust. But undifferentiated meteorites still retain the noble metals scattered through their mass, and the differentiated meteorites, the irons and stony-irons, contain concentrations of these elements. The rain of meteoritic dust must therefore contain noble metals in concentrations of from 1,000 to 10,000 times that found in Earth's crust. Could these elements be detected in the meteoritic materials mixed in the clay layer? Luis decided to look and selected iridium as the element of choice.

Consider the magnitude of the problem. One hundred thousand tons of meteoritic dust filtering down to Earth every year, when spread evenly over the entire surface, amounts to only about one 500-millionth of a gram per square inch. Typical meteoritic material contains about one-half to one part per million of iridium. Is it possible to detect iridium at this low concentration? Luis thought it was. The Alvarez team presented the problem to Berkeley's best nuclear chemist, Frank Asaro, a leading expert in neutron activation analysis. Asaro bombarded the Gubbio sample with neutrons, causing the material to become radioactive. Each element in the sample emitted gamma rays of a specific energy, each a signature of the elements present. Asaro found the expected iridium in the clay layer, but not the expected amount. Assuming that the rate of clay deposition and the rain of meteoritic material remained more or less constant, the ratio of clay to meteoritic material and therefore iridium should also remain constant, but it didn't. There was thirty times too much iridium.

Suddenly, the clay deposition rate seemed secondary. A more important issue had appeared. Where did the excess iridium come from? Serendipity had struck again.

The Wayward Asteroid

Supernovae manufacture heavy elements at the moment of explosion. The scientists briefly considered a nearby supernova as the culprit responsible for producing the iridium during its outburst. As a test, the

The Alvarez team. From left: *Helen Michel, Frank Asaro, Walter Alvarez, and Luis Alvarez.* –Lawrence Berkeley Laboratory, University of California

Berkeley team looked for a rare isotope of plutonium that should also be present in the Gubbio layer as the signature of such an event. It wasn't found–the supernova failed the test and the idea was cast aside.

Finally, after several other ideas failed stringent tests imposed on them, the Alvarez team decided an asteroid or possibly a comet must have been the impactor–the iridium was star stuff mixed into the body of an asteroid or comet and transported to Earth. From the amount of meteoritic material in the clay (about 10 percent meteoritic, 90 percent Earth material), based on the iridium content, Luis Alvarez estimated that the object had been about 6 miles in diameter. The Berkeley team announced their astounding discovery in a paper entitled "Extraterrestrial Cause for the Cretaceous-Tertiary Extinction," published in *Science,* June 6, 1980. It sent shock waves around the world. The implications were staggering.

Worldwide Iridium

If an asteroid had really struck the Earth and spread its dust worldwide, then iridium should be found wherever the K-T boundary was preserved. Geologists were skeptical of the conclusions the Alvarezes had reached in their paper, and at first reluctantly, then more enthusiastically, they began to sample known K-T boundaries in search of the telltale iridium. They were not disappointed; rather, they were astonished. Iridium was everywhere. Even higher concentrations showed up. In Texas it was 43 times more than in the surrounding rock; in New Zealand, 120 times; in Montana, 127 times; in Haiti, 300 times. A hole dug in the mid–North Pacific brought up 330 times more iridium, and Denmark topped the list with 340 times. By 1986, the number of sites reporting high iridium levels reached eighty. The current number is over one hundred.

Through the years, many hypotheses have been presented to explain the death of the dinosaurs. These ranged from small mammals that ate dinosaur eggs to a plague of exotic diseases. But none of the ideas could be applied to the worldwide extinction that took not just the dinosaurs but two-thirds of all living plants and animals. Any hypothesis must provide a means of testing its validity, or, more precisely, the hypothesis must be able to predict consequences that are testable. This is where the asteroid-impact idea shines. Iridium concentrations worldwide are predictable and testable. The tests were performed and the asteroid-impact hypothesis received an A+.

Geologists and paleontologists were slow to accept the theory of a worldwide catastrophic event, especially one initiated extraterrestrially. Most scientists feel much safer considering orderly Earthbound processes working gradually and predictably over geologic time. Catastrophism smacks of biblical ideas like Noah's flood and Revelations or the bizarre theories of Immanuel Velikovsky. Astronomers see the world differently. They don't mind the notion of terrestrial impacts and exploding supernovae. That is all in a day's work for them. It took a nuclear physicist and a geologist to seriously introduce the idea of a catastrophic astronomical event, one which began to change the way evolutionary biologists and paleontologists think about the world. Perhaps we humans need to be reminded every so often of our catastrophic origins: The same forces are still at work out there.

Dead Comets or Wet Asteroids?

For years, astronomers considered comets to be a slurry of ice, dust, and small rocks—not much in terms of solid substance. The planets, especially Jupiter and Saturn, could push them around, changing their orbits when they came too close. So they couldn't possibly be any more massive than the asteroids. When Giotto gave us humans our first look at the solid body of a comet, we learned it was just that, a solid hunk of ice and rock, not the loose conglomerate of gravel and ice we had envisioned before. Now, in the context of an Earth impactor, they have to be taken seriously.

A comet loses tons of water to space when it rounds the Sun, as much as a million tons per day. A typical comet losing water at that rate will cease to be a comet after about a thousand returns to the Sun. When it runs out, the "dead" comet continues to travel on the same elliptical orbit that sent it to the outer planets and inward to the inner planets, often crossing Earth's orbit. The distinction between an asteroid and a comet becomes blurred at this point. Is the core of a typical comet a solid stony body? We know from the Giotto pictures that the surface is covered with charcoal-like carbon compounds. The remaining stony material may be quite similar to a carbonaceous chondrite meteorite.

In 1977, an object was discovered moving in an orbit between Saturn and Uranus. The object was given the asteroid number 2,060 and the name Chiron. Its diameter, about 100 miles, places it in the asteroid class, and its surface is dark like that of carbonaceous asteroids. No comet nucleus is known to be that large. But what would an asteroid be doing out there, so far away from the asteroid belt? As if to mock our ignorance, Chiron suddenly doubled in brightness in 1988 and formed a gaseous halo around itself. This "asteroid" was acting suspiciously like a comet. Today we believe it is, in fact, a comet of huge proportions that may someday venture into the inner solar system, if Saturn gives it a tug.

There are other "asteroids" that have orbits typical of short-period comets. These may be dead comets or, like Chiron, a new comet just waiting for a chance to enter the inner solar system. All of these bodies orbit near Jupiter, Saturn, or Uranus, which means their orbits are not stable. They will one day be accelerated out of the solar system or possibly inward toward Earth.

The Oort Cloud of Comets

Most comets newly discovered every year have extremely long orbits that carry them as far as 50,000 AU from the Sun. From a study of these comets, astronomer Jan Oort in 1950 deduced that a huge spherical cloud of comets must exist surrounding the Sun but at an enormous distance. The cloud has an inner radius of 20,000 AU. Its outer limits may vary from 100,000 AU to 200,000 AU. This would place the outer edge of the cloud in interstellar space, more than halfway to the nearest star. The Oort cloud may contain as many as a trillion comets.

At a distance of 50,000 AU, the Sun's gravitational influence is weak, and passing stars or great molecular gas clouds in orbit around the Milky Way Galaxy can exert an equally strong gravitational force on the comets, causing them to change their orbits. Eventually, some of the disturbed comets leave the comet cloud altogether to wander forever among the stars. Others fall inward toward the Sun, taking millions of years to reach the inner solar system, where we can observe them.

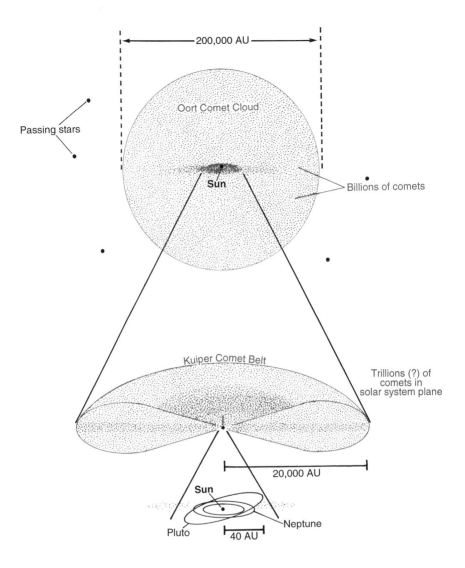

The Oort comet cloud and the Kuiper comet belt. The Kuiper belt may be the source of comets in the Oort cloud.

The Oort cloud may be in a state of equilibrium whereby as it loses comets, it also gains them. The constant supply comes from, of all places, the outer solar system. Some astronomers postulate that there exists yet another zone of comets, only very much closer to the Sun. This enormous zone, called the Kuiper belt, after the American planetary astronomer Gerard Kuiper, extends from the orbit of Uranus or Neptune to the innermost boundaries of the Oort cloud. It is where all the short-period comets probably originate; Comet Halley has an aphelion within this belt. There may be as many as ten trillion comets in the Kuiper belt, a number equivalent in mass to one thousand Earths.

The Kuiper belt can be pictured as analogous to the asteroid belt but composed of icy bodies that formed along with the asteroids, only much farther out in space. Comets in the Oort cloud did not originally form that far from the Sun. The Kuiper belt of comets probably supplied the original comets to the Oort cloud. They were accelerated by the outer planets' gravity into increasingly longer orbits, which eventually gathered them into the Oort cloud, where their orbits were gradually circularized by passing stars.

Today, comets and asteroids can no longer be considered as completely separate classes. Their difference in composition depends entirely upon where they formed in the early solar nebula. Some asteroids may be simply spent comets that have lost their water, suggesting that some comets actually metamorphose into asteroids or that comets may be asteroids with an abundance of ice.

Comet or Asteroid?

Was the object that did in the dinosaurs an asteroid or comet? Scientists involved in this research today refer to the asteroid *or* comet, not specifically one or the other. There may be as many as six thousand

Comet Swift-Tuttle in November 1992. Star trails indicate the comet's motion and show small wiggles produced by slight tracking errors at the telescope.

Earth-crossing asteroids, but the Kuiper belt, with 10 trillion comets, far outnumbers the asteroids: Each represents a potential impactor.

The potential for cometary impact drew national attention when the November 23, 1992, issue of *Newsweek* magazine hit the newsstands. Its startling cover showed Comet Swift-Tuttle, the comet responsible for the yearly Perseid meteor shower, approaching Earth. The comet hadn't been seen since President Lincoln was in the White House, and its whereabouts had been in question until an amateur astronomer in Japan spotted it on September 26. On November 7, 1992, the comet passed within 110 million miles of Earth.

Comets have the maddening habit of being unpredictable. Most, the long-period comets, suddenly appear in the sky, intruding into the inner solar system from somewhere beyond. Short-period comets, which remain within the confines of the solar system, have the potential to change their orbits in unpredictable ways. Comet Swift-Tuttle takes about 130 years to orbit the Sun. Within that long period, nongravitational effects can change its orbit so that its next perihelion passage may be off by several days. The geyserlike jets we have seen on Comet Halley and other comets as they approach the Sun can nudge them enough to make these unpredictable changes.

Comet Swift-Tuttle didn't even come close to Earth in 1992, but in 2126 it will be a spectacular sight in the sky. If predictions are correct,

Comet IRAS on March 11, 1983, during its close approach to Earth. The long star trails produced while tracking the comet attest to its rapid motion among the stars. –National Optical Astronomy Observatory

it will come within about 14 million miles, sixty times farther than the Moon and nine times closer than Comet Halley came in 1986. That's close for a comet, but most Earthlings weren't aware of the comet that passed within 3 million miles of Earth in 1983. I vividly remember Comet IRAS-Araki-Alcock (named for a spacecraft and two human observers) as a hazy cloud three times the Moon's apparent size at its closest point to Earth. What was astonishing about the comet was the speed at which it was traveling. Usually, comets seem to the naked eye to remain relatively fixed in the sky; their motion can be detected only from night to night. But not Comet IRAS. On May 11, this comet was so close to Earth that it moved across the northern sky in a single night.

Other considerations speak favorably of a comet as the impactor in question. Of the several comet nuclei that have been measured, at least half fall within the size constraints of the Alvarez hypothesis. Comet Halley measures 9.6 miles in its longest dimension. Comet IRAS, also an irregular body, has a mean diameter of about 5.5 miles. Swift-Tuttle also falls into this range. Apollo asteroids are much smaller, having diameters under 2 miles, and most are a few hundred feet to a half mile in diameter.

Comets travel at much higher speeds than asteroids. On the average, they would impact Earth at more than three times the average speed of an asteroid. Higher velocities mean greater impact energies, which translates into larger craters for a given impactor size.

One unknown when comparing comets and asteroids is the composition of comets. We can measure the iridium content of meteorites and be reasonably certain that the result is consistent with the parent asteroids. But we have yet to sample directly the icy nucleus of a comet. We're not certain of the proportions of ice to rock. We don't know if a comet has a rocky core or is icy throughout. Comets are fragile, and can break apart by themselves without striking anything.

Impact of Comet Shoemaker-Levy with Jupiter

Recently, the world was treated to a spectacular example of a comet breakup. On March 24, 1993, at Palomar Observatory, comet hunters Shoemaker and Levy discovered a curious "squashed" comet shaped like a bar with a long tail. A day later, using the Spacewatch telescope, James Scotti resolved the bar into a dozen or more individual bodies arranged in a row like beads on a string. Calculations showed that in July 1992 the wayward comet had passed only 26,000 miles above Jupiter's cloud tops, and the planet's strong gravity had torn it to pieces. The pieces continued to travel in the same orbit, each producing a dusty tail that blended together with the others to form a bar.

Months later, after astronomers had a chance to observe the motion of the comet string, they came to a startling conclusion. No longer did the comet orbit the Sun. Apparently, it had been orbiting Jupiter as a

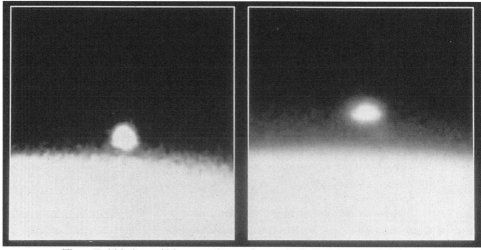

These Hubble Space Telescope photos show a fireball rising to a height of 1,860 miles above the impact point of fragment G. −NASA

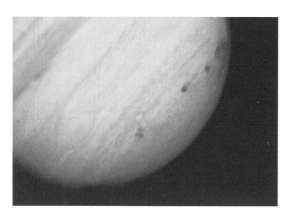

Jupiter shows numerous impact scars left by at least fifteen impacting pieces from Comet Shoemaker-Levy 9 between July 16 and July 22, 1994. −NASA

satellite for perhaps twenty years, and this extremely close passage had changed its orbit into a long, skinny path that took it 30 million miles beyond Jupiter. Now it was returning to the giant planet, to encounter it one last time—this time fatally. Over a period of six days, chunks of ice and rock were predicted to slam into Jupiter's atmosphere at 36 miles per second, releasing the energy of an unimaginable nuclear barrage. At least twenty-one pieces were detected (see color plate VIII), the largest perhaps between 1 and 2 miles across. Traveling at such a high speed, a 2-mile-diameter chunk of ice would release more than a million megatons of energy into Jupiter's atmosphere, to be preceded and followed by the other, smaller, pieces.

Before the predicted impacts, theoreticians had a field day. The most optimistic predicted the impacts would produce spectacular fireballs and huge scars in Jupiter's atmosphere. Most astronomers were more pessimistic, some expecting the giant planet to swallow the fragile icy chunks, leaving little visible evidence of the encounter. This event was unprecedented in human history; no one had ever witnessed a collision between planet and comet. But nature wasn't cooperative. A geometric twist of fate placed the collisions on the side of Jupiter facing away from Earth. We would not see any impacts directly from Earth. But the impact positions were tantalizingly close, and within minutes after impact, they became visible as the far side of Jupiter rotated into view.

On July 16, 1994, virtually every telescope on Earth was pointed at Jupiter to record the historic event. The first impact occurred at 12:12 A.M., Pacific standard time. There was one viewer with a ringside seat. The *Galileo* spacecraft, on its way to Jupiter, could see around the limb of the planet. It photographed the fireball rising from the upper atmosphere. Within minutes of the first collision, the impact point rotated around to face Earth, and the Hubble Space Telescope took the first picture from Earth. There was a huge, oval dark stain at 45 degrees south latitude, the site of impact.

Over the next six days, twenty-one comet pieces slammed into Jupiter at the rate of about one every seven hours. On July 18, fragment G, largest of the pieces, smacked into Jupiter and the Hubble Space Telescope photographed a huge fireball as it rose above Jupiter's horizon. The fireball rocketed skyward, climbing at over 10 miles per second, reaching nearly 2,000 miles above the cloud tops. This expanding mass of superheated gas carried a temperature of over 18,000 degrees Fahrenheit, hotter than the Sun's surface. Scientists estimated the energy released by fragment G was about 6 trillion tons of TNT. Though not all of the fragments created visible scars, fragment G left behind a brown circular feature, twice the size of Earth, on the cloud deck of Jupiter.

Gene Shoemaker, the codiscoverer of Comet Shoemaker-Levy, calculated the frequency of a comet striking Jupiter to be only once in 2,000 years. He describes how remarkable it really was: "What are the odds that this would happen in our lifetimes—with Hubble fixed, *Galileo* nearing Jupiter, infrared detectors having come of age, and before the money has run out? Folks, I think we've been privileged to witness a bloody miracle."

When scientists began the chemical analyses of the cooling fireballs, they found something unexpected—elements not usually associated with Jupiter. Among these were magnesium, sulfur, silicon, and iron—the major constituents of meteorites.

Even with the data from Comet Shoemaker-Levy, many questions remain about the stony content of comets, and no one can say how much iridium exists in those icy bodies. Comet densities are estimated to range from 0.5 to 1.2 times the density of water, which doesn't leave much room for stony material. Assuming the fraction of iridium in this stony material is the same as that in stony asteroids, a comet about twice the size of a typical asteroid would be needed to provide the same amount of iridium.

Doomsday

Whether the impactor was a comet or asteroid is perhaps a moot point. Whatever it was, there is now no question that it hit Earth 65 million years ago. Was there any warning that day when the impactor struck? Probably, but there was no creature intelligent enough to take notice. An asteroid 6 miles in diameter would be visible, if one knew where to look, by the time it passed inside the Moon's orbit. It would be three hours from impact and counting. The object would appear starlike, increasing noticeably in brightness every few minutes. Within an hour of impact it would rival the brightness of Venus and would cast a shadow. Fifteen minutes before impact it would be seen as an irregular mass growing rapidly in size. Within three seconds of impact, it would enter Earth's atmosphere, but the atmosphere would be no shield against this intruder. The ocean would provide only a momentary cushion to the asteroid. It would hit the ocean floor before it was half submerged.

Nothing in human experience compares with that moment of impact. It struck Earth at 20 miles per second (or more if it was a comet) and blasted a crater 100 miles wide and 5 miles deep. The asteroid was instantly vaporized on impact, and several times its mass in Earth rock met the same fate. Earth trembled as the shock wave spread outward from ground zero. Instantaneously, energy equivalent to that of a billion-megaton nuclear bomb was released, blowing away part of the atmosphere above ground zero and sending hundreds of cubic miles of dust and water vapor into and above the stratosphere. Molten rock debris that escaped vaporization was ejected from the growing crater. Some of this material was driven through the hole in the atmosphere and into space, where it became ballistic; some perhaps even orbited Earth before eventually reentering the atmosphere and falling as glassy blobs back to Earth, adding to the ejecta blanket rapidly building around the new crater. The ejected rock extended for hundreds of miles, thickest near the crater and thinning to an inch or less at the edge.

The expanding shock wave created enormous tidal waves several thousand feet high, which roared across the ocean at hundreds of miles per hour, carrying with them great masses of oceanic crust displaced by the explosion. These waves reached the nearest landmasses, where they dumped their rocky load on the continental shelf and continued hundreds of miles inland before dying out and returning to the sea. Accompanying these waves were winds that reached speeds of hundreds of miles per hour, blowing away from the impact point.

Other hurricane-force winds rushed back into the void left by the passage of the asteroid. It would take weeks for the atmosphere and oceans to stabilize.

The atmosphere suffered major changes that must have affected the biosphere. Under normal conditions atmospheric nitrogen and oxygen do not react together, but the enormous amount of energy released within the atmosphere caused nitrogen and oxygen to form nitrous and nitric oxides. These in turn reacted with water to form nitrous and nitric acids, which rained down upon Earth, making today's acid rains insignificant by comparison. The limestone covering the seabed was vaporized, releasing enormous quantities of carbon dioxide locked in the rock, subjecting Earth to global warming for hundreds of years.

Intense heat generated by the energy of impact would instantly have ignited forest fires over large areas of Earth's surface. Some of these fires were quickly extinguished by the blast of atmospheric shock waves extending from the impact point, but fires farther away would continue to burn, perhaps for months. As much as half of the world's forests may have burned. Hundreds of cubic miles of ash and smoke particles filled the atmosphere and remained suspended for years, adding to the dust already ejected high into the stratosphere. Sooty material found in the boundary layer is direct evidence of this worldwide conflagration.

The Cloud. While all this was going on below, the ejected cloud of dust and vapor over ground zero formed a huge, rapidly expanding fireball, much like the mushroom cloud of a nuclear explosion, but on an enormous scale. The cloud quickly reached into the stratosphere and, catching high-altitude winds, began to spread a mantle of black iridium-laden dust over Earth. As the cloud cooled, more dust condensed. The largest particles, about one-tenth of a millimeter in diameter, rained down within a few days, much like the ash from an erupting volcano. The smallest particles, about one-thousandth of a millimeter, remained suspended for months or years before falling back to Earth, gradually building up the famous clay layer.

An incredible 400 cubic miles of dust spread out over Earth—so much dust that it was thick enough to blot out the Sun completely. For months or possibly years, Earth was enveloped in darkness, blacker than night. Temperatures dropped by tens of degrees, falling below freezing in most places, even on the equator. Any dim light filtering through the gloom was not sufficient to allow photosynthesis. Plants on land and in the sea withered and died. Phytoplankton in the oceans, the bottom of the food chain, perished within the first ten to one hundred days. The food chain collapsed and with it the world's ecosystem. This was the beginning of the Great Dying.

The Crater. One glaring weak point exists in this scenario: There was no crater. A 100-mile-diameter crater is not an insignificant geological feature. Some scientists argued that it may be hidden in the ocean and now lies covered by hundreds of feet of limestone. Worse, others suggested that it may have been subducted beneath the continents through plate tectonic processes. Clearly, about half of the ocean floor has been lost to this process. Much has happened to Earth in 65 million years.

Many impact craters are known around the world, but they are either too old or too young or not the correct size to fit the catastrophic event. To find the crater (if it still exists), we have to start at square one. A glance at lunar impact craters shows that there is much more to a crater than a round hole. The most significant part of a crater is its ejecta blanket, which may extend for thousands of miles beyond the crater itself. If one could locate the ejecta, it would lead back to the crater. But where to begin looking?

The best place to begin is with the boundary-layer clay, the only material known to come from the impact site. In the mid-1980s, Alan Hildebrand, then a graduate student at the University of Arizona, decided to take the challenge. He and his advisor William Boynton tested the boundary clay for chemical signatures that would tell them if it came from oceanic or continental crust. The answer was both, with oceanic crust dominating. This suggested that the impact was on the edge of a continent or continental shelf. Therefore, there should be evidence of wave action from the enormous waves that made their way toward the continent.

Geologists began to find curious deposits of broken and mixed rocks resting on top of late Cretaceous beds. Several sites along the southern United States, most a few hundred miles from the Gulf of Mexico, contained these jumbled rock deposits. They obviously did not belong there but were transported by wave action. The Caribbean Basin was strewn with thick deposits, especially in Cuba and Haiti. They were also found along the east coast of Mexico. All of these deposits appeared to be ringing the impact zone.

While the wave deposits were being investigated, evidence of ejecta was found at several locations in North America. This ejecta was beneath the boundary clay, since it was deposited while the fine clay particles were still suspended in the atmosphere. It was unique to North America and the Caribbean Islands. In Haiti, it was much thicker than elsewhere. Since the ejecta should be thickest nearest the impact point, Hildebrand assumed that the crater must be somewhere in the ocean off the southwest coast of Haiti. Two additional discoveries intensified the search in this area. A close examination of the ejecta revealed quartz grains with telltale signs of impact shock. Glassy beads called tektites were found with aerodynamic shapes that formed when the molten rock was ejected high into the atmosphere. Their flight through the atmosphere at supersonic speeds while still molten shaped them into teardrops, flattened spheres, and dumbbells. Tektites are found in abundance around known impact sites and, like shatter cones and shocked quartz, are considered diagnostic for impact features.

In 1990, Hildebrand's search centered on a semicircular structure that had been detected through a seismic survey in the Colombian Basin deep in the Caribbean Sea about 50 miles north of Colombia and about 350 miles south of Haiti. Further seismic studies needed to be made to see if the other half of the structure existed, but interest waned when the tektite glass found in Haiti showed a composition more compatible with continental rather than deep oceanic crust.

The Smoking Gun. While this rash of discoveries was stirring up scientific debate among the world's geologists, another drama was quietly unfolding on the Yucatan Peninsula. It began a dozen years earlier, about the time Walter Alvarez was sampling the Gubbio boundary clays. Pemex, Mexico's national oil company, had hired a geophysi-

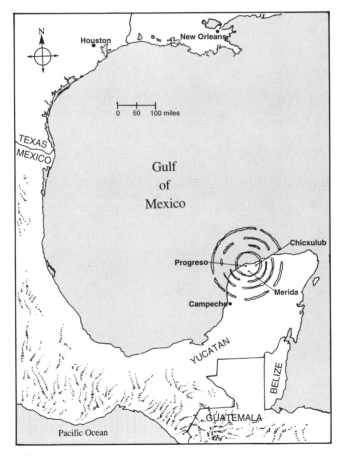

The Chicxulub Crater, 112 miles in diameter, overlies the Yucatan Peninsula. Recent gravity measurements indicate a multiringed structure typical of large impact craters on the Moon and other terrestrial planets.

Typical tektites from around the world. Tektites look like volcanic obsidian but differ in composition and structure. Their names come from the area in which they are found. Only one (b) of the four shown is associated with a known impact crater.

a: *Indochinites from Thailand with typical aerodynamically sculptured shapes.* From left to right: *elliptical; dumbbell; teardrop; disk. The dumbbell is 3.25 inches long.*

b: *These olive-green moldavites (about 1 inch across) are glass ejecta from the Ries Crater in Germany. They come from a strewn field extending from the crater eastward into the Czech Republic and Austria.*

c: *This large rizalite (2.75 inches in diameter) is from the Rizal Province, Philippines. Natural chemical etching produced the star-shaped, deeply grooved pattern and pitted surface.*

d: *A lens-shaped australite (1 inch across) from Australia shows a series of concentric rings and a flange around its edge caused by heating and melting as the stone reentered Earth's atmosphere at near orbital velocity.* –Robert A. Haag collection

cal company from Houston, Texas, to conduct an airborne magnetic survey of the Yucatan Peninsula. Every day an airplane towing a magnetometer traced a grid beginning in the Gulf of Mexico and crossing the shallow waters off the north Yucatan coast. Pemex was looking for thick, sedimentary beds that could be oil bearing. Glen Penfield, a young geophysicist just out of college, was sent to Merida to monitor the progress. Every day he pieced together strip charts made the day before, which revealed more and more sediments on the gulf floor.

Before long, the charts began to show a magnetic disturbance. It grew day by day until it formed a huge semicircular arc with cusps facing south. This curious structure was buried beneath more than a mile of limestone sediments, typical of the ocean floor in the area. The structure didn't seem to belong there, so Penfield decided to investigate further. In the 1960s, Pemex had made a gravity survey primarily covering the peninsula. Using the borrowed gravity map of the area, Penfield located another arclike structure partially over the Yucatan Peninsula, but this time facing north with its cusps projecting into the gulf. He placed the two tracings together. They fit perfectly. A huge crater more than 100 miles in diameter appeared, centered on the small coastal village of Puerto Chicxulub. Penfield had apparently found the "smoking gun."

Pemex was not interested in impact craters and refused to release the magnetic and gravity data to the world, but they did allow Penfield and their field geologist to announce the discovery at a meeting of the Society of Exploration Geophysicists in 1981, where they received an "underwhelming" reception—nobody at the meeting showed the least bit of interest. They were looking for oil, not impact craters. The incredible discovery stalled: wrong place, wrong time, wrong meeting.

Penfield tried to gather further evidence to support the find. What he needed were core samples to verify the structure's impact origin. Pemex had drilled several test wells on the peninsula dating back to the 1950s, but when Penfield tried to locate the core samples, he learned they had been destroyed in a warehouse fire in 1979.

What a surprise it must have been when a *Houston Chronicle* reporter told Hildebrand of his conversation with a geophysicist in Houston back in 1981. The geophysicist claimed he had found a huge crater on the Yucatan Peninsula. It didn't take Hildebrand long to find Glen Penfield, and in 1990 they located a core sample taken from just outside the crater rim. This sample was housed at the University of New Orleans. It revealed shocked quartz, confirming the crater's extraterrestrial origin.

Scientists are a cautious lot. Their constant cry is for more data, more evidence. They want to drill new core samples to see if they reveal impact-melted and shattered rock. The center of the structure seems to be raised. Is this a rebound feature typical of impact craters or is it

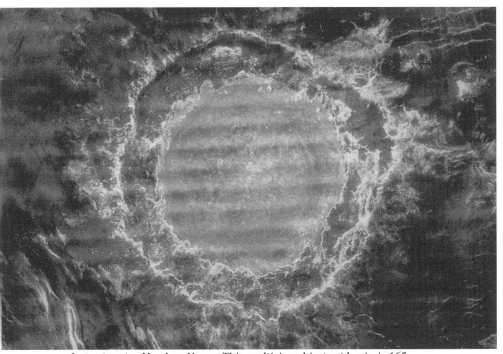

Impact crater Mead on Venus. This multiringed impact basin is 165 miles in diameter, similar in structure and size to the Chicxulub Crater. Its floor is flooded with basalt. –NASA

volcanic? Earlier drilling brought up igneous rock from about a mile down, signaling a molten origin. Was this igneous process induced by the impact of a 6-mile-wide asteroid?

Today, impact geologists no longer doubt the occurrence of a devastating impact 65 million years ago on the Yucatan Peninsula. They argue instead about the size of the crater and the total energy of the impact. In late 1993, Mexican and U.S. geologists working together announced the discovery of what appears to be a subtle fourth ring partially encircling the crater. If this measurement is confirmed, the crater's known diameter will increase to about 180 miles. Virgil L. Sharpton, of the Lunar and Planetary Institute in Houston and the leader of the team, has estimated that the total energy required to form this enlarged crater would be five times the original estimate, or equivalent to a 5-billion-megaton blast. This is probably the largest impact delivered to the inner solar system since life originated on Earth; it compares favorably in size with the 165-mile-diameter crater Mead, the largest impact crater on Venus.

Multiple Impacts

Images of the asteroids Gaspra and Ida show irregularly shaped masses, obviously pieces of larger bodies. Some of these pieces may travel together in twos and threes. In August 1989, a near-Earth asteroid

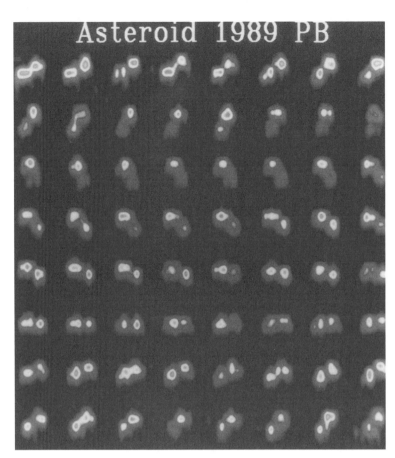

Asteroid 1989 PB

Several dozen "intensity" images of asteroid 1989PB made from a radar "motion picture" to be read left to right and down the page. Two distinct half-mile-diameter lobes appear to be in contact, both rotating around an axis that points toward the viewer. —NASA

was discovered when it was 3.4 million miles from Earth. The asteroid was given the temporary designation 1989PB. Two weeks later, using the world's largest radio telescope, in Puerto Rico, astronomers bounced radar waves off the asteroid and from the echoes constructed the first radar images of an asteroid.

As if this wasn't enough, another near-Earth asteroid, 4179 Toutatis, passed within 2.2 million miles of Earth on December 8, 1992. This time, astronomers were prepared to produce real images of the asteroid using radar echoes. The results were spectacular. The images reveal two irregularly shaped cratered objects about 2.5 and 1.6 miles in average diameter. The four pictures, taken on December 8, 9, 10, and

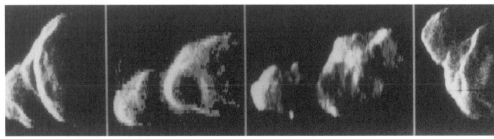

Radar images of asteroid 4179 Toutatis, showing its contact binary structure. —NASA

13, each show a different orientation as the asteroid rotated. Like 1989PB, Toutatis is a contact binary asteroid. In both cases, these asteroids were probably separate bodies that collided "gently" and stuck together.

The newest photos of Ida show a 1-mile-diameter moon orbiting 30 miles from the asteroid's surface. Some astronomers suspect that Gaspra may even be a fused binary. Contact binaries, first regarded as curious rarities, are turning out to be common. This information adds a new element to the evolving story of Earth impacts.

If a contact binary asteroid were to enter Earth's atmosphere, it would probably break apart at the joining point and both bodies would impact Earth close together, forming a double crater. Scientists believe this has happened many times. The best examples of double-impact craters are the 300-million-year-old East and West Clearwater Lakes in Quebec.

Multiple asteroid strikes may be only part of the story. Many researchers, including Hildebrand and Alvarez, believe the K-T impactor was a comet. Comets are fragile bodies, subject to breakup when they pass close to the Sun and likely to break apart at Earth's distance from the Sun. The pieces would drift apart sufficiently to strike Earth at different times, producing multiple craters separated by hundreds or thousands of miles. This raises the question of a multiple impact at the end of the Cretaceous.

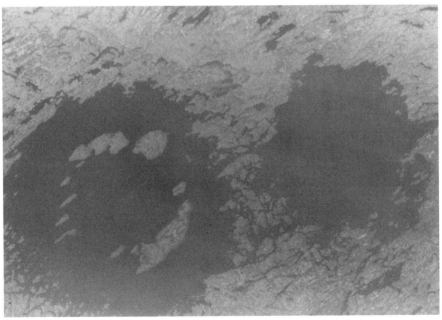

Clearwater Lakes near Hudson Bay in Quebec. The larger West Clearwater Lake, 21.6 miles across, appears to have a rebound ring. East Clearwater Lake is 15.6 miles in diameter and has a central uplift below its rim and within the lake. Both are associated with shock metamorphism. –NASA

There is at least one crater that dates near the K-T boundary. It is the largest suspected impact structure in the United States, located about 170 miles due north of Kansas City near the town of Manson in central Iowa. This is the crater Hildebrand and others studied first, because it was the candidate closest to the ejecta now associated with the Yucatan site. The 20-mile-diameter Manson structure is completely buried in glacial till. What is known about it comes from core drilling. It has a raised central zone of crystalline rock surrounded by a depressed area filled with brecciated Lower Cretaceous sediments. Shock features found in early core samples suggested an impact origin, although no shatter cones have been found.

New core samples drilled in 1992 brought up rock that had been melted by the impact. The radioactive clock in this material was reset when the melted rock liberated all its argon gas, accumulated through millions of years of radioactive decay of potassium-40. Since the impact, argon again began to accumulate at the same fixed rate. The amount of argon in the rock indicated an age for the Manson structure of 65.4 million years, virtually identical to that of the Chicxulub Crater. This seemed to settle the question. The dinosaurs had been hit by a double whammy!

But the latest dating of the drill core samples from Manson (1993) conflict with the 1992 estimates and place its age at 74 million years, about 10 million years too early. Regardless, the idea of multiple K-T impacts is still being pursued. A dual layer of sediments near Trinidad, Colorado, at the K-T boundary, suggests both a major and minor impact event but not simultaneous ones. Debris from the larger event (Chicxulub) was deposited first, producing a thick clay layer. There it remained long enough for plants to root themselves before the smaller impact event deposited a thinner layer above, destroying the growing plants. Evidence of this root system still survives. The two events must have taken place within a few years of each other. If the minor impact event did not produce the Manson Crater, then what did it produce? At the present time, there is no other candidate. The Sierra Madera crater in Pecos County, Texas, lies not too far away, but is too old to be considered a K-T candidate.

In 1997, Walter Alvarez came up with a solution to the missing second impact crater. There wasn't any. Instead, there were two fireballs from the single impact at Chicxulub. The first fireball ejected shocked quartz and other impact materials at a steep angle of 70 degrees. That fireball was caused by vaporization of limestone from the ocean floor, which released a huge hot cloud of carbon dioxide. The nearly simultaneous second fireball blew northwestward at a low angle of 45 degrees. That hot vapor cloud carrying glassy melt droplets arrived first in the northwestern states, followed shortly by ejecta from the second cloud, accounting for the double layer of impact debris.

The Rest of the Story

The importance of the impact scenario is its implied effects on life. Without impacts to change the course of evolution, life on Earth today would be very different. There is every reason to believe that the dinosaurs or their descendants would have continued to dominate the animal kingdom. That the course of evolution and our very existence were governed by an extraterrestrial event is a shocking realization. It is catastrophism at its best—totally uncontrollable. We seem to be at the mercy of the planetary laws of motion. Could it happen again? The answer is most certainly, yes! Did it happen at any other time in the past? The answer is again, yes! It did happen at other times in the past.

Extinctions Are a Fact of Life. The idea of global disaster is nothing new. It has been a main theme in the world's religions. Maybe it's not such a bad idea to wipe the slate clean once in a while and start all over again. In all recorded history, we humans have never experienced a global disaster like the K-T event. But other life-forms have—many times. Evolutionary biologists and their cousins the paleontologists know that life on this planet has not been a smooth progression of uninterrupted evolution of species. There have been numerous global mass extinctions throughout the 600 million years of multicellular life on this planet. Every time a mass extinction occurs, evolution accelerates rapidly. New species of animals and plants appear to fill the vacated niches left by past life. As life evolves and diversifies, the world's biota changes, and for the better, we'd like to think.

Everybody accepts past extinctions as a fact and many people, even some scientists, have finally accepted the growing evidence of an extraterrestrial cause for the dinosaurs' demise. Scientists have identified five major, or mass, extinctions—meaning global in extent—with a significant loss of families and genera of both plants and animals. This quite naturally generates uneasy thoughts. What about other extinctions . . . in the future?

The Alvarez team has found iridium anomalies for an additional extinction between the Eocene and Oligocene epochs in the Cenozoic era. This extinction event occurred between 35 and 39 million years ago. The boundary layer was not marked by an obvious clay layer like the K-T boundary, but by a layer of microtektites. There, among the tektites, the Alvarez team found iridium. The iridium concentration was much lower than in the K-T clay but still well above the expected background. The tektites were particularly interesting. There were two distinct layers found, suggesting two different impact events. The iridium in the layer also showed several peaks; apparently, the E-O boundary had an impact and several encores.

At least three impact craters are known with ages falling within this period. The Popigai Crater in Siberia is the largest, with a 60-mile

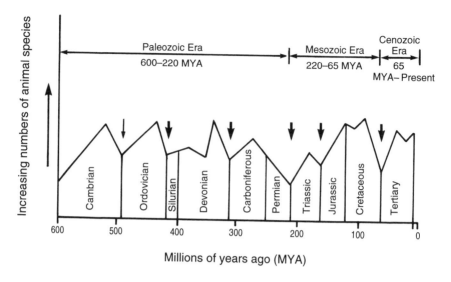

Extinction record over the past 600 million years of geologic history. Bold arrows indicate the five extinctions that most paleontologists consider as worldwide mass extinctions.

diameter. The other two are smaller: the Mistastin structure in Quebec is 16 miles across, and a newly discovered crater in the Atlantic Ocean, about 80 miles east of Atlantic City, New Jersey, measures 9 by 15 miles.

Now that geologists' minds are open to the idea of impact-induced extinctions, so are their eyes. Discoveries are being announced at an unprecedented rate. Researchers at the Royal Institute of Natural Sciences in Belgium found microtektites in 367-million-year-old rocks, much like the material in Haiti. This places it near the time of a known extinction in the middle Devonian (between the Frasnian and Fammenian epochs). The 30-mile-diameter Siljan Crater in Sweden is suspected ground zero, with an age of 365 million years. Nearly simultaneously, French and Australian scientists found an iridium anomaly in the same layers, strongly supporting the idea of yet another killer asteroid.

The greatest extinction in Earth's history occurred at the end of the Permian period, which also marked the end of the Paleozoic era. More than 90 percent of all living species disappeared. Throughout the world at the end of the Permian, conditions were primarily conducive to erosion rather than deposition. As a result, relatively little of the Permian-Triassic boundary has survived intact. Nevertheless, this was a prime target of the Alvarez team. But they were disappointed. In several locations, they found a distinct clay layer but no iridium excess.

Periodic Extinctions—The Unthinkable. Just when the dust was settling on the iridium issue and the world was actually beginning to like the exotic notion of dinosaur deaths by celestial hammer, two

400

paleontologists bravely stepped forward and dropped a bomb, stirring the dust all over again. David Raup and J. John Sepkoski from the University of Chicago published a paper proposing that the extinctions were periodic. That is, they occurred in cycles, and the cycle repeats every 26 million years. Sepkoski had been compiling lists of the scientific family names of all the life-forms that ever lived in the sea. Most, of course, had become extinct, and he noted the approximate time (epoch) when this had occurred, according to the fossil record. Before long he had compiled an impressive list, more complete than any before. Then he called in his colleague David Raup and together they analyzed the data. From the statistical analysis came this 26-million-year cycle.

The Chicago team plotted the percent extinction of families from one geologic epoch to the next, beginning 250 million years ago and continuing to the present, taking in the Mesozoic and Cenozoic eras and the very end of the Paleozoic. The accuracy of the extinction dates varies by a factor of ten from plus or minus 1 to 2 million years in the Tertiary period to plus or minus 10 to 20 million years in the Triassic. This reflects the margin of error in dating the rocks where the fossils are found, which increases with the age of the samples. Four peaks appear on the plot. These are the major extinctions, with the greatest

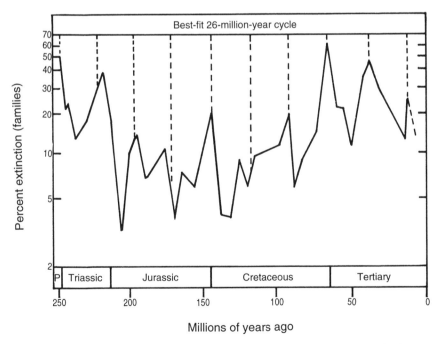

The original Raup-Sepkoski plot of twelve extinctions of families from the late Permian to the middle Miocene compared with a 26-million-year extinction cycle. Of the four highest peaks, the late Permian, late Triassic, and late Cretaceous peaks are well known as major mass extinctions. –Reproduced by permission of David Raup, University of Chicago **401**

loss of families. Four smaller peaks appear, less devastating but still significant. Four peaks of still lesser significance bring the total to twelve.

When the peaks are compared with a 26-million-year periodic cycle, one cannot help but be impressed. Over the 250-million-year interval, ten extinctions are predicted. Eight of the ten fall within the dating margin of error, and all are significant extinctions. The remaining four are considered minor extinctions with peaks not significantly above background.

When Raup and Sepkoski announced their results, they were predictably met with interest but few supporters. At first, even Luis Alvarez thought they had a crazy idea, but he later changed his mind and became a strong supporter. Results derived from statistical games simply don't match good, solid "iridium-tainted" measurements. Scientists correctly insisted on obtaining a method to test the idea.

If asteroids or comets impacted Earth every 26 million years, there ought to be a record of it among what was left of the impact craters scattered over Earth. When Raup and Sepkoski published their work in February 1984, there were eighty-eight known impact craters, with ages determined radiometrically to within a few million years. Richard Muller from Berkeley's Department of Physics joined Walter Alvarez in a statistical study of those craters known to have formed between 5 million and 250 million years before the present—the same time period that Raup and Sepkoski had selected in their extinction study. Of the craters falling within the specified age criteria, only the largest were chosen. From this reduced number (thirteen) Alvarez and Muller found a periodicity of 28.4 million years, close to but not an exact correlation with the 26-million-year cycle of extinctions. But the discrepancy of 2.4 million years fell well within the uncertainties of the age determinations of the boundaries, especially for those ages over 100 million years.

Periodicity was found primarily among the largest craters. All the major extinctions seemed to be accompanied by one or more craters. Between the cycles were undoubtedly smaller, low-energy impacts, producing small craters that did not result in worldwide extinctions. These impacts were essentially random and were produced by small asteroidal bodies. Most of these craters would not have survived erosion over millions of years, and many of the impactors would have fallen into the sea.

Other scientists performed the magic of statistical analysis on the same crater data and came up with different periodicities in the range of 30 to 32 million years. Still others came up with different hypotheses to explain the results, some terrestrial, some extraterrestrial. Some scientists simply threw up their hands and couldn't accept any of the data or the conclusions. Dating of the paleontological boundaries was not accurate enough, they said. The crater dates were equally inaccurate. To many, the crater record was not complete enough to draw conclusions about periodicities. Yet like the iridium anomalies, the largest craters remain as plain as ever at the major extinction boundaries, enticing us to look again. The debate continues.

Nemesis—The Death Star. As the controversies raged on, the Berkeley group took a bold next step. It didn't matter that their hypothesis of a periodic bombardment of comets or asteroids left the geological world in a turmoil. They were convinced that periodic extinctions were real, but there was no known geological process that could account for a 26- or 28-million-year cycle. These numbers are truly astronomical—and that's where the answer must lie.

About half the stars in the sky are really two or more stars orbiting around each other. Typically it takes them hundreds or thousands of years to complete an orbit.

Astronomers have often wondered why the Sun seems to be a loner, without a companion star traveling with it around the galaxy. Roughly half the stars in the Milky Way Galaxy are red dwarf stars with very low luminosities. In fact, they are so dim that only those closest to the Sun can be seen, even with powerful telescopes. What if the Sun had a red

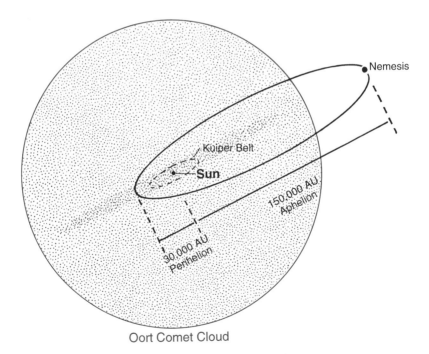

Eccentric orbit of a hypothetical red dwarf star around the Sun. The star has a perihelion near the outer edge of the Kuiper comet belt, where the comet population is highest, and its aphelion carries it beyond the outer edges of the Oort comet cloud.

dwarf companion star so far away that it was not noticeable from Earth? Could it have such a large orbit that it would take 26 to 28 million years to round the distant Sun?

A star with an orbital period of 26 million years would orbit with an average distance of about 88,000 AU if its orbit were circular. This would place it in the outer part of the Oort cloud, where the comets are sparsely distributed. But, as in most double-star systems with dwarf components, the dwarf star would probably not have a circular orbit but one with considerable eccentricity. At its closest point to the Sun, it could come within about 30,000 AU and at its farthest distance about 150,000 AU. During its perihelion passage, the star would pass into the more populated inner region of the Oort cloud, encountering swarms of comets. The star's gravity would perturb the comets, sending millions inward toward the Sun.

The dwarf star would take about a million years to pass through the cloud, and the comets, once on their way to the Sun, would take about a million years to arrive. Thus, the rain of comets through the inner solar system would be more or less continuous for a million years. After passing through the cloud, the star would move slowly outward to its aphelion point, more than 2 light years from the Sun, taking 13 million years to reach it. During this time, there would be a hiatus in comet activity. Then, after another 13 million years, the star would once again enter the Oort cloud, harvesting yet another population of comets to bring Armageddon to the solar system and death to life on Earth.

The outer planets from Jupiter to Neptune are Earth's only line of defense. They would send many of the invaders back to the Oort cloud, but there would be too many—some would get through. For a million years the skies of Earth would be filled with long-tailed comets, a sight that humans have never seen.

This incredible scenario is the work of Richard Muller and Marc Davis at Berkeley and Piet Hut at Princeton University. They published their ideas in the scientific journal *Nature* in a paper entitled "Extinction of Species by Periodic Comet Showers" in April 1984, creating yet another storm of controversy. In a footnote to this paper, Muller gave the hypothetical star the name Nemesis: "If and when the companion is found, we suggest it be named Nemesis, after the Greek goddess who relentlessly persecutes the excessively rich, proud and powerful. We worry that if the companion is not found, this paper will be our nemesis."

The press had a great time with the story. Nemesis became the "Death Star." Everyone reading the stories of Nemesis wondered when it would strike again. To calm the worldwide jitters, the authors pointed out that we are currently at midcycle. Nemesis is at its aphelion point. That means we are 13 million years after an extinction occurred in the Miocene epoch and 13 million years before another bout with Oort's comets.

The Future

Nemesis, though speculative, is respectable science because, like periodic mass extinctions, it is testable. The test is to find the star. Soon after the Nemesis idea was announced, Muller assembled a group at Berkeley to search for the elusive star. At first, it seemed simple enough. Just look for a reddish star of the 15th magnitude (about the brightness of Pluto) that shows a relatively large yearly shift in position due to its proximity to the Sun. What was believed would take only a few months has stretched to years. The star has yet to be found. Today the periodic-extinctions hypothesis is incomplete without the discovery of Nemesis, much as the Alvarez Cretaceous-extinction hypothesis was incomplete without the discovery of the K-T crater. Yet, as the years go by, evidence piles up for periodic extinctions. What began as one, then two iridium anomalies at two mass extinction boundaries has grown to five. The theory of periodic mass extinctions cries for an explanation, and the only really plausible explanation awaits confirmation of Nemesis.

The tedious search continues. Muller's group is using automated telescopes to survey thousands of faint, red dwarf stars at intervals of six months to see if any show telltale parallactic shifts in their position due to Earth's motion. These shifts would demonstrate the star distance from Earth. Of the 3,100 stars selected as possible candidates, more than half have already been eliminated, and every clear night reduces the number by ten. We should know in the near future if Nemesis does exist.

The latter half of the twentieth century has seen the rise of the new science of impact geology. It is truly a hybrid science requiring a free exchange of ideas from a diverse group of established sciences. It has profoundly shaken the foundations of paleontology and evolutionary biology and is changing our thinking about our relationship to the cosmos. All this started so modestly. Flashing trails of light in the sky. Stones pelting Earth from space. A gaping crater in the northern Arizona desert.

Epilogue

For one hundred fifty years, scientists have studied meteorites with the knowledge that they are rocks from space. These small bodies have led some of us on a journey to know ourselves. They have been responsible, in part, for our origins and, in part, for the demise of our animal ancestors. They are why we are here. We are born of star stuff, and our legacy is out there among the asteroids and comets and stars.

In late 1984, I gave a small iron meteorite to astronaut Dick Scobee in commemoration of his visit to the University of Arizona, where he had graduated in engineering. Scobee was scheduled to command the space shuttle early the next year. I suggested, half in jest, that he return the meteorite to its home in space on his next voyage. In a way, it would be a historic event—a meteorite returning to space! A few weeks later a letter arrived from Scobee. He thanked me for the meteorite, calling it a "nice little treasure" and assured me that he would, indeed, return it "at least temporarily to the environment from which it came." On January 28, 1986, the meteorite began its historic return journey but never made it home. Nor did the crew of seven. Their ship was the *Challenger*.

I frequently look upon this sad story in a philosophical way. Here, symbolized in the space shuttle, was the culmination of evolution on Earth. We humans have made it to the top of the ladder. Through millions of years of evolution we evolved the most complex organic entity in the universe, the human brain. We learned to use this unique tool to develop our sciences and engineer the machines that would someday extend our reach into the cosmos. These machines went before us into space, giving us tantalizing views of the asteroids and comets, our forebears. Like lollipops to a child, the meteorites tantalize us, beckoning us into space. Now it is ours to return to the environment whence we came.

Meteorite Verification Laboratories

Several museums and laboratories in the United States are equipped to examine and authenticate any suspected meteorite specimen sent to them. This requires that a small sample, if not the entire specimen, be sent to them for analysis. Usually the service is performed free of charge; however, if the specimen proves to be a meteorite, the institution may request that the finder consider donating the specimen to the institute, or they may offer to purchase part or all of the specimen.

The American Museum
of Natural History
Central Park West
at 79th Street
New York, NY 10024

Center for Meteorite Studies
Arizona State University
Tempe, AZ 85281

The Field Museum
of Natural History
S. Lake Shore Drive
Chicago, IL 60605

Institute of Geophysics
and Planetary Sciences
University of California
Los Angeles, CA 90024

Institute of Meteoritics
Department of Geology
University of New Mexico
Albuquerque, NM 87131

Lunar and Planetary Laboratory
Space Sciences Building
University of Arizona
Tucson, AZ 85721

National Museum
of Natural History
Department of Mineral Sciences
Smithsonian Institution
Washington, DC 20560

New England
Meteoritical Services
P.O. Box 440-A
Mendon, MA 01756

Commercial Meteorite Dealers

Meteorites may be purchased or traded from several commercial businesses. The market for meteorites remains fluid. With the exception of the most common meteorites, for which the supply is plentiful, the availability of specimens fluctuates. Relatively rare meteorites become available to the commercial market through museum trades, private collections, and occasional new finds. These disappear quickly into university, museum, or private collections. Request that your name be placed on a "call me when something new comes in" file if you wish to be kept aware of the latest acquisitions. Commercial collectors are very accommodating and will make every effort to acquire the specimens you are seeking.

Michael Blood
6106 Kerch Street
San Diego, CA 92115

Alain Carion
92, rue Saint-Louis en l'Isle
Paris (4 ème) France

Michael I. Casper,
Meteorites, Inc.
Post Office Drawer J
Ithaca, NY 14851

Direct Line Resources
P.O. Box 982
Lynnwood, WA 98046

Excalibur Mineral Company
1000 North Division Street
Peekskill, NY 10566

Robert A. Haag
P.O. Box 27527
Tucson, AZ 85726

Killgore Southwest
P.O. Box 95
Payson, AZ 85547

Allan Langheinrich
290 Brewer Rd.
Ilion, NY 13357

Macovich Collection
of Meteorites
1501 Broadway, Suite 1304
New York, NY 10036

Magic Mountain Gems
3906 W. Ina, #285
Tucson, AZ 85741

Mare Meteoritics
P.O. Box 19041
Oakland, CA 94619

Meteorite Market
P.O. Box 33873
Juneau, AK 99803

Meteorites
577 Hunterdale Road
Evans, GA 30809

Mile High Meteorites
1710 Robb Street
Lakewood, CO 80215

Mineralogical Research Co.
15840 E. Alta Vista Way
San Jose, CA 95127

Mitterling Meteorites
5245 S. Country Club Rd.
Warsaw, IN 46580

Montana Meteorite Laboratory
P.O. Box 1063
Malta, MT 59538

New England
Meteoritical Services
P.O. Box 440-A
Mendon, MA 01756

Blaine Reed Meteorites
907 County Road 207 #1
Durango, CO 81301

S. A. R. L.
Labenne Meteorites
16 Boulevard Gambetta
02700 Tergnier, France

Swiss Meteorite Laboratory
P.O. Box 126
CH-8750 Glarus, Switzerland

Edwin Thompson Meteorites
5150 SW Dawn Street
Lake Oswego, OR 97035

Walter Zeitschel
Meteorite Collection
P.O. Box 2340
D-63413 Hanau
Germany

Additional dealer information is on the Internet at the following sites:
www.meteoritecentral.com • www.meteorite.com

Selected List of Authenticated Impact Craters Worldwide

More than one hundred fifty impact craters have been identified worldwide and newly identified craters continue to be added to the list each year. Almost all the structures exhibit a variety of evidence pointing to an impact origin. Most are visible on the surface. The sixty-three craters listed here were selected because they show strong impact evidence or are among the largest on the planet. The type of evidence is tabulated as follows: (1) shock metamorphism; (2) coesite or stishovite; (3) shatter cones; (4) raised rims or central uplift; (5) ring structure; (6) shattered rock breccia or a breccia lens; (7) impact melt or impactite glass; (8) meteorites or meteoritic oxide.

Name and Location	Diameter (miles)	Evidence (type)	Approx. Age (millions of years)	Visible on Surface
Acraman South Australia, Australia	55.9	6	>450	yes
Aouelloul Crater Mauritania	0.24	7, 8	3.1 ± 0.3	yes
Araguainha Dome Mato Grosso, Brazil	24.8	1, 3, 4	247 ± 5.5	yes
Boxhole Crater Northern Territory, Australia	0.11	8	0.03 ± 0.0005	yes
Brent Crater Ontario, Canada	2.36	6, 7	450 ± 30	yes
Campo del Cielo Argentina	0.03	8	<0.004	yes
Carswell Lake Saskatchewan, Canada	24.2	1, 6	115 ± 10	yes
Charlevoix Structure Quebec, Canada	33.5	1, 3, 4	357 ± 15	yes
Chesapeake Bay Virginia, U.S.A.	52.8	1	35.5 ± 0.6	no
Chicxulub Yucatan, Mexico	112	1, 6	64.98 ± 0.05	no
Clearwater Lakes Quebec, Canada	15.6, 21.6	7	290 ± 20	yes
Crooked Creek Structure Missouri, U.S.A.	4.35	1, 3, 6	320 ± 80	yes

Name and Location	Diameter (miles)	Evidence (type)	Approx. Age (millions of years)	Visible on Surface
Dalgaranga Crater Western Australia	0.01	8	0.027	yes
Decaturville Dome Missouri, U.S.A.	3.73	3, 6	<300	yes
Deep Bay, Reindeer Lake Saskatchewan, Canada	8.07	6	100 ± 50	yes
Flynn Creek Disturbance Tennessee, U.S.A.	2.20	3, 6	360 ± 20	yes
Gosses Bluff Northern Territories, Australia	13.7	3	142.5 ± 0.8	yes
Gow Lake Saskatchewan, Canada	3.11	4, 6	<250	yes
Haughton Dome Northwest Territories, Canada	14.9	1, 3, 4	23 ± 1	yes
Haviland Crater Kansas, U.S.A.	0.01	8	<0.001	yes
Henbury Craters Northern Territory, Australia	0.01 (largest of 13 craters)	7, 8	<0.005	yes
Holleford Crater Ontario, Canada	1.46	2, 6	550 ± 100	no
Ile Rouleau Structure Quebec,Canada	2.48	3, 6	<300	yes
Ilyinets Structure Ukraine	2.79	6, 7	395 ± 5	no
Janisjarvi Structure Karelia, Russia	8.70	1, 3, 6, 7	698 ± 22	yes
Kaalijarv Crater Estonia	0.07	3, 6, 8	0.004 ± 0.001	yes
Kaluga Structure Russia	9.32	6, 8	380 ± 10	no
Kara-Kul Tajikistan	40.4	2, 4, 6, 7	<5	no
Kentland Disturbance Indiana, U.S.A.	8.08	2, 3, 6	<97	yes
Kursk Structure Russia	3.42	4, 6, 7	250 ± 80	no
Logoysk Crater Belarus	10.6	2, 6, 7	40 ± 5	no
Lonar Crater Maharashtra, India	1.14	4, 6	0.052 ± 0.006	yes
Manicouagan Lake Quebec, Ontario	62.1	3, 5	214 ± 1	yes
Manson Structure Iowa, U.S.A.	21.7	4, 6, 7	73.8 ± 0.3	no
Meteor Crater Arizona, U.S.A.	0.74	1, 2, 3, 6	0.049 ± 0.003	yes
Mien Sweden	5.59	2, 7	121.0 ± 2.3	yes
Mistastin Labrador, Canada	17.4	4, 7	38 ± 4	yes
Mizarai Lithuania	3.11	3, 6, 7	570 ± 50	no

Name and Location	Diameter (miles)	Evidence (type)	Approx. Age (millions of years)	Visible on Surface
Monturaqui Crater Antofagasta, Chile	0.28	7, 8	<1	yes
New Quebec Crater Quebec, Canada	2.14	7	1.4 ± 0.1	yes
Nicholson Lake Northwest Territories, Canada	7.77	3, 6	<400	yes
Nördlinger Ries Bayern, Germany	14.9	2, 3, 6, 7	15 ± 1	yes
Odessa Crater Texas, U.S.A.	0.10	8	<0.05	yes
Popigai Basin Siberia, Russia	62.1	2, 6, 7	35 ± 5	yes
Red Wing Creek North Dakota, U.S.A.	5.59	3, 4	200 ± 25	no
Ries Basin Bayern, Germany	14.8	1, 2, 3, 4, 6, 7	15 ± 1	yes
Rio Cuarto Argentina	2.80 (largest of 10 total)	2, 7, 8(?)	<0.1	yes
Rochechouart France	14.3	1, 6, 7	186 ± 8	yes
Rotmistrovka Basin Ukraine	1.68	1, 6, 7	140 ± 20	no
Serpent Mound Ohio, U.S.A.	4.97	2, 3	<320	yes
Sierra Madera Texas, U.S.A.	8.08	3, 6	<100	yes
Sikhote-Alin Russia	0.02 (largest of many craters)	6, 7, 8	0.027	yes
Siljan Ring Crater Sweden	32.3	1, 3, 4	368.0 ± 1.1	yes
Slate Islands Ontario, Canada	18.6	1, 3, 4	~450	yes
Steinheim Basin Bayern, Germany	2.17	3	15 ± 1	yes
Strangways Crater Northern Territories, Australia	15.5	1, 3, 4	<470	yes
Sudbury Structure Ontario, Canada	155	1, 3	1850 ± 3	yes
Talemzane Crater Algeria	1.09	2, 3	<3	yes
Vepriai Lithuania	4.97	3, 6	>160 ± 30	no
Vredefort Ring South Africa	186	2, 3, 6	2023 ± 4	yes
Wabar Saudi Arabia	0.07	7, 8	0.006 ± 0.002	yes
Wells Creek Tennessee, U.S.A.	7.46	3, 6	200 ± 100	yes
Wolfe Creek Western Australia, Australia	0.54	4, 8	<0.3	yes

a:

b:

c:

Three pictures of a slab of a medium octahedrite from Bear Creek, Jefferson County, Colorado: a: before, b: after two minutes, c: after complete etching to reveal the Widmanstätten figures. The etching procedure described in Appendix D was closely followed. To fully etch the meteorite took about five minutes. The specimen measures 1.375 inches long.

APPENDIX D

Etching an Iron Meteorite

The etching procedure is suitable only for known meteorites intended for display. If you have found a new iron meteorite, it is best to leave the preparation to scientists, since many preparation and coating materials can contaminate the specimen and destroy the trace-element information. New specimens should be studied thoroughly before proceeding with display preparations.

To etch an iron meteorite properly, the specimen must be cut and the flat face ground with coarse, medium, and fine abrasive. Number 600 grit produces a semipolish that is adequate for etching purposes.

The etching solution is a mixture of concentrated nitric acid and 95 percent ethanol (nitol). Working with concentrated acids is dangerous; they must be handled with extreme care. When mixing the nitol solution, always pour the acid into the alcohol—NEVER the reverse. You will need the following materials:

1. Two 250-ml glass beakers or glass jars
2. One petri dish or glass dish
3. One small (half-inch), fine-bristle paintbrush
4. Glass stirring rod
5. Concentrated nitric acid, 15 ml
6. 95 percent ethanol, 500 ml
7. One can polyurethane spray
8. Latex gloves and goggles

The recipe makes 265 ml of etching solution, much more than you would need for a single specimen. Unless you are etching many specimens, you need to prepare only one-tenth of this amount.

Before proceeding with the etching, you should don the latex gloves and protective goggles. Etching should be done in a porcelain sink with running water available. Begin by carefully pouring the nitric acid into 250 ml of ethanol. Thoroughly stir with the glass stirring rod. The meteorite specimen may have an irregular shape, which means the prepared surface will not rest in a horizontal position. You will either have to hold the specimen with one hand or place a lump of modeling clay under the specimen to prop it into a horizontal position.

Place the meteorite in the petri dish or shallow glass dish to catch the nitol solution. With the paintbrush, evenly spread nitol solution over the surface with a constant sweeping motion.

This constant motion of the brush over the surface maintains an even flow of nitol. At this acid concentration, you will begin to see the Widmanstätten figures appear in one or two minutes. Refer to the photographs to judge the progress of the etching. As the etching proceeds, the kamacite dissolves, leaving the taenite border. Look for the appearance of distinct plates that have a satinlike sheen. As you tilt the specimen, the plates will appear alternately bright and dark depending upon your viewing angle. The taenite will look silvery and bright, while the kamacite will have a more satin finish. When you have reached this point, place the specimen, still in the dish, under slowly running tap water. Be careful not to splash the nitol. Wash the specimen for five minutes. Wipe dry and then immerse the entire meteorite in the remaining 250 ml of unused ethanol. This will act as a drying agent, eliminating any water that may have found its way between the kamacite plates.

Leave the meteorite in the alcohol bath for about an hour. Then remove the specimen and place it in an oven at the lowest temperature (150 degrees Fahrenheit) for an additional hour. Let it cool to room temperature. Finally, prop the meteorite into a horizontal position on a block of wood and spray the surface with a thorough coating of polyurethane. To keep the specimen dust free while drying, cover it with a deep pan lid. The polyurethane gives the meteorite a beautiful sheen and protects it from contact with outside moisture and fingerprints. If you get fingerprints on the face before spraying, they must be removed immediately. Acids in natural body oils will etch into the surface, leaving your permanent "signature."

Even after these elaborate drying and coating procedures, you may notice the appearance of brownish rust between the plates after a few months. This may be due to trapped moisture or, more probably, "lawrencite disease," meaning that the meteorite is being attacked from within. The only remedy is to repolish the specimen and go through the etching process again. If it is a small specimen, it may be helpful to place it in an airtight container, like a zipper bag, with some silica gel—a moisture absorbent—or place it facedown in a storage box.

Testing a Meteorite for Nickel

All iron meteorites contain nickel, making a test for nickel diagnostic for meteorite authentication. This even works with chondrites, since they contain a sufficient amount of metal to test positive for nickel. A common chemical test for the presence of nickel can be made in a high-school chemistry lab.

Wearing protective rubber gloves and goggles, dissolve a small piece (less than 1 gram) of the suspected meteorite in heated, concentrated hydrochloric acid.

When it's dissolved, add to this solution a few drops of nitric acid to oxidize the iron in solution. Then add a few drops of citric acid to prevent the iron from precipitating. Neutralize the solution with ammonium hydroxide. The solution should be clear of precipitate. If not, filter the solution to obtain a clear liquid. To test for nickel, add a few drops of dimethyl gloxine to the solution. If nickel is present, even in minute quantities, the solution will turn bright cherry red.

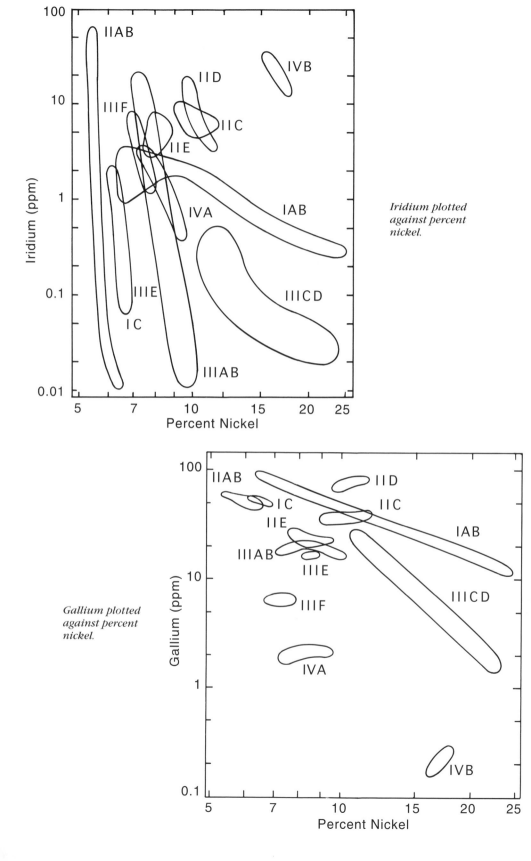

Iridium plotted against percent nickel.

Gallium plotted against percent nickel.

The Chemical Classification of Iron Meteorites

Before the mid-1960s, scientists classified iron meteorites by structure. The researchers polished and etched the irons and noted any Widmanstätten structure. Chapter 12 discusses that structural classification and how it relates to the amount of nickel alloyed with the iron. It was a crude chemical classification and the most diversified irons, the octahedrites, could be subdivided into groups depending upon the width of the Widmanstätten bands. Scientists did not know how or if these different irons were related to each other nor how many parent bodies they represented.

Recently, researchers devised a more refined method of classification using techniques that determine the trace element compositions in meteorites. There are certain elements, the siderophiles, that have an affinity for iron—their atoms can readily bond with iron atoms as the iron crystallizes in a melt. These elements include nickel as the major element and several trace elements such as iridium, gallium, and germanium.

Let's see how trace elements can be used to subdivide iron meteorites into distinct chemical groups. In a slowly cooling iron-nickel melt, as the iron begins to crystallize out it contains less nickel than the melt itself. This means that the first solid metal produced is relatively low in nickel. As the crystallization process continues, the magma becomes richer in nickel, causing the metal to become richer in nickel, though still below the nickel concentration in the melt. This process, called fractional crystallization, results in iron metal with various percentages of nickel. The trace elements have varying affinities for iron as a metal solid versus iron as a liquid. For example, iridium prefers to combine with the first crystallizing solid iron which, in the beginning, is low in nickel. As the percent nickel increases in the melt, the amount of iridium in the resulting metal decreases, since most of it was tied up in the first metal to crystallize. If we plot the percentage nickel against iridium abundance in the metal, we see the relationship between nickel and iridium abundance. As percent nickel goes down, the iridium (in parts per million) goes up dramatically. For a given initial magma composition, well-defined fields appear into which certain iron meteorites can be grouped. Other fields appear representing magmas with different initial compositions (low or high nickel) and where other iron meteorites are grouped. These different fields represent different parent bodies.

A modern plot of percent nickel to parts per million iridium is based on the work of John Wasson and his colleagues at the University of California at Los Angeles, who analyzed about 500 iron meteorites. They distinguished thirteen fields, each representing a separate chemical group. They gave each group a Roman numeral I–IV, a convention that began several years earlier when only four groups were recognized. The letter designations A through F were added to further subdivide the four original groups. Some groups were closely related, and in some cases actually graded into each other, so they were later combined. Thus, group IA and IB are now IAB; groups IIA and IIB are now IIAB.

Plots were also made using the trace elements germanium and gallium. Here the fields are noticeably different in the range over which the fields extend compared with iridium, but the same chemical groupings appear. The small, distinct fields of the gallium and germanium plots are most useful for classifying a new iron meteorite. The iridium plot with its extensive fields gives meteoriticists the resolution necessary to distinguish meteorites within the same group.

This classification scheme tells us that the iron meteorites in collections worldwide apparently came from at least thirteen separate parent bodies. In addition, about 13 percent of the iron meteorites do not fall neatly into any of the fields. By convention, there must be at least five specimens of the same chemical affinity to constitute a chemical group. No meteorites in that 13 percent pass the minimum number rule so they are ungrouped. They may represent several dozen more parent bodies.

As unwieldy as this chemical scheme seems, the fields are related in a general way with the structural classification. For example, it is relatively easy to distinguish low-nickel hexahedrites and high-nickel ataxites by looking at the iridium plots. They are IIAB and IVB respectively. The table of structural classification of iron meteorites in chapter 12 relates the structure of iron meteorites with nickel content and the chemical classification.

GLOSSARY

ablation. The removal of material by heating and vaporization as a meteorite passes through Earth's atmosphere.

accretion. The gradual accumulation of material through the collision of particles in the solar nebula, forming a subplanetary or planetary body.

achondrite. A class of stony meteorite formed by igneous processes; they lack the spherical inclusions called chondrules.

albedo. The percentage of incident sunlight reflected off the surface of a planetary body.

Amor asteroid. An asteroid whose perihelion distance lies just outside Earth's orbit, defined as between 1.017 AU and 1.3 AU from the Sun.

amphoterite. A type of chondritic meteorite with low metal and low iron; an LL-chondrite.

anhedral. Irregular-shaped mineral crystals not bounded by their typical crystal faces.

aphelion. A position on an elliptical orbit where a celestial body is farthest from the Sun. See *perihelion.*

Apollo asteroid. An asteroid that passes inside Earth's orbit; defined as having its mean distance from the Sun equal to or greater than 1 AU and a perihelion distance of less than Earth's aphelion distance.

asteroid. A rocky or metallic orbiting body of subplanetary size showing no cometary activity; usually but not necessarily confined to the asteroid belt.

asteroid belt. A region between the orbits of Mars and Jupiter, between 2.2 and 4.0 AU from the Sun, where most of the asteroids are located.

astrobleme. An ancient, deeply eroded crater formed by explosion of a meteorite, asteroid, or comet upon impacting Earth; usually contains evidence of impact shock. Literally means "star wound."

astronomical unit (AU). The mean distance between Earth and the Sun; a standard unit of distance usually given as 92,960,116 miles.

ataxite. An iron meteorite with a nickel content of greater than 13 percent composed of nearly pure taenite and showing no macroscopic structure.

Aten asteroid. An asteroid whose entire orbit lies inside Earth's orbit; defined as having an aphelion distance of greater than 0.983 AU and a mean distance of less than 1 AU.

aubrite. A stony meteorite formed by igneous processes containing enstatite as its primary mineral; also called an enstatite achondrite.

Barringer Crater. Meteor Crater, Arizona; named after Daniel M. Barringer, the mining engineer who investigated the crater in the first three decades of the twentieth century.

basalt. A common volcanic igneous rock usually erupted onto Earth's surface from vents; mineral content is primarily pyroxenes and plagioclase feldspars.

breccia. A rock made up of cemented angular, broken pieces of rock.

bronzite. A magnesium-rich orthopyroxene $(Mg,Fe)SiO_3$; found in stony meteorites.

calcium-aluminum inclusions (CAIs). Irregular, white inclusions occurring in carbonaceous chondrites probably representing material condensed first from the original solar nebula.

carbonaceous chondrite. A primitive class of chondritic stony meteorite that contains highly oxidized minerals and organic compounds.

celestial sphere. The apparent sphere of the sky centered on Earth.

Ceres. 1 Ceres; the largest asteroid and the first discovered (1801).

chassignite. A rare achondrite meteorite primarily made of olivine; it is one member of the SNC group of meteorites.

chondrite. A common type of stony meteorite containing spherical bodies called chondrules.

chondrules. Spherical, millimeter- or submillimeter-sized bodies found in stony chondrite meteorites formed by remelting of mineral grains in the solar nebula.

coesite. A rare, high-pressure polymorph of quartz; often found associated with impact craters.

cohenite. An accessory mineral found in iron meteorites; an iron-nickel carbide $(Fe,Ni,Co)_3C$.

commensurate orbits. Asteroid orbits whose periods are simple multiples or fractions of Jupiter's orbital period.

cosmic dust. Microscopic silicate or melted iron particles probably produced from comet tails or by collisions of asteroid bodies, or may also be shed by massive stars.

cosmic-ray exposure age. The length of time a meteoroid has been exposed to cosmic rays in space; determined by the presence of specific isotopes produced by cosmic ray bombardment.

cosmic velocity. The velocity of an orbiting body while in space.

crystallization. The formation of mineral crystals with an ordered atomic crystalline structure.

crypto-explosion crater. See *astrobleme.*

crystal lattice. An orderly arrangement of the atoms of a mineral forming a particular structure, usually expressed in the crystal form of the mineral.

cumulate. An igneous rock made of crystals that have formed within and settled out of a magma.

daughter isotopes. Secondary isotopes formed by the decay of the original, or parent, isotopes.

differentiation. A process in which a homogeneous planetary body liquefies and separates into layers of different density and composition; the body usually separates into a core, mantle, and crust.

diogenite. An achondrite meteorite of igneous origin primarily composed of orthopyroxene crystals as a cumulate; usually of hypersthene, bronzite, and ferrosilite.

distribution ellipse. An elliptical area covering several square miles over which meteorites tend to fall, with the more massive meteorites distributed on the far end of the ellipse.

eccentricity. Referring to an elliptical orbit, the degree to which an orbit is elongated; technically the ratio between the foci of the ellipse and its major axis.

E-chondrite. Enstatite chondrite.

ecliptic. The plane of the solar system; Earth's orbital plane projected against the stars is the reference plane from which all orbits in the solar system are measured.

end-member. One of the two or more simple compounds (minerals) that make up a solid solution.

enstatite. The magnesium end-member of orthopyroxenes; made of magnesium silicate ($Mg_2Si_2O_6$).

enstatite chondrite. A chondritic meteorite containing the pyroxene enstatite and metal; the most reduced of the chondrites.

equilibrium. A state in which the mineral phases in a rock do not undergo further changes in time as long as the conditions remain the same.

eucrite. The most common achondrite meteorite; of igneous origin, similar in composition to terrestrial and lunar basalts.

euhedral. Mineral crystals fully bounded by well-formed typical crystal faces.

fall. An observed fall of a meteorite that is subsequently recovered; see *find.*

fayalite. An iron-rich olivine (Fe_2SiO_4); the iron end-member of olivine.

feldspar. A general term for aluminum silicate minerals with various amounts of sodium, calcium, and potassium.

find. A meteorite that was found but not observed to fall; see *fall.*

forsterite. A magnesium-rich olivine (Mg_2SiO_4); the magnesium end-member of olivine.

fractionation. Fractional crystallization; a process in which minerals crystallize out of a magma at specific temperatures, thereby changing the composition of the magma.

fragment. A piece of a larger single meteorite that has been torn free of the main mass and usually shows fragmented characteristics.

fusion crust. A dark, glassy coating that forms on stony meteorites as they ablate in the atmosphere; usually made of glass and iron oxide.

gegenschein. A faint, diffuse, glowing region on the ecliptic opposite the Sun, produced by sunlight reflecting off interplanetary dust particles.

gneiss. A coarse-grained metamorphic rock formed from igneous rock such as granite; exhibits a banded structure.

graphite. The amorphous state of carbon; often found as nodular inclusions in iron meteorites.

H-chondrite. A group of common chondritic meteorites with high total iron content.

half-life. The interval of time needed for half of the remaining atoms of a radioactive isotope to decay into its daughter products.

hexahedrite. An iron meteorite containing less than 5 percent nickel; usually occurs as nearly pure kamacite with Neumann lines across the crystal faces.

Hirayama family. A group of asteroids with similar orbits, formed by the breakup of a larger asteroid.

howardite. A brecciated achondrite composed of eucrite and diogenite fragments; thought to be the soil of a parent body.

hypersthene. A magnesium-rich orthopyroxene ($(Mg,Fe)SiO_3$); a common pyroxene in chondritic meteorites.

igneous rock. A rock composed of interlocking mineral crystals formed from a melt or magma.

impact shock. Shock effects produced by an impacting meteorite; shock effects are usually seen in quartz minerals.

impactites. Small, roughly spherical glassy masses, melted as a result of heat generated by an impacting meteorite.

individual. A single meteorite not fragmented off a larger meteoroid, which has reached Earth intact.

inertia. The property of a body that resists a change in its state of motion; the momentum of the body is a measure of its inertia.

interstellar grains. Solid grains of material produced and ejected by massive stars; probably composed of silicate grains surrounded by methane ice.

ion. An atom with an electrical charge produced by a gain or loss of one or more electrons.

irons. Meteorites composed primarily of iron, alloyed with a small percentage of nickel and cobalt.

isotope. Atoms that have the same number of protons but a different number of neutrons; atoms of the same element but with different atomic masses, some of which are unstable.

kamacite. An alloy of nickel-iron with not more than 7.5 percent nickel; an important alloy in nickel-iron meteorites.

kinetic energy. The energy of a moving body; defined as the product of one-half the mass times the square of the velocity.

Kirkwood gaps. Gaps in the asteroid belt produced by orbital resonance between the asteroids and Jupiter.

L-chondrite. A class of chondritic meteorites with metal (iron) and combined iron amounts between the H- and LL-chondrites.

LL-chondrite. A class of chondritic meteorites with the lowest amounts of metal (iron) and combined iron.

lodranite. A rare type of stony-iron meteorite with equal amounts of bronzite and olivine in a nickel-iron aggregate; classified with the mesosiderites.

magma. Molten rock containing many dissolved minerals and gasses, which crystallize out to form igneous rock.

magnetite. An iron oxide (Fe_3O_4), usually black, found in carbonaceous chondrites and in the crusts of most meteorites.

mantle. A zone in a differentiated parent body between the core and crust.

maria. Large impact basins on the Moon filled with basalts.

melt. See *magma.*

mesosiderite. A class of stony-iron meteorites composed of iron-nickel alloy and angular rock fragments of eucrite and diogenite composition.

metal. A general term for uncombined metal in meteorites, usually the iron alloy kamacite or taenite, depending upon the nickel content.

metamorphism. Recrystallization of minerals in a meteorite or terrestrial rock as a result of heat and pressure on the solid rock, which results in the destruction of the original texture.

meteor. The light produced when a meteoroid enters Earth's atmosphere and burns due to friction.

meteor shower. A rain of tiny cometary dust particles that enters Earth's atmosphere annually from a specific direction and is usually associated with specific comets.

meteorite. A meteoroid, probably from a fragmented asteroid, that passes through Earth's atmosphere and arrives safely on Earth.

meteoriticist. A scientist trained in chemistry and mineralogy who studies meteorites.

meteoritics. The study of meteorites and the origin of the solar system.

meteoroid. A rock of subasteroid diameter that orbits the Sun near Earth's orbit and represents fragments of asteroid parent bodies.

micrometeoroid. A meteoroid of submillimeter size from either comets or disintegrated asteroids.

molecular cloud. Large, cold, and relatively dense interstellar clouds of hydrogen gas, solid dust grains, and a predominance of molecules; they are involved with the birth of stars.

momentum. The product of the mass and velocity of a body; a measure of the inertia of a moving body.

monomict breccia. A rock made of angular fragments of similar composition.

nakhlite. A rare class of achondritic meteorite composed of the green clinopyroxene augite; a class of SNC meteorite.

Neumann lines. A fine, linear structure representing twinning boundaries in the nickel-iron alloy kamacite, produced by impact shock; found in hexahedrite meteorites.

noble metals. Precious and rare metals such as platinum, gold, iridium, osmium; the platinum group.

noctilucent clouds. Ice clouds condensed around meteoritic dust about 50 miles above Earth, high enough to be illuminated by sunlight at night.

octahedrite. A subclass of iron meteorite containing the nickel-iron alloys kamacite and taenite arranged in plates on the faces of an octahedron.

olivine. A silicate of either magnesium or iron or both in varying amounts $(Mg,Fe)_2SiO_4$; common in stony and stony-iron meteorites.

Oort cloud. A hypothetical cloud of comets found in a zone between 20,000 and 150,000 AU from the Sun.

ordinary chondrite. The most common class of stony meteorites, which contain chondrules; the class includes E-, H-, L-, and LL-chondrites.

oriented meteorite. Meteorites that show a preferred orientation of fall, which produces a conical shape for the meteorite as it ablates in the atmosphere.

orthoferrosilite. Orthopyroxene end-member containing iron ($FeSiO_3$).

orthopyroxene. A pyroxene series containing either iron or magnesium or both in varying amounts; a solid solution between enstatite and orthoferrosilite.

pallasite. A class of stony-iron meteorite containing approximately equal amounts of metal and olivine; the metal makes up a continuous network with isolated grains of olivine.

parent body. Astronomical bodies of subplanetary to planetary size that fragment by collision to produce meteorites.

penetration hole. The hole made by a meteorite that impacts Earth but does not explode.

perihelion. A position on an elliptical orbit where a celestial body is closest to the Sun. See *aphelion.*

perturbation. A small gravitational force on an orbiting body exerted by another, more massive body, often resulting in the gradual change of the orbit.

petrologic type. A scale used to denote the texture of chondrite meteorites; denotes increasing metamorphism in chondrules.

phase diagram. A diagram that illustrates the stability of minerals at certain temperatures, pressure, and compositions; phase diagrams for metal alloys in iron meteorites show continuing recrystallization in the solid state.

piezoglypt. See *regmaglypt.*

plagioclase. A class of feldspar minerals in solid solution with sodium and calcium ions; plagioclase ranges from albite or all-sodium plagioclase to anorthite or all-calcium plagioclase.

planetesimal. Small bodies up to a few hundred miles in diameter that formed from the first solid grains to condense out of the solar nebula; planetesimals accreted to each other to form the planets of the solar system.

plasma. Hot ionized gas; the gas that surrounds a bright fireball as it passes through the atmosphere.

platinum group. See *noble metals.*

plessite. A fine-grained mixture of kamacite and taenite formed late in the diffusion process at low temperatures; usually found filling spaces between the Widmanstätten patterns in octahedrite meteorites.

polymict breccia. A rock made up of angular fragments from other rocks of different compositions.

polymorph. A mineral that has several crystal forms; for example, graphite is the amorphous form, while diamond is the crystal form, both being polymorphs of carbon.

protoplanets. The next stage of growth for a solar-system body after the planetesimal stage, leading to planets.

pyroxene. An important silicate mineral found in all stony meteorites; different pyroxenes form by solid solution with magnesium, iron, and calcium.

pyrrhotite. A terrestrial iron sulfide (FeS); troilite is the equivalent mineral in meteorites.

radiant point. The point in the sky from which meteors in a meteor shower seem to diverge; an illusion of perspective.

reflection spectrum. An absorption spectrum produced as light is partially absorbed by mineral crystals on the surface of an asteroid and then reflected back into space.

refractory elements. Elements that are the first to condense out of a cooling gas at relatively high temperatures; elements that have high vaporization temperatures.

regmaglypt. A deep pit or cavity on the exterior of meteorites produced by ablation of certain minerals in the meteorite as it passes through Earth's atmosphere. Synonym: *piezoglypt.*

resonance. Gravitational perturbations on a body orbiting the Sun that arise from a third body whose orbital period is a simple multiple or fraction of the other body's orbital period; resonance between Jupiter and the asteroids changes the asteroids' orbits.

retardation point. The point in the flight of a meteoroid through Earth's atmosphere at which the meteoroid loses its cosmic velocity and falls freely due to Earth's gravity.

schreibersite. A nickel-iron cobalt phosphide $(Fe,Ni,CO)_3P$; an accessory mineral in iron meteorites; not found on Earth.

serpentine. A hydrous magnesium silicate mineral, $Mg_6Si_4O_{10}(OH)_8$, commonly found in carbonaceous chondrites.

shatter cone. A conical fracturing of rock as shock waves generated by an impacting meteorite are propagated through the bedrock.

shergottite. One of the three rare SNC achondritic meteorites; of basaltic composition with plagioclase and pyroxene as major components.

shock metamorphism. Changes in rocks and minerals due to shock generated by an impacting body; may result in changes in crystal structure, especially in quartz after melting and recrystallization.

siderophile. An "iron-loving" element; an element that readily forms an alloy with iron.

silicated iron. An iron meteorite that contains inclusions of silicate minerals.

SNC meteorites. Rare achondritic meteorites that have young isotopic ages; may have originated on Mars; SNC is an abbreviation for shergottite, nakhlite, and chassignite.

solar nebula. The gas and dust cloud thought to have given birth to the solar system.

solar wind. A constant outflow of charged particles from the Sun.

solid solution. Substitution of ions, usually metals, of similar size, producing variations in the composition of minerals; usually forms a solid-solution series of related minerals.

spectrophotometer. An instrument that measures the reflection spectrum of an asteroid.

sporadic meteor. An unpredictable, isolated meteor not associated with a periodic meteor shower.

stellar cocoon. A fragment of a molecular cloud that collapses to produce a star.

stishovite. A high-pressure form of quartz formed by impact shock.

stones. Meteorites of stony composition.

stony-iron meteorites. A class of meteorites that contain both silicate minerals and nickel-iron in approximately equal proportions; pallasites and mesosiderites are subclasses.

strewn field. An elliptical field defining the fall path of a multiple meteorite fall.

subhedral. Mineral crystals bounded by some of their typical crystal faces. Intermediate between anhedral and euhedral crystals.

supernova. A star of several solar masses that explodes due to instabilities in its core; produces heavy elements in the process.

taenite. A nickel-iron alloy with at least 20 percent nickel; commonly found in iron meteorites.

thermonuclear reactions. Nuclear reactions at high temperatures in the cores of stars that produce energy to sustain the stars.

train. A trail of ionized gas and meteoritic dust left behind a burning meteoroid as it passes through Earth's atmosphere.

troilite. A bronze-colored magnetic iron sulfide found in meteorites similar but not identical to terrestrial pyrrhotite.

ureilite. A rare type of achondrite consisting of pyroxene crystals in a carbon-rich matrix.

volatile elements. Elements that are the last to condense out of a cooling gas; elements that have low vaporization temperatures.

Widmanstätten figures. Plates of kamacite bordered by taenite that grow on the octahedral faces of octahedrite meteorites.

zenith. A point on the celestial sphere directly overhead from any location on Earth.

zodiacal light. Light from the Sun scattered by interplanetary dust particles along the ecliptic plane and between Earth and the Sun.

REFERENCES

Two magazines specifically dedicated to meteorites should interest the meteorite researcher and collector.

Meteorite!
Pallasite Press
P.O. Box 33-1218
Takapuna, Auckland, New Zealand

Meteoritics and Planetary Sciences
Journal of the Meteoritical Society
Department of Chemistry and Biochemistry
University of Arkansas
Fayetteville, Arkansas 72701

Meteorite! is a young magazine; the first issue was published in February 1995. Well-written articles by both collectors and researchers are aimed at the general reader with an interest in meteorites. The magazine is published quarterly in New Zealand.

Meteoritics and Planetary Sciences, published bimonthly, is the professional journal of the Meteoritical Society. Intended for scientists and informed amateurs, this research journal covers a wide range of topics: asteroids, comets, meteors, impact craters, interplanetary dust, tektites, and the origin of the solar system. Readers should have a grasp of basic mineralogy and petrology to get the most out of this journal. Subscribers become members of the Meteoritical Society.

General References

Bagnall, Philip M. 1991. *The Meteorite and Tektite Collector's Handbook.* Richmond, Va.: Willmann-Bell.

Beatty, J. Kelly, and Andrew Chaiken. 1990. *The New Solar System.* 2nd ed. Cambridge, Mass.: Sky Publishing.

Chapman, Clark R. 1982. *Planets of Rock and Ice.* New York: Scribner.

Corliss, William R. 1986. *The Sun and Solar System Debris.* Glen Arm, Md.: Sourcebook Project.

Cunningham, Clifford J. 1988. *Introduction to Asteroids.* Richmond, Va.: Willmann-Bell.

Davies, John K. 1986. *Cosmic Impact.* London: Fourth Estate.

Goldsmith, Donald. 1985. *Nemesis: The Death Star and Other Theories of Mass Extinction.* New York: Walker and Co.

Hartmann, William K., and Ron Miller. 1991. *The History of Earth.* New York: Workman Publishing.

Heide, Fritz. 1964. *Meteorites.* Chicago: University of Chicago Press.

Hoyt, William Graves. 1987. *Coon Mountain Controversies.* Tucson: University of Arizona Press.

Hutchison, Robert. 1983. *The Search for Our Beginnings.* British Museum. London: Oxford University Press.

Hutchison, Robert, and Andrew Graham. 1993. *Meteorites.* New York: Sterling Publishing Co., Inc.

Krinov, E. L. 1966. *Giant Meteorites.* New York: Pergamon Press.

LaPaz, Lincoln, and Jean LaPaz. 1961. *Space Nomads.* New York: Holiday House.

LeMaire, T. R. 1980. *Stones from the Stars.* New Jersey: Prentice-Hall, Inc.

Mark, Kathleen. 1987. *Meteorite Craters.* Tucson: University of Arizona Press.

Mason, Brian. 1962. *Meteorites.* New York: Wiley.

Muller, Richard. 1988. *Nemesis.* New York: Weidenfeld and Nicolson.

Nininger, H. H. 1933. *Our Stone-Pelted Planet.* New York: Houghton Mifflin.

——. 1956. *Arizona's Meteorite Crater.* Denver: American Meteorite Museum.

——. 1959. *Out of the Sky.* New York: Dover Publications.

——. 1972. *Find a Falling Star.* New York: Paul S. Eriksson, Inc.

O'Keefe, John A. 1963. *Tektites.* Chicago: University of Chicago Press.

Pearl, Richard M. 1975. *Fallen from Heaven: Meteorites and Man.* Colorado Springs: Earth Science Publishing.

Povenmire, Harold R. 1980. *Fireballs, Meteors, and Meteorites.* Indian Harbor Beach, Fla.: JSB Enterprises.

Raup, David M. 1991. *Extinction: Bad Genes or Bad Luck?* New York: W. W. Norton and Co.

Sagan, Carl, and Ann Druyan. 1985. *Comet.* New York: Random House.

Sears, D. W. 1988. *Thunderstones: A Study of Meteorites Based on Falls and Finds in Arkansas.* University Museum Special Publication. Fayetteville: University of Arkansas.

Whipple, F. L. 1985. *The Mystery of Comets.* Washington, D.C.: Smithsonian Institution Press.

Willey, R. R. 1987. *The Tucson Meteorites.* Washington, D.C.: Smithsonian Institution Press.

Advanced References

Buckwald, V. F. 1975. *Handbook of Iron Meteorites.* Vols. 1–3. Los Angeles: University of California and Arizona State University presses.

Burke, John G. 1986. *Cosmic Debris: Meteorites in History.* Berkeley: University of California Press.

Clarke, Roy S., Jr. 1971. *The Allende, Mexico, Meteorite Shower.* Washington, D.C.: Smithsonian Institution Press.

Clarke, Roy S., Jr., ed. 1993. *The Port Orford Oregon Meteorite Mystery.* Smithsonian Contributions to the Earth Sciences, No. 31. Washington, D.C.: Smithsonian Institution Press.

Delsemme, A. H. 1977. *Comets, Asteroids, Meteorites: Interrelationships, Evolution and Origins.* Toledo: University of Toledo Press.

Dodd, R. T. 1981. *Meteorites: A Petrologic-Chemical Synthesis.* New York: Cambridge University Press.

———. 1986. *Thunderstones and Shooting Stars.* Cambridge: Harvard University Press.

Farrington, Oliver C. 1915. *Meteorites.* Chicago: Published by the author.

Gehrels, T., ed. 1988. *Asteroids.* Tucson: University of Arizona Press.

Graham, A. L., A. W. R. Bevan, and R. Hutchison. 1985. *Catalogue of Meteorites.* Tucson: University of Arizona Press.

Kerridge, John F., and Mildred Shapley Matthews, eds. 1979. *Meteorites and the Early Solar System.* Tucson: University of Arizona Press.

Klein, Hurlbut. 1977. *Manual of Mineralogy.* New York: Wiley.

McCall, G. J. H., ed. 1979. *Astroblemes—Cryptoexplosion Structures. Benchmark Papers in Geology.* Vol. 50. Stroudsbury, Pa.: Dowden, Hutchison and Ross.

McSween, Harry Y., Jr., 1987. *Meteorites and Their Parent Planets.* New York: Cambridge University Press.

Nininger, H. H. 1977. *Meteorites: A Photographic Study of Surface Features. Part 1: Shapes.* Center for Meteorite Studies. Tempe: Arizona State University.

Sears, D. W. 1978. *The Nature and Origin of Meteorites.* Bristol, England: Adam Hilger, Ltd.

Wasson, John T. 1985. *Meteorites: Their Record of Early Solar System History.* New York: W. H. Freeman.

Wilkening, L., ed. 1982. *Comets.* Tucson: University of Arizona Press.

Articles and Technical Papers

Articles and technical papers for the general reader are included in this list. An asterisk (*) indicates technical papers.

*Ahrens, Thomas J., and Alan W. Harris. 1992. Deflection and Fragmentation of Near-Earth Asteroids. *Nature* (December):429-33.

*Alvarez, Luis, Walter Alvarez, Frank Asaro, and Helen V. Michel. 1980. Extraterrestrial Cause for the Cretaceous-Tertiary Extinction. *Science* (June):1095-1107.

*Alvarez, Walter, and Richard A. Muller. 1984. Evidence from Crater Ages for Periodic Impacts on the Earth. *Nature* (April):718-20.

Beatty, J. Kelly. 1991. Killer Crater in the Yucatan? *Sky and Telescope* (July):38-40.

Bilkadi, Zayn. 1986. The Wabar Meteorite. *Aramco World Magazine* (November/December):26-33.

*Bradley, J. P., and D. E. Brownlee. 1991. An Interplanetary Dust Particle Linked Directly to Type CM Meteorites and an Asteroidal Origin. *Science* (February):549-52.

Brownlee, Donald E. 1982. Cosmic Dust. *Natural History* (April):72-77.

———. Cosmic Dust. 1984. *The Planetary Report* (March/April):9-11.

Burnham, Robert. 1991. Arizona's Meteor Crater. *Earth* (January):50-58.

Chaikin, Andrew. 1983. A Stone's Throw from the Planets. *Sky and Telescope* (February):122-23.

———. 1984. Target: Tunguska. *Sky and Telescope* (January):18-21.

Chyba, Christopher. 1993. Death from the Sky. *Astronomy* (December):38-45.

Clark, Arthur. 1986. The Camel's Hump. *Aramco World Magazine* (November/December):32-33.

Cowen, R. 1993. Near-Earth Asteroids: Class Consciousness. *Science News* (February):117.

*Davis, Marc, Piet Hut, and Richard A. Muller. 1984. Extinction of Species by Periodic Comet Showers. *Nature* (April):715-17.

di Cicco, Dennis. 1983. Wethersfield Meteorite: The Odds Were Astronomical. *Sky and Telescope* (February):118-19.

———. 1993. New York's Cosmic Car Conker. *Sky and Telescope* (February):118.

*Dietz, Robert S. 1959. Shatter Cones in Cryptoexplosion Structures. *Journal of Geology* (67):502-3.

———. 1991. Are We Mining an Asteroid? *Earth* (January):36-41.

———. 1991. Demise of the Dinosaurs: A Mystery Solved? *Astronomy* (July):30-37.

Farkelmann, K. A. 1993. Tunguska: The Explosion of a Stony Meteorite. *Science News* (January):23.

Federer, Charles A., ed. 1962. Leonid Meteors Give Unexpected Display. *Sky and Telescope* (February):61-66.

———. 1967. Great Leonid Meteor Shower of 1966. *Sky and Telescope* (January):4-10.

Flam, Faye. 1991. Seeing Stars in a Handful of Dust. *Science.* Research News (July):380-81.

Hildebrand, Alan R., and William V. Boynton. 1991. Cretaceous Ground Zero. *Natural History* (June):47-53.

Hovey, Edmund Otis. 1907. The Meteorites in the Collection of the American Museum of Natural History. Guide Leaflet No. 26.

Jacchia, Luigi G. 1974. A Meteorite that Missed the Earth. *Sky and Telescope* (July):4-9.

Kerr, Richard A. 1993. Second Crater Points to Killer Comets. *Science Research News* (March):1543.

Krinov, E. L. 1969. New Studies of the Sikhote-Alin Meteorite Shower. *Sky and Telescope* (February):87-90.

Lange, Erwin F. 1958. Oregon Meteorites. *Oregon Historical Quarterly* 59(3).

Marvin, Ursula B. 1981. The Search for Antarctic Meteorites. *Sky and Telescope* (November):423-27.

——. 1986. Meteorites, the Moon, and the History of Geology. *Journal of Geological Education* 34:140-65.

*——. 1992. The Meteorite of Ensisheim: 1492 to 1992. *Meteoritics* (March): 28-70.

——. 1996. E. F. F. Chladni and the Origins of Modern Meteorite Research. *Meteoritics and Planetary Sciences* (September) 545-88.

Mason, Brian. 1981. A Lode of Meteorites. *Natural History* (April):62-64.

Ming, Chou. 1965. China's Largest Meteorite. *Sky and Telescope* (December):347.

Monastersky, Richard. 1992. Closing in on the Killer. *Science News* (January):56-58.

——. 1992. Anti-Impactors Have Their Day in K-T Court. *Science News* (November):310.

Morrison, David, and Clark R. Chapman. 1990. Target Earth: It Will Happen. *Sky and Telescope* (March):261-65.

Norton, O. Richard. 1959. The Barringer Meteorite Crater. *Griffith Observer* (May):62-73.

——. 1995. The Great Meteorite Caper. *Lapidary Journal.* (December):55-60.

——. 1996. There Was an Old Woman . . . , Part I. *Lapidary Journal.* (December):48-54.

——. 1997. There Was an Old Woman . . . , Part II. *Lapidary Journal.* (January):60-68.

——. 1997. Exploring Meteorite Thin Sections, Part I. *Meteorite!* (February):16-21.

——. 1997. Exploring Meteorite Thin Sections, Part II. *Meteorite!* (August):19-22.

Oliver, Robert W. 1988. Port Orford Meteorite. *Oregon Coast* (August/September):22-25.

Plotkin, Howard. 1993. The Port Orford Meteorite Hoax. *Sky and Telescope* (September):35-38.

*Raup, David M., and J. John Sepkoski Jr. 1984. Periodicity of Extinctions in the Geologic Past. *Proceedings of the National Academy of Sciences* (February):801-5.

*Rawcliffe, R. D. 1974. Meteor of August 10, 1972. *Nature* 247:449-50.

Reck, Kathy. 1977. Legality of Mining Claim Ignored in Old Woman Meteorite Case. *California Mining Journal* (November):4, 5, 17.

Robinson, Leif J., ed. 1993. Tunguska: An Asteroid. News Notes. *Sky and Telescope* (March):15.

Rubin, Alan E. 1997. Mineralogy of Meteorite Groups. *Meteoritics and Planetary Science* (March):231-47.

Schultz, Peter H., and J. Kelly Beatty. 1992. Teardrops on the Pampas. *Sky and Telescope* (April):389-92.

*Sharpton, Virgil. L. 1993. Chicxulub Multiring Impact Basin: Size and Other Characteristics Derived from Gravity Analysis. *Science* (261):1564–67.

Shoemaker, Eugene M. 1959. *Impact Mechanics at Meteor Crater, Arizona.* U.S. Geological Survey.

Spratt, Christopher, and Sally Stephens. 1992. Meteorites That Have Struck Home. *Mercury* (Journal of the Astronomical Society of the Pacific) (March/April):50–56.

The True First-Hand Story of the Old Woman Meteorite. 1977. *California Mining Journal* (August):11–13.

Weissman, Paul R. 1993. Comets at the Solar System's Edge. *Sky and Telescope* (January):26–29.

Witze, Alexander M. 1992. The Great Stone of Ensisheim Turns 500. *Sky and Telescope* (November):502–3.

Watson, Traci. 1993. Uncovering a 10-ton Hoax. Random Samples. *Science* (February):1256.

Weaver, Kenneth F. 1986. Meteorites: Invaders from Space. *National Geographic* (September):390–418.

*Wilshire, H. G., T. W. Offield, K. A. Howard, and D. Cummings. 1972. Geology of the Sierra Madera Cryptoexplosion Structure, Pecos County, Texas. *U.S. Geological Survey Professional Paper 599H.*

Yeomans, Donald K. 1991. The Search for Near-Earth Asteroids. *The Planetary Report* (November/December):4–7.

INDEX—General

Abercrombie, Thomas, J., 144
achondrites, 201–13
  classification of, 203
Ahnighito meteorite. *See*
  Greenland meteorites
Ainsa, Augustin, 263–64
Alvarez, Luis, 376–79, 398, 402
Alvarez, Walter, 375–79, 402
American Meteorite Laboratory,
  287. *See also* Nininger, H. H.
amino acids, 196, 198
amphoterites, 187, 190, 346
Antarctica, 175, 235, 315–22
Antarctic meteorites
  age of, 319–21
  ALHA 80133, 322
  ALHA 81005, lunar rock
    discovery, 320–21
  ALHA 81006, Plate VII
  ALHA 81031, 319
  ALH 84001, vi, 199, 209-10,
    212-13
  Allan Hills, 209-10, 317–18
  cataloging, 322
  Derrick Peak, 319, 323
  glacial mechanism, 315–16
  meteorite "gold" rush, 318–19
  Yamato Mountains meteorite
    discovery, 315–18
Antiquities Act, 253
Apollo asteroids, 362–64, 368,
  369, 385
Asaro, Frank, 378–79
asteroid belt, 207, 327
  early asteroid population, 337–38
  Jupiter perturbation of, 338,
    342–43, 358
  Kirkwood gaps, 342–43, 360

asteroid impacts, predictions of,
  366–68
asteroids
  discovery of first, 338–41
  families, 359–60
  near-Earth, 362–66
  reflection spectra, 352–54
  types of, 354–56
asteroids (names)
  2101 Adonis, 362
  132 Aethra, 361
  1221 Amor, 361–62
  1862 Apollo, 362
  5 Astraea, 341
  2062 Aten, 363–64
  1 Ceres, 338, 341, 354, 358
  2060 Chiron, 381
  433 Eros, 354, 361
  951 Gaspra, 239–41, 326, 395, 396
  Hermes, 363
  243 Ida, 359–60, 395, 397
  3 Juno, 341
  253 Mathilde, xii, 1–2
  2 Pallas, 341, 358
  1989PB, 396
  16 Psyche, 354
  4279 Toutatis, 395–97
  4 Vesta, 207, 341, 354, 356–57
asteroid surveys
  Palomar Earth-Crossing, 364–66
  Spacewatch, 363, 365–66
astroblemes, 136–37
ataxites, 216, 227–28
aubrites, 204

bacteria, fossil, 212–13
Barringer, Daniel M., 95, 116–22,
  278

O. Richard Norton and Dorothy Sigler Norton.

O. Richard Norton is former director of the Grace Flandrau Planetarium and Science Center at the University of Arizona and the Fleischmann Planetarium at the University of Nevada at Reno. He studied meteoritics under world-renowned meteoriticist Frederick C. Leonard. Mr. Norton currently teaches astronomy at Central Oregon Community College and serves as president of Science Graphics, a company that produces instructional science slides for use in college and university classrooms.

Dorothy Sigler Norton is an artist and scientific illustrator specializing in astronomy, geology, and paleontology. Dorothy's artwork includes colorful paintings and drawings that illustrate books, journals, and magazines. She has published illustrations in *National Geographic,* and her large paintings of extinct animals grace the walls of the National Geological Museums in Japan. Her scientific illustrations in slide form are being used as teaching aids in colleges and universities throughout the world.

Check for our books at your local bookstore. Most stores will be happy to order any that they do not stock. Or order directly from us, either by mail, using the enclosed order form, or our toll-free number, 1-800-234-5308, and put your order on your Mastercard or Visa charge card. We will gladly send you a complete catalog upon request.

Some other Mountain Press titles:

____ROADSIDE GEOLOGY OF ALASKA	$15.00
____ROADSIDE GEOLOGY OF ARIZONA	$15.00
____ROADSIDE GEOLOGY OF COLORADO	$16.00
____ROADSIDE GEOLOGY OF HAWAII	$20.00
____ROADSIDE GEOLOGY OF IDAHO	$15.00
____ROADSIDE GEOLOGY OF LOUISIANA	$15.00
____ROADSIDE GEOLOGY OF MAINE	$18.00
____ROADSIDE GEOLOGY OF MONTANA	$18.00
____ROADSIDE GEOLOGY OF NEW MEXICO	$15.00
____ROADSIDE GEOLOGY OF NEW YORK	$20.00
____ROADSIDE GEOLOGY OF NORTHERN CALIFORNIA	$15.00
____ROADSIDE GEOLOGY OF OREGON	$15.00
____ROADSIDE GEOLOGY OF SOUTH DAKOTA	$20.00
____ROADSIDE GEOLOGY OF TEXAS	$16.00
____ROADSIDE GEOLOGY OF UTAH	$16.00
____ROADSIDE GEOLOGY OF VERMONT & NEW HAMPSHIRE	$10.00
____ROADSIDE GEOLOGY OF VIRGINIA	$12.00
____ROADSIDE GEOLOGY OF WASHINGTON	$17.00
____ROADSIDE GEOLOGY OF WYOMING	$15.00
____ROADSIDE GEOLOGY OF THE YELLOWSTONE COUNTRY	$12.00
____COLORADO ROCKHOUNDING	$18.00
____NEW MEXICO ROCKHOUNDING	$20.00
____FIRE MOUNTAINS OF THE WEST	$20.00
____GEOLOGY UNDERFOOT IN SOUTHERN CALIFORNIA	$14.00
____GEOLOGY UNDERFOOT IN ILLINOIS	$15.00
____GEOLOGY UNDERFOOT IN DEATH VALLEY AND OWENS VALLEY	$16.00
____NORTHWEST EXPOSURES	$24.00
____ROCKS FROM SPACE	$30.00

Please include $3.00 per order to cover postage and handling.

Please send the books marked above. I have enclosed $_____

Name_____

Address_____

City_____State_____Zip_____

☐ Payment enclosed (check or money order in U.S. funds)   **OR**   Bill my:

☐ VISA ☐ MC Expiration Date:_____ Daytime Phone_____

Card No._____

Signature_____

MOUNTAIN PRESS PUBLISHING COMPANY
P.O. Box 2399 • Missoula, MT 59806 • Order Toll-Free 1-800-234-5308
E-mail: mtnpress@montana.com • Website: www.mtnpress.com